Basics of

Web Design

HTML5 & CSS

Sixth Edition

Basics of

Web Design

HTML5 & CSS

Sixth Edition

Terry Ann Felke-Morris, Ed.D.
Professor Emerita
Harper College

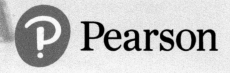

Content Development: Tracy Johnson
Content Management: Dawn Murrin, Tracy Johnson
Content Production: Carole Snyder

Product Management: Holly Stark
Product Marketing: Wayne Stevens
Rights and Permissions: Anjali Singh

Please contact https://support.pearson.com/getsupport/s/ with any queries on this content

Library of Congress Cataloging-in-Publication Data

Names: Felke-Morris, Terry, author.
Title: Basics of web design : HTML5 & CSS / Terry Ann Felke-Morris, Ed.D., Professor Emerita.
Description: Sixth edition. | Hoboken : Pearson, 2022.
Identifiers: LCCN 2020030388 | ISBN 9780137313266 (Print Offer) | ISBN 9780137313211 (Rental)
Subjects: LCSH: HTML (Document markup language) | Cascading style sheets. | Web site development—Computer programs. | Web sites—Design.
Classification: LCC QA76.76.H94 F455 2020 | DDC 006.7/4—dc23
LC record available at https://lccn.loc.gov/2020030388

1 2020

Rental
ISBN 10:   0-13-731321-7
ISBN 13: 978-0-13-731321-1

Print Offer
ISBN-10:   0-13-731326-8
ISBN-13: 978-0-13-731326-6

Preface

Basics of Web Design: HTML5 & CSS is intended for use in a beginning web design or web development course. Topics are introduced in two-page sections that focus on key points and often include a hands-on practice exercise. The text covers the basics that web designers need to develop their skills:

- Introductory Internet and World Wide Web concepts
- Creating web pages with HTML5
- Configuring text, color, and page layout with Cascading Style Sheets
- Configuring images and multimedia on web pages
- Exploring CSS Flexbox and CSS Grid layout systems
- Web design best practices
- Creating responsive web pages that display well on both desktop and mobile devices
- Accessibility, usability, and search engine optimization considerations
- Obtaining a domain name and a web host
- Publishing to the Web

Student files include solutions to the Hands-On Practice exercises, starter files for the Hands-On Practice exercises, and the starter files for the Case Study. The eText offers student file downloads by chapter (where used) wiithin each chapter introduction. Student files are also available for download from the companion website for this book at www.pearson.com/felke-morris.

Building on this textbook's successful fifth edition, the sixth edition features:

- Additional Hands-On Practice exercises
- Updated code samples, case studies, and web resources
- Updates for HTML5 elements and attributes
- Expanded treatment of page layout design and responsive web design techniques
- Expanded treatment of CSS Flexible Layout Module (Flexbox) and CSS Grid Layout systems
- Expanded coverage of responsive image techniques including lazy loading
- Updated reference sections for HTML5 and CSS

Features of the Text

Design for Today and Tomorrow. The textbook prepares students to design web pages that work today in addition to being ready to take advantage of new HTML5 and CSS coding techniques of the future.

Well-Rounded Selection of Topics. This text includes both "hard" skills such as HTML5 and Cascading Style Sheets (Chapters 1–2 and 4–11) and "soft" skills such as web design

(Chapter 3) and publishing to the Web (Chapter 12). This well-rounded foundation will help students as they pursue careers as web professionals. Students and instructors will find classes more interesting because they can discuss, integrate, and apply both hard and soft skills as students create web pages and websites. The topics in each chapter are typically-introduced on concise two-page sections that are intended to provide quick overviews and timely practice with the topic.

Two-Page Topic Sections. Most topics are introduced in a concise, two-page section. Many sections also include immediate hands-on practice of the new skill or concept. This approach is intended to appeal to your busy students—especially the millennial multitaskers—who need to drill down to the important concepts right away.

Hands-On Practice. Web design is a skill, and skills are best learned by hands-on practice. This text emphasizes hands-on practice through practice exercises within the chapters, end-of-chapter exercises, and the development of a website through ongoing real-world case studies. This variety provides instructors with a choice of assignments for a particular course or semester.

Website Case Study. There are case studies that continue throughout most of the text (beginning at Chapter 2). The case studies serve to reinforce skills discussed in each chapter. Sample solutions to the case study exercises are available on the Instructor Resource Center available through https://pearsonhighered.com/felke.

Focus on Web Design. Every chapter offers an additional activity that explores web design topics related to the chapter. These activities can be used to reinforce, extend, and enhance the course topics.

FAQs. In her web design courses, the author is frequently asked similar questions by students. They are included in the book and are marked with the identifying FAQ icon.

Focus on Accessibility. Developing accessible websites is more important than ever, and this text is infused with accessibility techniques throughout. The special icon shown here makes accessibility information easy to find.

Focus on Ethics. Ethical issues related to web development are highlighted throughout the text with the special ethics icon shown here.

Quick Tips. Quick tips, which provide useful background information, or help with productivity, are indicated with this Quick Tip icon.

Explore Further. The special icon identifies enrichment topics along with web resources useful for delving deeper into a concept introduced in book.

Reference Materials. The appendices offer reference material, including an HTML5 reference, a Cascading Style Sheets reference, a WCAG 2.1 Quick Reference, an overview of ARIA Landmark Roles and a Web Safe Color Palette.

VideoNotes. VideoNotes are Pearson's visual tool designed for teaching students key programming concepts and techniques. These short step-by-step videos demonstrate how to solve problems from design through coding. VideoNotes allow for self-placed instruction with easy navigation including the ability to select, play, rewind, fast-forward, and stop within each VideoNote exercise. Margin icons in your textbook let you know when a VideoNote video is available for a particular concept or hands-on practice.

Supplemental Materials

Student Resources. Student resources provide both reinforcement and practice of new concepts and skills include:

- ▶ VideoNotes
- ▶ Student Files containing the following:
  - ▶ Hands-On Practice starter files
  - ▶ Hands-On Practice solutions
  - ▶ Case Study starter files

Author's Website. In addition to the publisher's companion website for this book, the author maintains a website at https://www.webdevbasics.net. This website contains a page for each chapter with additional resources and updates. This website is not supported by the publisher.

Acknowledgments

Very special thanks go to the people at Pearson, including Tracy Johnson, Carole Snyder, Scott Disanno, and Erin Sullivan.

Most of all, I would like to thank my family for their patience and encouragement. My wonderful husband, Greg Morris, has been a constant source of love, understanding, support, and encouragement. Thank you, Greg! A big shout-out to my children, James and Karen, who grew up thinking that everyone's Mom had their own website. Thank you both for your understanding, patience, and timely suggestions. Finally, a very special dedication to the memory of my father who will be greatly missed.

About the Author

Dr. Terry Ann Felke-Morris is a Professor Emerita at Harper College in Palatine, Illinois. She holds a Doctor of Education degree, a Master of Science degree in information systems, and numerous certifications, including Adobe Certified Dreamweaver 8 Developer, WOW Certified Associate Webmaster, Microsoft Certified Professional, Master CIW Designer, and CIW Certified Instructor.

Dr. Felke-Morris received the Blackboard Greenhouse Exemplary Online Course Award in 2006 for use of Internet technology in the academic environment. She is the recipient of two international awards: the Instructional Technology Council's Outstanding e-Learning Faculty Award for Excellence and the MERLOT Award for Exemplary Online Learning Resources—MERLOT Business Classics.

With more than 25 years of information technology experience in business and industry, Dr. Felke-Morris published her first website in 1996 and has been working with the Web ever since. A long-time promoter of web standards, she was a member of the Web Standards Project Education Task Force. Dr. Felke-Morris is the author of the popular textbook *Web Development and Design Foundations with HTML5*, currently in its tenth edition. She was instrumental in developing the Web Development degree and certificate programs at Harper College. For more information about Dr. Terry Ann Felke-Morris, visit https://terrymorris.net.

CONTENTS

VideoNotes

Locations of **VideoNotes**
www.pearson.com/felke-morris

Internet and Web Basics

The Internet and the Web are parts of our daily lives. How did they begin? What networking protocols and programming languages work behind the scenes to display a web page? This chapter provides an introduction to some of these topics and is a foundation for the information that web developers need to know. This chapter also gets you started with your very first web page. You'll be introduced to Hypertext Markup Language (HTML), the language used to create web pages.

You'll learn how to...

- Describe the evolution of the Internet and the Web
- Explain the need for web standards
- Describe universal design
- Identify benefits of accessible web design
- Identify reliable resources of information on the Web
- Identify ethical uses of the Web

- Describe the purpose of web browsers and web servers
- Identify Internet protocols
- Define URIs and domain names
- Describe HTML, XHTML, and HTML5
- Create your first web page
- Use the body, head, title, and meta elements
- Name, save, and test a web page

The Internet and the Web

The Internet

The **Internet**, the interconnected network of computer networks, seems to be everywhere today. You can't watch television or listen to the radio without being urged to visit a website. Even newspapers and magazines have their place on the Internet. It is possible that you may be reading an electronic copy of this book that you downloaded over the Internet. With the increased use of mobile devices, such as tablets and smartphones, being connected to the Internet has become part of our daily lives.

The Birth of the Internet

The Internet began as a network to connect computers at research facilities and universities. Messages in this network would travel to their destinations by multiple routes or paths, allowing the network to function even if parts of it were broken or destroyed. The message would be rerouted through a functioning portion of the network while traveling to its destination. This network was developed by the Advanced Research Projects Agency (ARPA)—and the ARPAnet was born. Four computers (located at University of California, Los Angeles; Stanford Research Institute; University of California, Santa Barbara; and the University of Utah) were connected by the end of 1969.

Growth of the Internet

As time went on, other networks, such as the National Science Foundation's NSFnet, were created and connected with the ARPAnet. Use of this interconnected network, or Internet, was originally limited to government, research, and educational purposes. The ban on commercial use of the Internet was lifted in 1991.

The growth of the Internet continues—Internet World Stats[1] reported that over 4.8 billion users, about 62% of the world's population, were using the Internet by 2020.

When the restriction on commercial use of the Internet was lifted, the stage was set for future electronic commerce. However, while businesses were no longer banned, the Internet was still text based and not easy to use. The further developments addressed this issue.

The Birth of the Web

VideoNote
Evolution of the Web

While working at CERN, a research facility in Switzerland, **Tim Berners-Lee** envisioned a means of communication for scientists by which they could easily "hyperlink" to another research paper or article and immediately view it. For this purpose, Berners-Lee created the World Wide Web. In 1991, Berners-Lee posted the code in a newsgroup and made it freely available. This version of the World Wide Web used **Hypertext Transfer Protocol (HTTP)** to communicate between the client computer and the web server, and it was text based, employing **Hypertext Markup Language (HTML)** to format the documents.

The First Graphical Browser

In 1993, Mosaic, the first graphical web browser, became available. Marc Andreessen and graduate students working at the National Center for Supercomputing Applications (NCSA) at the University of Illinois Urbana–Champaign developed Mosaic. Some individuals in this group later created another well-known web browser, Netscape Navigator, which is an ancestor of today's Mozilla Firefox.

Convergence of Technologies

By the early 1990s, personal computers with easy-to-use graphical operating systems (such as Microsoft's Windows, IBM's OS/2, and Apple's Macintosh OS) were increasingly available and affordable. Online service providers such as CompuServe, AOL, and Prodigy offered low-cost connections to the Internet. Figure 1.1 depicts this convergence of available computer hardware, easy-to-use operating systems, low-cost Internet connectivity, the HTTP protocol and HTML language, and a graphical browser that made information on the Internet much easier to access. The **World Wide Web**—the graphical user interface providing access to information stored on web servers connected to the Internet—had arrived!

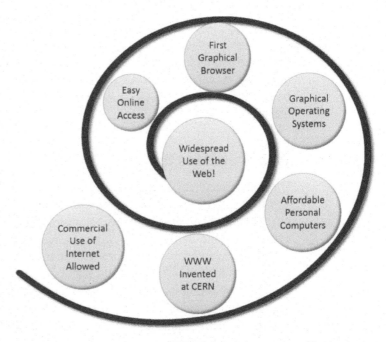

FIGURE 1.1 *Convergence of technologies.*

Web Standards and Accessibility

You are probably aware that no single person or group runs the World Wide Web. However, the **World Wide Web Consortium**, commonly referred to as the **W3C**, takes a proactive role in developing recommendations and prototype technologies related to the Web.[2] Topics that the W3C addresses include web architecture, standards for web design, and accessibility. In an effort to standardize web technologies, the W3C produces guidelines called recommendations.

W3C Recommendations

The W3C Recommendations are created in working groups with input from many major corporations involved in building web technologies. These recommendations are not rules; they are guidelines. Major software companies that build web browsers do not always follow the W3C Recommendations. This makes life challenging for web developers because not all web browsers will display a web page in exactly the same way. The good news is that there is a trend toward conforming to the W3C Recommendations in new versions of major web browsers. You'll follow W3C Recommendations as you code web pages in this book. Following the W3C Recommendations is the first step toward creating a website that is accessible.

Web Standards and Accessibility

The Web Accessibility Initiative, referred to as the WAI, is a major area of work by the W3C.[3] The Web can present barriers to individuals with visual, auditory, physical, and neurological disabilities. An **accessible website** provides accommodations that help individuals overcome these barriers. The WAI has developed the **Web Content Accessibility Guidelines (WCAG)** for web content developers, web authoring tool developers, and web browser developers to facilitate use of the Web by those with special needs.[4] The most recent version of WCAG is WCAG 2.1, which extends WCAG 2.0 and introduces additional success criteria including requirements for increased support of mobile device accessibility, low vision accessibility, and cognitive and learning disability accessibility.

Accessibility and the Law

The **Americans with Disabilities Act (ADA)** of 1990 is a federal civil rights law that prohibits discrimination against people with disabilities. The ADA requires that business, federal, and state services are accessible to individuals with disabilities.

Section 508 of the Federal Rehabilitation Act was amended in 1998 to require that U.S. government agencies give individuals with disabilities access to information technology that is comparable to the access available to others. This law requires developers creating information technology (including web pages) for use by the federal government to provide for accessibility. The GSA Government-wide IT Accessibility Initiative provides accessibility requirement resources for information technology developers.[5] As the Web and Internet technologies developed, it became necessary to revise the original Section 508 requirements. In 2017, an update to Section 508 Standards became official which requires meeting the requirements of WCAG 2.0 Level A & AA Success Criteria. This textbook focuses on WCAG 2.0 and 2.1 guidelines to provide for accessibility.

In recent years, state governments have also begun to encourage and promote web accessibility. The Illinois Information Technology Accessibility Act (IITAA) guidelines are an example of this trend.[6]

Putting It All Together: Universal Design for the Web

Universal design is a "strategy for making products, environments, operational systems, and services welcoming and usable to the most diverse range of people possible".[7] Examples of universal design are all around us. The cutouts in sidewalk curbs providing for wheelchair accessibility also benefit a person pushing a stroller or riding a Segway Personal Transporter (Figure 1.2). Doors that open automatically improve accessibility and also benefit people carrying packages. A ramp is useful for a person dragging a rolling backpack or carry-on bag.

Awareness of universal design by web developers has been steadily increasing. Forward-thinking web developers design with accessibility in mind because it is the right thing to do. Providing access for visitors with visual, auditory, and other challenges should be an integral part of web design rather than an afterthought.

FIGURE 1.2 *A smooth ride is a benefit of universal design.*

A person with visual difficulties may not be able to use graphical navigation buttons and may use a screen reader device to provide an audible description of the web page. By making a few simple changes, such as providing text descriptions for the images and perhaps providing a text navigation area at the bottom of the page, web developers can make the page accessible. Often, providing for accessibility increases the usability of the website for all visitors.

Accessible websites with alternate text for images, headings used in an organized manner, and captions or transcriptions for multimedia are more easily used not only by visitors with disabilities but also by visitors using a mobile browser. Finally, accessible websites may be more thoroughly indexed by search engines, which can be helpful in bringing new visitors to a site. As this book introduces web development and design techniques, corresponding web accessibility and usability issues are discussed.

Web Browsers and Web Servers

Network Overview

A **network** consists of two or more computers connected for the purpose of communicating and sharing resources. A diagram of a small network is shown in Figure 1.3. Common components of a network include the following:

- Server computer(s)
- Client computer(s)
- Shared devices such as printers
- Networking devices (routers, hubs, and switches) and the media that connect them

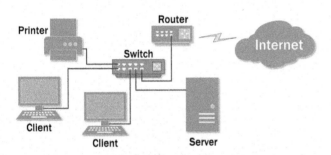

FIGURE 1.3 *Common components of a network.*

The **clients** are the computer workstations used by individuals, such as a PC on a desk. The **server** receives requests from clients for resources, such as a file. Computers used as servers are usually kept in a protected, secure area and are only accessed by network administrators. Networking devices such as hubs and switches provide network connections for computers, and routers direct information from one network to another. The **media** connecting the clients, servers, peripherals, and networking devices may consist of copper cables, fiber optic cables, or wireless technologies.

The Client/Server Model

The term **client/server** dates from the 1980s and refers to computers joined by a network. Client/server can also describe a relationship between two computer programs—the client and the server. The client requests some type of service (such as a file or database access) from the server. The server fulfills the request and transmits the results to the client over a network. While both the client and the server programs can reside on the same computer, typically they run on different computers (Figure 1.4). It is common for a server to handle requests from multiple clients.

The Internet is a great example of client/server architecture at work. Consider the following scenario: A person is at a computer using a web browser client to access the Internet. The person uses the web browser to visit a website, say http://www.google.com. The server is the web server program running on the computer with an IP address that corresponds to google.com. The web server is contacted, it locates the web page and related resources that

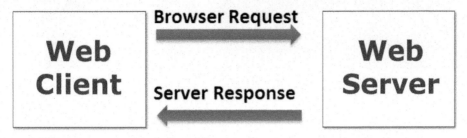

FIGURE 1.4 *Web client and web server.*

were requested, and it responds by sending them. Here's how to distinguish between web clients and web servers:

Web Client

- ▶ Connected to the Internet when needed
- ▶ Usually runs web browser (client) software such as Edge or Firefox
- ▶ Uses HTTP
- ▶ Requests web page from a web server
- ▶ Receives web page and associated files from a web server

Web Server

- ▶ Continually connected to the Internet
- ▶ Runs web server software (such as Apache or Microsoft Internet Information Server)
- ▶ Uses HTTP
- ▶ Receives a request for the web page
- ▶ Responds to the request and transmits the status code, web page, and associated files

When clients and servers exchange files, they often need to indicate the type of file that is being transferred; this is done through the use of a **Multi-Purpose Internet Mail Extensions (MIME) type,** which is a standard for the exchange of multimedia documents among different computer systems. MIME was initially intended to extend the original Internet e-mail protocol, but it is also used by HTTP. MIME provides for the exchange of seven different media types on the Internet: audio, video, image, application, message, multipart, and text. MIME also uses subtypes to further describe the data. The MIME type of a web page is text/html. MIME types of gif and jpeg images are image/gif and image/jpeg, respectively.

A web server determines the MIME type of a file before it is transmitted to the web browser. The MIME type is sent along with the document. The web browser uses the MIME type to determine how to display the document.

How does information get transferred from the web server to the web browser? Clients (such as web browsers) and servers (such as web servers) exchange information through the use of communication protocols such as HTTP, TCP, and IP, which are introduced in the next section.

Internet Protocols

Protocols are rules that describe how clients and servers communicate with each other over a network. There is no single protocol that makes the Internet and the Web work—a number of protocols with specific functions are needed.

E-Mail Protocols

Most of us take e-mail for granted, but there are two servers involved in its smooth functioning—an incoming mail server and an outgoing mail server. When you send e-mail to others, **Simple Mail Transfer Protocol (SMTP)** is used. When you receive e-mail, **Post Office Protocol** (POP; currently **POP3**) and **Internet Message Access Protocol (IMAP)** can be used.

Hypertext Transfer Protocol

Hypertext Transfer Protocol (HTTP) is a set of rules for exchanging files such as text, graphic images, sound, video, and other multimedia files on the Web. Web browsers and web servers usually use this protocol. When the user of a web browser requests a file by typing a website address or clicking a hyperlink, the browser builds an HTTP request and sends it to the server. The web server in the destination machine receives the request, does any necessary processing, and responds with the requested file and any associated media files.

Hypertext Transfer Protocol Secure (HTTPS)

Hypertext Transfer Protocol Secure (HTTPS) combines HTTP with a security and encryption protocol. Using HTTPS provides a more secure transaction because the information passed between the browser and the web server is encrypted. See Chapter 12 for more information on HTTPS.

File Transfer Protocol

File Transfer Protocol (FTP) is a set of rules that allows files to be exchanged between computers on the Internet. Unlike HTTP, which is used by web browsers to request web pages and their associated files in order to display a web page, FTP is used simply to move files from one computer to another. Web developers commonly use FTP to transfer web page files from their computers to web servers.

Transmission Control Protocol/Internet Protocol

Transmission Control Protocol/Internet Protocol (TCP/IP) has been adopted as the official communication protocol of the Internet. TCP and IP have different functions that work together to ensure reliable communication over the Internet.

TCP. The purpose of TCP is to ensure the integrity of network communication. TCP starts by breaking files and messages into individual units called **packets**. These packets (see Figure 1.5) contain information such as the destination, source, sequence number, and checksum values used to verify the integrity of the data.

FIGURE 1.5 *TCP packet.*

TCP is used together with IP to transmit files efficiently over the Internet. IP takes over after TCP creates the packets, using IP addressing to send each packet over the Internet using the best path at the particular time. When the destination address is reached, TCP verifies the integrity of each packet using the checksum, requests resend if a packet is damaged, and reassembles the file or message from the multiple packets.

IP. Working in harmony with TCP, IP is a set of rules that controls how data are sent between computers on the Internet. IP routes a packet to the correct destination address. Once sent, the packet gets successively forwarded to the next closest router (a hardware device designed to move network traffic) until it reaches its destination.

IP Addresses

Each device connected to the Internet has a unique numeric **IP address**. These addresses consist of a set of four groups of numbers called octets. The current widely used version of IP, **IPv4**, uses 32-bit (binary digit) addressing. This results in a decimal number in the format of xxx.xxx.xxx.xxx, where each xxx is a value from 0 to 255. Theoretically, this system allows for at most 4 billion possible IP addresses (although many potential addresses are reserved for special uses). However, even this many addresses will not be enough to meet the needs of all of the devices expected to be connected to the Internet in upcoming years.

IPv6, Internet Protocol Version 6, intended to replace IPv4, was designed as an evolutionary set of improvements and is backwardly compatible with IPv4. Service providers and Internet users can update to IPv6 independently without having to coordinate with each other. IPv6 provides for more Internet addresses because the IP address is lengthened from 32 bits to 128 bits. This means that there are potentially 2^{128} unique IP addresses possible, or 340, 282, 366, 920, 938, 463, 463, 347, 607, 431, 768, 211, 456.

The IP address of a device may correspond to a domain name. The **Domain Name System (DNS)** associates these IP addresses with the text-based URLs and domain names you type into a web browser address box (more on this later). For example, at the time this was written an IP address for Google was 216.58.194.46.

You can enter this number in the address text box in a web browser (as shown in Figure 1.6), press Enter, and the Google home page will be displayed. Of course, it's much easier to type "google.com," which is why domain names such as google.com were created in the first place! Since long strings of numbers are difficult for humans to remember, the DNS was introduced as a way to associate text-based names with numeric IP addresses.

FIGURE 1.6 *Entering an IP address in a web browser.*

Uniform Resource Identifiers and Domain Names

URIs and URLs

A **Uniform Resource Identifier (URI)** identifies a resource on the Internet. A **Uniform Resource Locator (URL)** is a type of URI that represents the network location of a resource such as a web page, a graphic file, or an MP3 file. The URL consists of the protocol, the domain name, and the hierarchical location of the file on the web server.

The URL http://www.webdevbasics.net, as shown in Figure 1.7, denotes the use of HTTP protocol and the web server named www at the domain name of webdevbasics.net. In this case, the root file (which is usually index.html or index.htm) of the 6e directory will be displayed.

FIGURE 1.7
URL describing a file within a folder.

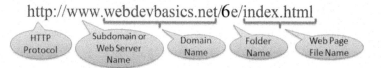

Domain Names

A **domain name** locates an organization or other entity on the Internet. A domain name is associated with a unique numeric IP address assigned to a device. This association is stored in the DNS database.

Let's consider the domain name www.google.com. The .com is the top-level domain name. The portion google.com is the domain name that is registered to Google and is considered a second-level domain name. The www is the name of the web server (sometimes called a **host**) at the google.com domain.

A **subdomain** can be configured to house a separate website located at the same domain. For example, Google's Gmail can be accessed by using the subdomain "gmail" in the domain name (gmail.google.com). Google Maps can be accessed at maps.google.com and Google News Search is available at news.google.com. The combination of host or subdomain, second-level domain, and top-level domain name (such as www.google.com or gmail.google.com) is called a **Fully Qualified Domain Name (FQDN)**.

Top-Level Domain Names

A **top-level domain (TLD)** identifies the rightmost part of the domain name. A TLD is either a **generic top-level domain (gTLD)**, such as com for commercial, or a country-code top-level domain, such as fr for France. The Internet Assigned Numbers Authority (IANA) website has a complete list of TLDs.[8]

Generic Top-Level Domain (gTLD) Names

The Internet Corporation for Assigned Names and Numbers (ICANN) administers gTLDs.[9] The .com, .org, and .net gTLD designations are currently used on the honor system, which means that an individual who owns a shoe store (not related to networking) can register shoes.net. Table 1.1 shows a collection of gTLDs and their intended use.

TABLE 1.1 *Generic Top-Level Domains*

gTLD	Intended for Use By
.aero	Air-transport industry
.bank	Banks and other financial institutions
.biz	Businesses
.club	Clubs and groups with common interests
.com	Commercial entities
.coop	Cooperative
.edu	Restricted to accredited degree-granting institutions of higher education
.gov	Restricted to government use
.info	Unrestricted use
.int	International organization (rarely used)
.jobs	Human resource management community
.mil	Restricted to military use
.museum	Museums
.name	Individuals
.net	Entities associated with network support of the Internet, usually Internet service providers or telecommunication companies
.org	Nonprofit entities
.pro	Accountants, physicians, and lawyers
.tel	Contact information for individuals and businesses
.travel	Travel industry

Expect the number and variety of gTLD names to increase. As of 2020, there were over 1,900 gTLD names applications submitted with over 1200 new gTLD names approved.[10] The new gTLDs include place names (.quebec, .vegas, .moscow, and .amsterdam), retail terms (.blackfriday, .sale, .shop, and .shopping), financial terms (.cash, .trade, .loans, and .mortgage), technology terms (.systems, .technology, .digital, and .app), company names (.apple, .guardian, .toshiba, and .volkswagen), professions (.doctor, .accountant, .dentist, and .attorney), sports (.baseball, .football, and .hocky), and even whimsical fun terms (.ninja, .buzz, and .cool). ICANN has set a schedule to periodically launch new gTLDs.

Country-Code Top-Level Domain Names

Two-character country codes have also been assigned as TLD names. The country-code TLD (ccTLD) names were originally intended to designate the geographical location of the individual or organization that registered the name. The IANA website has a complete list of country-code TLDs.[11]

Domain names with country codes are often used for municipalities, schools, and community colleges in the United States. For example, the domain name www.harper.cc.il.us denotes, from right to left, the United States, Illinois, community college, Harper, and the web server named "www" as the website for Harper College in Illinois.

Although the original intent of country-code TLD names were to designate websites in a particular geographical location, some country codes have become available for commercial use. Examples of nongeographical use of ccTLDs include domain names such as mediaqueri.es, livestre.am, webteacher.ws, youtu.be, who.is, amc.tv, and bit.ly. Table 1.2 lists some popular country codes used on the Web.

TABLE 1.2 *Country-Code TLDs*

ccTLD	Country
.am	Armenia
.au	Australia
.be	Belgium
.dj	Dijbouti
.ch	Switzerland
.es	Spain
.in	India
.is	Iceland
.it	Italy
.lv	Latvia
.ly	Libya
.me	Montenegro
.tm	Turkmenistan
.tv	Tuvalu
.us	United States
.ws	Samoa

Domain Name System (DNS)

The DNS associates domain names with IP addresses. As shown in Figure 1.8, the following happens each time a new URL is typed into a web browser:

1. The DNS is accessed.
2. The corresponding IP address is obtained and returned to the web browser.
3. The web browser sends an HTTP request to the destination computer with the corresponding IP address.
4. The HTTP request is received by the web server.
5. The necessary files are located and sent by HTTP responses to the web browser.
6. The web browser renders and displays the web page and associated files.

We all get impatient sometimes when we need to view a web page. The next time you wonder why it's taking so long to display a web page, consider all of the processing that goes on behind the scenes before the web browser receives the files needed to display the web page.

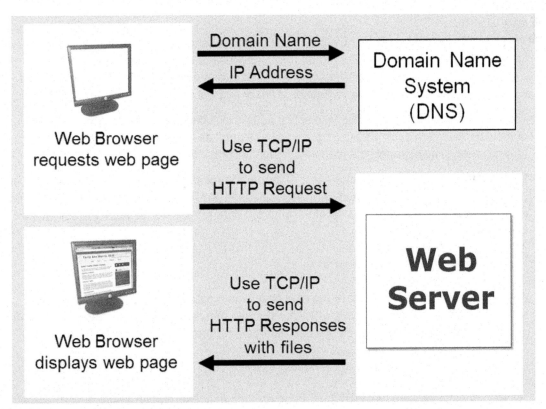

FIGURE 1.8 *Accessing a web page.*

Information on the Web

FIGURE 1.9 *Who really updated that web page you are viewing?*

These days anyone can publish just about anything on the Web. In this section, we'll explore how you can tell if the information you've found is reliable and also how you can use that information. There are many websites—but which ones are reliable sources of information? When visiting websites to find information, it is important not to take everything at face value (Figure 1.9).

▶ Is the organization credible?

Anyone can post anything on the Web! Choose your information sources wisely.

First, evaluate the credibility of the website itself. Does it have its own domain name, such as http://mywebsite.com, or is it a free website consisting of just a folder of files hosted on a free web hosting site (such as weebly.com, awardspace.com, or 000webhost.com)?

The URL of a site hosted on a free web server usually includes part of the free web host's domain name. Information obtained from a website that has its own domain name will usually (but not always) be more reliable than information obtained from a free website.

Evaluate the type of domain name—is it a nonprofit organization (.org), a business (.com or .biz), or an educational institution (.edu)? Businesses may provide information in a way that gives them an advantage, so be careful. Nonprofit organizations or schools will sometimes treat a subject more objectively.

▶ How recent is the information?

Another item to look at is the date the web page was created or last updated. Although some information is timeless, very often a web page that has not been updated for several years is outdated and may not be the best source of information.

▶ Are there links to additional resources?

Hyperlinks indicate websites with supporting or additional information that can be helpful to you in your research as you explore a topic. Look for these types of hyperlinks to aid you in your studies.

▶ Is it Wikipedia?

Wikipedia[12] is a good place to begin research, but don't accept what you read there for fact, and avoid using Wikipedia as a resource for academic assignments. Why? Well, except for a few protected topics, anyone can update Wikipedia with anything!

Usually it all gets sorted out eventually—but be aware that the information you read may not be valid.

Feel free to use Wikipedia to begin exploring a topic, but scroll down to the bottom of the Wikipedia web page and look for "References"—and explore those websites and others that you may find. As you gather information on these sites, also consider the other criteria: credibility, domain name, timeliness, and links to additional resources.

Ethical Use of Information on the Web

The wonderful technology called the World Wide Web provides us with information, graphics, music, and video—all virtually free (after you pay your Internet service provider, of course). Let's consider the following issues relating to the ethical use of this information:

- ▶ Is it acceptable to copy someone's graphic to use on your website?
- ▶ Is it acceptable to copy someone's music or video to use on your website?
- ▶ Is it acceptable to copy someone's website design to use on your site or on a client's site?
- ▶ Is it acceptable to copy an essay that appears on a web page and use it or parts of it as your writing?
- ▶ Is it acceptable to insult someone on your website or link to another website in a derogatory manner?

The answer to all these questions is no. Using a person's graphic, music, or video without permission is the same as stealing it. In fact, if you link to it, you are actually using up some of his or her bandwidth and may be costing him or her money. Copying the website design of another person or company is also a form of stealing. Any text or graphic on a website is automatically copyrighted in the United States whether or not a copyright symbol appears on the site. Insulting a person or company on your website or linking to another website in a derogatory manner could be considered a form of defamation.

Issues related to intellectual property, copyright, and freedom of speech are regularly discussed and decided in courts of law. Good web etiquette requires that you ask permission before using others' work, give credit for what you use as a student ("fair use" in the U.S. copyright law), and exercise your freedom of speech in a manner that is not harmful to others. The **World Intellectual Property Organization (WIPO)**[13] is dedicated to protecting intellectual property rights internationally.

What if you'd like to retain ownership but make it easy for others to use or adapt your work? **Creative Commons** is a nonprofit organization that provides free services that allow authors and artists to register a type of a copyright license called a Creative Commons license.[14] There are several licenses to choose from, depending on the rights you wish to grant. The Creative Commons license informs others exactly what they can and cannot do with your creative work.

HTML Overview

Markup languages consist of sets of directions that tell web browser software (and other user agent software that retrieves and renders web content) how to display and manage a web document. These directions are usually called tags and perform functions such as displaying graphics, formatting text, and referencing hyperlinks.

The World Wide Web is composed of files containing Hypertext Markup Language (HTML) and other markup languages that describe web pages. Tim Berners-Lee developed HTML using **Standard Generalized Markup Language (SGML)**. SGML prescribes a standard format for embedding descriptive markup within a document and for describing the structure of a document. SGML is not in itself a document language, but rather a description of how to specify one and create a document type definition. The W3C sets the standards for HTML and its related languages. HTML (like the Web itself) is in a constant state of change.

What Is HTML?

HTML (Hypertext Markup Language) is the set of markup symbols or codes placed in a file that is intended for display on a web page. These markup symbols and codes identify structural elements such as paragraphs, headings, and lists. HTML can also be used to place media (such as graphics, video, and audio) on a web page and describe fill-in forms. The web browser interprets the markup code and renders the page. HTML permits the platform-independent display of information across a network. No matter what type of computer a web page was created on, any web browser running on any operating system can display the page.

Each individual markup code is referred to as an **element** or a **tag**. Each tag has a purpose. Tags are enclosed in angle brackets, the < and > symbols. Most tags come in pairs: an opening tag and a closing tag. These tags act as containers and are sometimes referred to as container tags. For example, when an HTML document is displayed by a web browser, the text that appears between the `<title>` and `</title>` tags would be displayed in the title bar on the browser window.

Some tags are used alone and are not part of a pair. For example, a `<hr>` tag that displays a horizontal line on a web page is a stand-alone or self-contained tag and does not have a closing tag. You will become familiar with these as you use them. Most tags can be modified with **attributes** that further describe their purpose.

What Is XML?

XML (eXtensible Markup Language) was developed by the W3C to create common information formats and share the format and the information on the Web. It is a text-based syntax designed to describe, deliver, and exchange structured information, such as RSS (Rich Site Summary) web feeds. XML is not intended to replace HTML, but to extend the power of HTML by separating data from presentation. Using XML, developers can create any tags they need to describe their information.

What Is XHTML?

eXtensible Hypertext Markup Language (XHTML) uses the tags and attributes of HTML4 along with the more rigorous syntax of XML. XHTML was used on the Web for over a decade, and you'll find many web pages coded with this markup language. At one point, the W3C was working on a new version of XHTML, called XHTML 2.0. However, the W3C stopped development of XHTML 2.0 because it was not backward compatible with HTML4. Instead, the W3C decided to move forward with HTML5.

HTML5—The Newest Version of HTML

HTML5 is the successor to HTML and replaces XHTML. HTML5 incorporates features of both HTML and XHTML, adds new elements, provides new functionality such as form edits and native video, and is designed to be backward compatible.

The W3C approved HTML5 for final Recommendation status in late 2014 and continued to update the language. HTML 5.3 reached in working draft status in 2018.[15] A change in the procedure to update HTML5 occurred in 2019, when the W3C and the WHATWG (Web Hypertext Application Technology Working Group) agreed to collaborate on the development of HTML specifications.[16] The **WHATWG** is an open group founded by individuals working at many leading technology organizations including Apple, the Mozilla Foundation, and Opera Software. The WHATWG plans to prepare a Review Draft[17] of the HTML Living Standard every six months. Recent versions of popular browsers offer good support for HTML5, which you'll learn to use as you work through this textbook. HTML5 documentation is available at https://html.spec.whatwg.org.

Under the Hood of a Web Page

FIGURE 1.10 *It's what is under the hood that matters.*

You already know that the HTML markup language tells web browsers how to display information on a web page. Let's take a closer look at what's "under the hood" (Figure 1.10) of every web page you create.

Document Type Definition

Because multiple versions and types of HTML and XHTML exist, the W3C recommends identifying the markup language used in a web page document with a **Document Type Definition (DTD)**. The DTD identifies the version of HTML in the document. Web browsers and HTML code validators use the information in the DTD when processing the web page. The DTD statement, commonly called a **DOCTYPE** statement, is the first line of a web page document. The DTD for HTML5 is:

```
<!DOCTYPE html>
```

Web Page Template

Every single web page you create will include the html, head, title, meta, and body elements. We will follow the coding style to use lowercase letters and place quotes around attribute values. A basic HTML5 web page template (found in the student files at chapter1/template.html) is:

```
<!DOCTYPE html>
<html lang="en">
<head>
<title>Page Title Goes Here</title>
<meta charset="utf-8">
</head>
<body>
. . . body text and more HTML tags go here . . .
</body>
</html>
```

With the exception of the specific page title, the first seven lines will usually be the same on every web page that you create. Review the code above and notice the DTD statement has its own formatting, but that the HTML tags all use lowercase letters. Next, let's explore the purpose of the html, head, title, meta, and body elements.

HTML Element

The purpose of the **html element** is to indicate that the document is HTML formatted. The html element tells the web browser how to interpret the document. The opening `<html>` tag is placed on a line below the DTD. The closing `</html>` tag indicates the end of the web page and is placed after all other HTML elements in the document.

The html element also needs to indicate the spoken language, such as English, of the text in the document. This additional information is added to the `<html>` tag in the form of an **attribute**, which modifies or further describes the function of an element. The **lang attribute** specifies the spoken language of the document. For example, `lang="en"` indicates the English language. Search engines and screen readers may access this attribute.

The html element contains the two sections of a web page: the head and the body. The **head section** contains information that describes the web page document. The **body section** contains the actual tags, text, images, and other objects that are displayed by the web browser as a web page.

Head Section

Elements that are located in the **head section** include the title of the web page, meta tags that describe the document (such as the character encoding used and information that may be accessed by search engines), and references to scripts and styles. Many of these do not show directly on the web page.

Head Element. The **head element** contains the head section, which begins with the `<head>` tag and ends with the `</head>` tag. You'll always code at least two other elements in the head section: a title element and a meta element.

Title Element. The first element in the head section, the **title element**, configures the text that will appear in the title bar of the browser window. The text between the `<title>` and `</title>` tags is called the title of the web page. This title text is accessed when web pages are bookmarked and printed. Popular search engines, such as Google, use the title text to help determine keyword relevance and even display the title text on the results page of a search. A descriptive title that includes the website or organization name is a crucial component for establishing a brand or presence on the Web.

Meta Element. The **meta element** describes a characteristic of a web page, such as the character encoding. **Character encoding** is the internal representation of letters, numbers, and symbols in a file, such as a web page or other file, that is stored on a computer and may be transmitted over the Internet. There are many different character-encoding sets. A form of Unicode called utf-8 character encoding is typically used for web pages.[18] The meta tag is not used as a pair of opening and closing tags. It is a stand-alone self-contained tag (referred to as a **void** element in HTML5). The meta tag uses the **charset attribute** to indicate the character encoding. An example meta tag is:

```
<meta charset="utf-8">
```

Body Section

The **body section** contains text and elements that display directly on the web page within the browser window, also referred to as the browser viewport. The purpose of the body section is to configure the contents of the web page.

Body Element. The **body element** contains the body section, which begins with the `<body>` tag and ends with the `</body>` tag. You will spend most of your time writing code in the body of a web page. Text and elements typed between the opening and closing body tags will be displayed on the web page in the browser viewport.

Your First Web Page

No special software is needed to create a web page document—all you need is a text editor. The Notepad text editor is included with Microsoft Windows. TextEdit is distributed with the Mac OS operating system. An alternative to using a simple text editor or word processor is to use a commercial web authoring tool, such as Adobe Dreamweaver. There are also many free or shareware editors available, including Notepad++, Brackets, Visual Studio Code, and BBEdit. Regardless of the tool you use, having a solid foundation in HTML will be useful. The examples in this book use Notepad.

VideoNote

Your First Web Page

 Hands-On Practice 1.1 ————————————————

Now that you're familiar with the basic elements used on every web page, it's your turn to create your first web page, shown in Figure 1.11.

FIGURE 1.11

Your first web page.

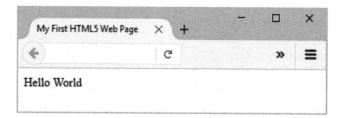

Create a Folder

You'll find it helpful to create folders to organize your files as you develop web pages and create your own websites. Use your operating system to create a new folder named chapter1 on your hard drive or on a portable flash drive.

To create a new folder on a Mac:

1. In the Finder, go to the location where you'd like to create the new folder.
2. Choose File > New Folder. An untitled folder is created.
3. To rename the folder with a new name: select the folder and click on the current name. Type a name for the folder and press the Return key.

To create a new folder with Windows:

1. Right-click on the Start Button and select File Explorer. Then, navigate to the location where you'd like to create the new folder, such as Documents, your C: drive, or an external USB drive.

2. Select the Home tab. Select New folder.

3. To rename the new folder: right-click on it, select Rename from the context-sensitive menu, type the new name, and press the Enter key.

Now, you are ready to create your first web page. Launch Notepad or another text editor. Type in the following code.

```
<!DOCTYPE html>
<html lang="en">
<head>
<title>My First HTML5 Web Page</title>
<meta charset="utf-8">
</head>
<body>
Hello World
</body>
</html>
```

Notice that the first line in the file contains the DTD. The HTML code begins with an opening `<html>` tag and ends with a closing `</html>` tag. The purpose of these tags is to indicate that the content between the tags makes up a web page.

The head section is delimited by `<head>` and `</head>` tags and contains a pair of title tags with the words "My First HTML5 Web Page" in between along with a `<meta>` tag to indicate the character encoding.

The body section is delimited by `<body>` and `</body>` tags. The words "Hello World" are typed on a line between the body tags. See Figure 1.12 for a screenshot of the code as it would appear in Notepad. You have just created the source code for a web page document.

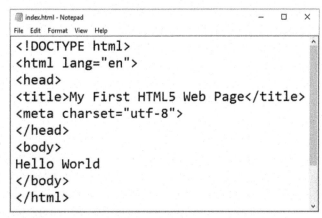

FIGURE 1.12 *Your web page source code displayed in Notepad.*

? FAQ   **Do I have to start each tag on its own line?**

No, you are not required to start each tag on a separate line. A web browser can display a page even if all the tags follow each other on one line with no spaces. Humans, however, find it easier to write and read web page code if line breaks and indentation are used.

Save Your File

Web pages use either an .htm or .html file extension. A common file name for the home page of a website is index.html or index.htm. The web pages in this book use the .html file extension.

You will save your file with the name of index.html.

1. Display your file in Notepad or another text editor.
2. Select File from the menu bar, and then select Save As.
3. The Save As dialog box appears. Using Figure 1.13 as an example, type the file name.
4. Click the Save button.

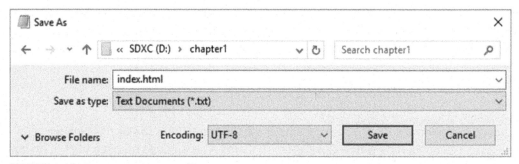

FIGURE 1.13 *Save and name your file.*

Sample solutions for Hands-On Practice exercises are available in the student files. If you would like, compare your work with the solution (chapter1/index.html) before you test your page.

Why does my file have a .txt file extension?

In some older versions of Windows, Notepad will automatically append a .txt file extension. If this happens, rename your file index.html.

Why should I create a folder, why not just use the desktop?

Folders will help you to organize your work. If you just used the desktop, it would quickly become cluttered and disorganized. It's also important to know that websites are organized on web servers within folders. By starting to use folders right away to organize related web pages, you are on your way to becoming a successful web designer.

Test Your Page

There are two ways to test your page:

1. In Windows Explorer (Windows) or the Finder (Mac), navigate to your index.html file. Double-click index.html. The default web browser will launch and will display your index.html page.

2. Launch a web browser. Select File > Open, and navigate to your index.html file. Double-click index.html and click OK. The browser will display your index.html page.

If you are using Microsoft Edge, your page should look similar to the one shown in Figure 1.14. A display of the page using Firefox is shown in Figure 1.11. Notice how the title text, "My First HTML5 Web Page" displays in the tab and the title bar of the browser window. Some search engines use the text surrounded by the `<title>` and `</title>` tags to help determine relevance of keyword searches, so make certain that your pages contain descriptive titles. The `<title>` tag is also used when viewers bookmark your page or add it to their Favorites. An engaging and descriptive page title may entice a visitor to revisit your page. If your web page is for a company or an organization, it's a good idea to include the name of the company or organization in the title.

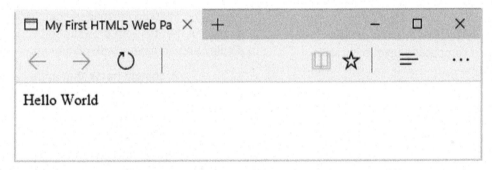

FIGURE 1.14 *Web page displayed by Microsoft Edge. Courtesy of Microsoft Corporation.*

 FAQ When I viewed my page in a web browser, the file name was index.html.html—why did this happen?

This usually happens when your operating system is configured to hide file extension names. You will correct the file name, using one of the following two methods:

▶ Use the operating system to rename the file from "index.html.html" to "index.html".

OR

▶ Open the index.html.html file in your text editor and save it with the name "index.html".

It's a good idea to change the settings in your operating system to show file extension names. Follow the steps at the resources below to show file extension names:

▶ *Windows:* http://www.file-extensions.org/article/show-and-hide-file-extensions-in-windows-10
▶ *Mac:* http://www.fileinfo.com/help/mac_show_extensions

Review and Apply

Review Questions

Multiple Choice. Choose the best answer for each question.

1. Select the item below that indicates the top-level domain name for the URL http://www.mozilla.com.

   a. http

   b. com

   c. mozilla

   d. www

2. What is a unique text-based Internet address corresponding to a computer's unique numeric IP address called?

   a. IP address

   b. domain name

   c. URL

   d. user name

3. The purpose of _____ is to ensure the integrity of the communication.

   a. IP

   b. TCP

   c. HTTP

   d. FTP

4. Choose the true statement:

   a. The title of the web page is displayed by the meta element.

   b. Information about the web page is contained in the body section.

   c. The content that displays in the browser viewport is contained in the head section.

   d. The content that displays in the browser viewport is contained in the body section.

True or False. Choose the best answer, true or false, for each question.

5. _____ Markup languages contain sets of directions that tell web browser software how to display and manage a web document.

6. _____ A domain name that ends in .net indicates that the website is for a networking company.

7. _____ The World Wide Web was developed to allow companies to conduct e-commerce over the Internet.

Fill in the Blanks.

8. _____ is the set of markup symbols or codes placed in a file intended for display on a web browser.

9. Web page documents typically use the _____ or _____ file extension.

10. The home page of a website is typically named _____ or _____.

Review Answers

1. b   2. b   3. b   4. d   5. True   6. False   7. False   8. HTML

9. .htm, .html   10. index.htm, index.html

Hands-On Exercises

1. A blog, or web log, is a journal that is available on the Web—it's a frequently updated page with a chronological list of ideas and links. Blog topics range from political journals to technical information to personal diaries. It's up to the person, called a blogger, who creates and maintains the blog.

   Create a blog to document your learning experiences as you study web design. Visit one of the many sites that offers free blogs, such as https://blogger.com, https://tumblr.com, or https://www.wordpress.com. Follow their instructions to establish your blog. Your blog could be a place to note websites that you find useful or interesting. You might report on websites that contain useful web design resources. You might describe sites that have interesting features, such as compelling graphics or easy-to-use navigation. Write a few sentences about the site that you find intriguing. After you begin to develop your sites, you could include the URLs and reasons for your design decisions. Share this blog with your fellow students and friends.

2. Twitter (https://www.twitter.com) is a social networking website for microblogging, or frequently communicating with a brief message (280 characters or less) called a tweet. Twitter users (referred to as twitterers) tweet to update a network of friends and followers about their daily activities, observations, and information related to topics of interest. A hashtag (the # symbol) can be placed in front of a word or term within a tweet to categorize the topic, such as typing the hashtag #SXSWi in all tweets about the SXSW Interactive Conference for the web design industry. The use of a hashtag makes it easy to search for tweets about a category or an event in Twitter.

   If you don't already use Twitter, sign up for a free account at https://www.twitter.com. Use your Twitter account to share information about websites that you find useful or interesting. Post at least three tweets. You might tweet about websites that contain useful web design resources. You might describe sites that have interesting features, such as compelling graphics or easy-to-use navigation. After you begin to develop your own websites, you can tweet about them, too!

   Your instructor may direct you to include a distinctive hashtag (for example, something like #CIS110) in your tweets that are related to your web design studies. Searching Twitter for the specified hashtag will make it easy to collect all the tweets posted by the students in your class.

Web Research

The World Wide Web Consortium creates standards for the Web. Visit its site at https://www.w3c.org and then answer the following questions:

a. How did the W3C get started?

b. Who can join the W3C? What does it cost to join?

c. The W3C develops standards for a variety of technologies. Explore the W3C's website, locate a technology that interests you, click its link, and read several of the associated pages. List three facts or issues you discover.

Focus on Web Design

Visit a website referenced in this chapter that interests you. Print the home page or one other pertinent page from the site. Write a one-page summary and your reaction to the site. Address the following topics:

a. What is the purpose of the site?
b. Who is the intended audience?
c. Do you think that the site reaches its intended audience? Why or why not?
d. Is the site useful to you? Why or why not?
e. List one interesting fact or issue that this site addresses.
f. Would you encourage others to visit this site?
g. How could this site be improved?

Endnotes

1. "World Internet Users Statistics and 2020 World Population Stats." *Internet World Stats*, 3 Mar. 2020, www.internetworldstats.com/stats.htm.
2. *W3C*, www.w3.org/.
3. w3c_wai. "Home Web Accessibility Initiative (WAI) W3C." *Web Accessibility Initiative (WAI)*, www.w3.org/WAI/.
4. w3c_wai. "WCAG 2.1 at a Glance." *Web Accessibility Initiative (WAI)*, 2019, www.w3.org/WAI/standards-guidelines/wcag/glance/.
5. "GSA Government-Wide IT Accessibility Program." *Section508.Gov | GSA Government-Wide IT Accessibility Program*, www.section508.gov/.
6. "IIITAA 2.0 Standards." *Illinois Department of Human Services*, www.dhs.state.il.us/page.aspx?item=96985.
7. "Universal Design." *Universal Design–Office of Disability Employment Policy–United States Department of Labor*, www.dol.gov/odep/topics/UniversalDesign.htm.
8. "Root Zone Database." *IANA*, www.iana.org/domains/root/db.
9. "By the Numbers & Survey Report." *ICANN*, www.icann.org/.
10. "Program Statistics." *Program Statistics | ICANN New GTLDs*, newgtlds.icann.org/en/program-status/statistics.
11. "Root Zone Database." *IANA*, www.iana.org/domains/root/db.
12. "Main Page." *Wikipedia*, Wikimedia Foundation, 5 Feb. 2020, www.wikipedia.org/.
13. *Inside WIPO*, www.wipo.int/about-wipo/en/.
14. "When We Share, Everyone Wins." *Creative Commons*, creativecommons.org/.
15. "HTML 5.3." Edited by Patricia Aas, et al., *W3C*, 18 Oct. 2018, www.w3.org/TR/html53/.
16. Jaffe, Jeff, et al. "W3C And WHATWG to Work Together to Advance the Open Web Platform." *W3C*, 28 May 2019, www.w3.org/blog/2019/05/w3c-and-whatwg-to-work- together-to-advance-the-open-web-platform/.
17. "HTML." *HTML Standard*, WHATWG, 1 Apr. 2020, html.spec.whatwg.org/.
18. "UTF-8, UTF-16, UTF-32 & BOM." *[Unicode]*, The Unicode Consortium, 24 Jan. 2020, www.unicode.org/faq/utf_bom.html#UTF8.

HTML Basics

In the previous chapter, you created your first web page using HTML5. You coded a web page and tested it in a browser. You used a Document Type Definition to identify the version of HTML being used along with the `<html>`, `<head>`, `<title>`, `<meta>`, and `<body>` tags. In this chapter, you will continue your study of HTML and configure the structure and formatting of text on a web page with HTML elements. You're also ready to explore hyperlinks, which make the World Wide Web into a web of interconnected information. In this chapter, you will configure the anchor element to connect web pages with hyperlinks. As you read this chapter, be sure to work through the examples. Coding a web page is a skill, and every skill improves with practice.

You'll learn how to...

- Configure the body of a web page with headings, paragraphs, divs, lists, and blockquotes
- Configure special entity characters, line breaks, and horizontal rules
- Configure text with phrase elements
- Test a web page for valid syntax
- Configure a web page with HTML5 structural elements: header, nav, main, footer, section, aside, and article
- Use the anchor element to link from page to page
- Configure absolute, relative, and e-mail hyperlinks

Heading Element

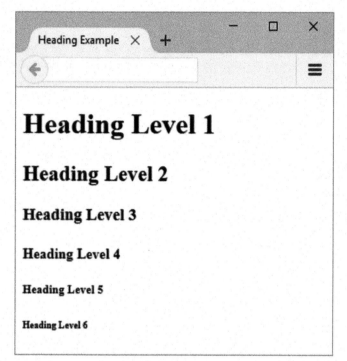

FIGURE 2.1 *Sample heading.html.*

Heading elements are organized into six levels: h1 through h6. The text within a heading element is rendered as a "block" of text by the browser (referred to as **block display**) and appears with empty space (sometimes called "white space" or "negative space") above and below. The size of the text is largest for **<h1>** (called a heading 1 tag) and smallest for **<h6>** (called a heading 6 tag). Depending on the font being used, the text within <h4>, <h5>, and <h6> tags may be displayed smaller than the default text size. All text within heading tags is displayed with bold font weight.

Figure 2.1 shows a web page document with six levels of headings.

 Hands-On Practice 2.1

To create the web page shown in Figure 2.1, launch a text editor and open the template.html file from the chapter2 folder in the student files. Modify the title element and add heading tags to the body section as indicated by the following code:

```
<!DOCTYPE html>
<html lang="en">
<head>
<title>Heading Example</title>
<meta charset="utf-8">
</head>
<body>
<h1>Heading Level 1</h1>
<h2>Heading Level 2</h2>
<h3>Heading Level 3</h3>
<h4>Heading Level 4</h4>
<h5>Heading Level 5</h5>
<h6>Heading Level 6</h6>
</body>
</html>
```

Save the document as heading.html on your hard drive or flash drive. Launch a browser such as Edge or Firefox to test your page. It should look similar to the page shown in Figure 2.1. You can compare your work with the solution found in the student files (chapter2/heading.html).

? FAQ Why doesn't the heading tag go in the head section?

It's common for students to try to code the heading tags in the head section of the document, but someone doing this won't be happy with the way the browser displays the web page. Even though "heading tag" and "head section" sound similar, always code heading tags in the body section of the web page document.

Accessibility and Headings

Heading tags can help to make your pages more accessible and usable. It is good coding practice to use heading tags to outline the structure of your web page content. To indicate areas within a page hierarchically, code heading tags numerically as appropriate (h1, h2, h3, and so on), and include page content in block display elements such as paragraphs

and lists. In Figure 2.2, the `<h1>` tag contains the name of the website in the logo header area at the top of the web page, the `<h2>` tag contains the major topic or name of the page in the content area, and other heading elements are coded in the content area as needed to identify subtopics.

Visually challenged visitors who are using a screen reader can configure the software to display a list of the headings used on a page in order to focus on the topics that interest them. Your well-organized page will be more usable for every visitor to your site, including those who are visually challenged.

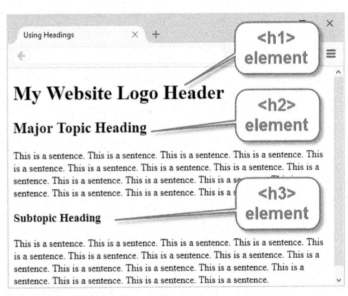

FIGURE 2.2 *Heading tags outline the page.*

More Heading Options in HTML5

You may have heard about the HTML5 header element. The header element offers additional options for configuring headings and typically contains an h1 element. We'll introduce the header element later in this chapter.

Paragraph Element

A **paragraph element** groups sentences and sections of text together. Text within **\<p\>** and **\</p\>** tags is rendered as block display with empty space above and below.

Figure 2.3 shows a web page document containing a paragraph after the first heading.

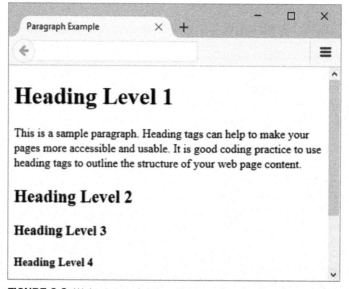

FIGURE 2.3 *Web page using headings and a paragraph.*

 Hands-On Practice 2.2 ——————————————————————

To create the web page shown in Figure 2.3, launch a text editor and open the heading.html file from the chapter2 folder in the student files. Modify the page title and add a paragraph of text to your page below the line with the \<h1\> tags and above the line with the \<h2\> tags:

```
<!DOCTYPE html>
<html lang="en">
<head>
<title>Paragraph Example</title>
<meta charset="utf-8">
</head>
<body>
<h1>Heading Level 1</h1>
<p>This is a sample paragraph. Heading tags can help to make your
pages more accessible and usable. It is good coding practice to use
heading tags to outline the structure of your web page content.
</p>
<h2>Heading Level 2</h2>
<h3>Heading Level 3</h3>
<h4>Heading Level 4</h4>
<h5>Heading Level 5</h5>
<h6>Heading Level 6</h6>
</body>
</html>
```

Save the document as paragraph.html on your hard drive or flash drive. Launch a browser to test your page. It should look similar to the page shown in Figure 2.3. You can compare your work with the solution found in the student files (chapter2/paragraph.html). Notice how the text in the paragraph wraps automatically as you resize your browser window.

Alignment

As you test your web pages, you may notice that the headings and text begin near the left margin. This is called **left alignment**, and it is the default alignment for web pages. There are times when you want a paragraph or heading to be centered or right aligned (justified). In previous versions of HTML, the align attribute can be used for this. However, the align attribute is **obsolete** in HTML5, which means that the attribute has been removed from the W3C HTML5 specification. You'll learn techniques to configure alignment with Cascading Style Sheets (CSS) in Chapters 6 through 8.

Quick TIP When writing for the Web, avoid long paragraphs. People tend to skim web pages rather than reading them word for word. Use heading tags to outline the page content along with short paragraphs (about three to five sentences each) and lists (which you'll learn about later in this chapter).

Line Break and Horizontal Rule

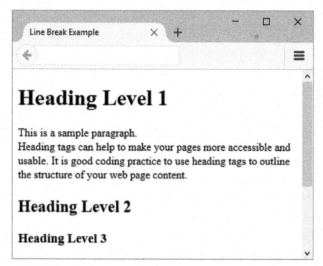

FIGURE 2.4 *Notice the line break after the first sentence.*

The Line Break Element

The **line break element** causes the browser to advance to the next line before displaying the next element or text on a web page. The line break tag is not coded as a pair of opening and closing tags. It is a void element and is coded as **
**. Figure 2.4 shows a web page document with a line break after the first sentence in the paragraph.

 Hands-On Practice 2.3

To create the web page shown in Figure 2.4, launch a text editor and open the paragraph.html file from the chapter2 folder in the student files. Modify the text between the title tags to be "Line Break Example." Place your cursor after the first sentence in the paragraph (after "This is a sample paragraph."). Press the Enter key. Save your file. Test your page in a browser and notice that even though your source code displayed the "This is a sample paragraph." sentence on its own line, the browser did not render it that way. A line break tag is needed to configure the browser to display the second sentence on a new line. Edit the file in a text editor and add a <br> tag after the first sentence in the paragraph as shown in the following code snippet:

```
<body>
<h1>Heading Level 1</h1>
<p>This is a sample paragraph. <br> Heading tags can help to make your
pages more accessible and usable. It is good coding practice to use
heading tags to outline the structure of your web page content.
</p>
<h2>Heading Level 2</h2>
<h3>Heading Level 3</h3>
<h4>Heading Level 4</h4>
<h5>Heading Level 5</h5>
<h6>Heading Level 6</h6>
</body>
```

Save your file as linebreak.html. Launch a browser to test your page. It should look similar to the page shown in Figure 2.4. You can compare your work with the solution found in the student files (chapter2/linebreak.html).

The Horizontal Rule Element

Web designers often use visual elements such as lines and borders to separate or define areas on web pages. The **horizontal rule** element, `<hr>`, configures a horizontal line across a web page. Since the horizontal rule element does not contain any text, it is coded as a void element and not in a pair of opening and closing tags. The horizontal rule element has an additional purpose in HTML5, it can be used to indicate a thematic break or change in the content. Figure 2.5 shows a web page document (also found in the student files at chapter2/hr.html) with a horizontal rule after the paragraph. In Chapter 6, you'll learn how to configure lines and borders on web page elements with Cascading Style Sheets (CSS).

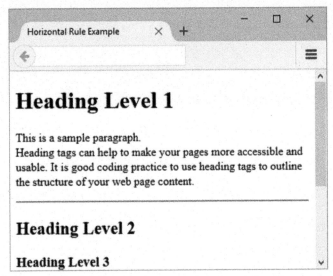

FIGURE 2.5 *The horizontal line is below the paragraph.*

 Hands-On Practice 2.4 ─────────────────

To create the web page shown in Figure 2.5, launch a text editor and open the linebreak.html file from the chapter2 folder in the student files. Modify the text between the title tags to be: Horizontal Rule Example. Place your cursor on a new line after the `</p>` tag. Code the `<hr>` tag on the new line as shown in the following code snippet:

```
<body>
<h1>Heading Level 1</h1>
<p>This is a sample paragraph. <br> Heading tags can help to make your
pages more accessible and usable. It is good coding practice to use
heading tags to outline the structure of your web page content.
</p>
<hr>
<h2>Heading Level 2</h2>
<h3>Heading Level 3</h3>
<h4>Heading Level 4</h4>
<h5>Heading Level 5</h5>
<h6>Heading Level 6</h6>
</body>
```

Save your file as hr.html. Launch a browser to test your page. It should look similar to the page shown in Figure 2.5. You can compare your work with the solution found in the student files (chapter2/hr.html).

 When you are tempted to use a horizontal rule on a web page, consider whether it is really needed. Usually, just leaving extra blank space on the page will serve to separate the content.

Blockquote Element

Besides organizing text in paragraphs and headings, sometimes you need to add a quotation to a web page. The **blockquote element** is used to display a block of quoted text in a special way—indented from both the left and right margins. A block of indented text begins with a **`<blockquote>`** tag and ends with a **`</blockquote>`** tag.

Figure 2.6 shows a web page document with a heading, a paragraph, and a blockquote.

FIGURE 2.6 *The text within the blockquote element is indented.*

> **Quick TIP**
>
> You've probably noticed how convenient the `<blockquote>` tag could be if you needed to indent an area of text on a web page. You may have wondered whether it would be OK to use `<blockquote>` anytime you'd like to indent text or whether the blockquote element is reserved only for long quotations. The semantically correct usage of the `<blockquote>` tag is to use it only when displaying large blocks of quoted text within a web page. Why should you be concerned about semantics? Consider the future of the Semantic Web, described in *Scientific American* as "A new form of Web content that is meaningful to computers will unleash a revolution of new possibilities." Using HTML in a semantic, structural manner is one step toward the Semantic Web. So, avoid using a `<blockquote>` just to indent text. You'll learn modern techniques to configure margins and padding on elements later in this book.

Hands-On Practice 2.5

To create the web page shown in Figure 2.6, launch a text editor and open the template.html file from the chapter2 folder in the student files. Modify the text in the title element. Add a heading tag, a paragraph tag, and a blockquote tag to the body section as indicated by the following code:

```
<!DOCTYPE html>
<html lang="en">
<head>
<title>Blockquote Example</title>
<meta charset="utf-8">
</head>
<body>
<h1>The Power of the Web</h1>
<p>According to Tim Berners-Lee, the inventor of the World Wide Web,
at https://www.w3.org/Press/IPO-announce:
</p>
<blockquote>
The power of the Web is in its universality. Access by everyone
regardless of disability is an essential aspect.
</blockquote>
</body>
</html>
```

Save the document as blockquote.html on your hard drive or flash drive. Launch a browser such as Edge or Firefox to test your page. It should look similar to the page shown in Figure 2.6. You can compare your work with the solution found in the student files (chapter2/blockquote.html).

 Why does my web page still look the same?

Often, students make changes to a web page but get frustrated because their browser shows an older version of the page. The following troubleshooting tips are helpful when you know you have modified your web page, but the changes do not show up in the browser:

1. Be sure to save your web page file after you make changes.

2. Verify the location where you save your file—the hard drive, a particular folder.

3. Verify the location from where your browser is requesting the file—the hard drive, a particular folder.

4. Be sure to click the web browser Refresh or Reload button.

Phrase Element

A **phrase element** indicates the context and meaning of the text between the container tags. It is up to each browser to interpret that style. Phrase elements are displayed directly in line with the text (referred to as inline display) and can apply to either a section of text or even a single character of text. For example, the **<strong>** element indicates that the text associated with it has strong importance and should be displayed in a "strong" manner in relation to normal text on the page.

Table 2.1 lists common phrase elements and examples of their use. Notice that some tags, such as <cite> and <dfn>, result in the same type of display (italics) as the <em> tag in today's browsers. These tags semantically describe the text as a citation or definition, but the physical display is usually italics in both cases.

TABLE 2.1   *Phrase Elements*

Element	Example	Usage
<abbr>	WIPO	Identifies text as an abbreviation
<b>	**bold** text	Has no extra importance but is styled in bold font
<cite>	*cite* text	Identifies a citation or reference; usually displayed in italics
<code>	code text	Identifies program code samples; usually a fixed-space font
<dfn>	*dfn* text	Identifies a definition of a word or term; usually displayed in italics
<em>	*emphasized* text	Causes text to be emphasized; usually displayed in italics
<i>	*italicized* text	Has no extra importance but is styled in italics
<kbd>	kbd text	Identifies keyboard input; usually has a fixed-space font
<mark>	mark text	Highlights text in order to be easily referenced
<q>	"quoted" text	Indicates a short quote; usually displays quotation marks
<samp>	samp text	Shows program sample output; usually a fixed-space font
<small>	small text	Defines a smaller text
<strong>	**strong** text	Strong importance; usually displayed in bold
<sub>	$_{sub}$text	Displays a subscript as small text below the baseline
<sup>	^{sup}text	Displays a superscript as small text above the baseline
<var>	*var* text	Identifies and displays a variable output; usually displayed in italics

Note that all phrase elements are container tags—both an opening and a closing tag is used. As shown in Table 2.1, the `<strong>` element indicates that the text associated with it has "strong" importance. Usually the browser (or other user agent) will display `<strong>` text in bold type. A screen reader, such as JAWS or Window-Eyes, might interpret `<strong>` text to indicate that the text should be more strongly spoken. In the following line, the phone number is displayed with strong importance:

Call for a free quote for your web development needs: **888.555.5555**

The code follows:

```
<p>Call for a free quote for your web development needs:
<strong>888.555.5555</strong></p>
```

Notice that the opening `<strong>` and closing `<strong>` tags are within the paragraph tags (`<p>` and `</p>`). This code is properly nested and is considered to be **well-formed**. When improperly nested, the `<p>` and `<strong>` tag pairs overlap each other instead of being nested within each other. Improperly nested code will not pass validation testing (see the HTML Syntax Validation section later in this chapter) and may cause display issues.

Figure 2.7 shows a web page document (also found in the student files at chapter2/em.html) that uses the `<em>` tag to display the emphasized phrase, "Access by everyone," in italics.

FIGURE 2.7 *The `<em>` tag in action.*

The code snippet follows:

```
<blockquote>
The power of the Web is in its universality.
<em>Access by everyone</em> regardless of disability is an essential
aspect.
</blockquote>
```

Ordered List

Lists are used on web pages to organize information. When writing for the Web, headings, short paragraphs, and lists can make your page more clear and easy to read. HTML can be used to create three types of lists—description lists, ordered lists, and unordered lists. All lists are rendered as block display with empty space above and below. This section focuses on the **ordered list**, which displays a numbering or lettering system to sequence the information in the list. An ordered list can be organized using numerals (the default), uppercase letters, lowercase letters, uppercase Roman numerals, and lowercase Roman numerals. See Figure 2.8 for a sample ordered list.

My Favorite Colors

1. Blue
2. Teal
3. Red

FIGURE 2.8 *Sample ordered list.*

Ordered lists begin with an **<ol>** tag and end with an **</ol>** tag. Each list item begins with an **<li>** tag and ends with an **</li>** tag. The code to configure the heading and ordered list shown in Figure 2.8 follows:

```
<h1>My Favorite Colors</h1>
<ol>
  <li>Blue</li>
  <li>Teal</li>
  <li>Red</li>
</ol>
```

The `type`, `start`, and `reversed` Attributes

The **type attribute** configures the symbol used for ordering the list. For example, to create an ordered list organized by uppercase letters, use `<ol type="A">`. Table 2.2 documents the type attribute and its values for ordered lists.

TABLE 2.2 *The type Attribute for Ordered Lists*

Value	Symbol
1	Numerals (the default)
A	Uppercase letters
a	Lowercase letters
I	Roman numerals
i	Lowercase Roman numerals

Another handy attribute that can be used with ordered lists is the **start attribute**, with which you can specify the start value for the list (for example, `start="10"`). Use the **reversed attribute** (`reversed="reversed"`) to indicate that a list is in descending order.

Hands-On Practice 2.6 ————————————————

In this Hands-On Practice, you will use a heading and an ordered list on the same page. To create the web page shown in Figure 2.9, launch a text editor and open the template.html file from the chapter2 folder in the student files. Modify the title element and add h1, h2, ol, and li tags to the body section, as indicated by the following code:

```
<!DOCTYPE html>
<html lang="en">
<head>
<title>Heading and List</title>
<meta charset="utf-8">
</head>
<body>
<h1>My Favorite Colors</h1>
<ol>
  <li>Blue</li>
  <li>Teal</li>
  <li>Red</li>
</ol>
</body>
</html>
```

FIGURE 2.9 *An ordered list.*

Save your file as ol.html. Launch a browser and test your page. It should look similar to the page shown in Figure 2.9. You can compare your work with the solution in the student files (chapter2/ol.html).

Take a few minutes to experiment with the `type` attribute. Configure the ordered list to use uppercase letters instead of numerals. Save your file as ola.html. Test your page in a browser. You can compare your work with the solution in the student files (chapter2/ola.html).

? FAQ Why is the web page code in the examples indented?

Actually, it doesn't matter to the browser if web page code is indented, but humans find it easier to read and maintain code when it is logically indented. This makes it easier for you or another web developer to understand the source code in the future. For example, it's common practice to indent <li> tags a few spaces in from the left margin because it makes it easier to "see" the list with a quick glance at the source code. There is no "rule" as to how many spaces to indent. However, your instructor or the organization you work for may have a standard for you to follow. Web pages with consistent indentation are easier to maintain.

Unordered List

An **unordered list** displays a bullet, or list marker, before each list entry. The default list marker is determined by the browser but is typically a disc, which is a filled-in circle. See Figure 2.10 for a sample unordered list.

My Favorite Colors

- Blue
- Teal
- Red

FIGURE 2.10 *Sample unordered list.*

Unordered lists begin with a **<ul>** tag and end with a **</ul>** tag. The ul element is a block display element and is rendered with empty space above and below. Each list item begins with an <li> tag and ends with an </li> tag. The code to configure the heading and unordered list shown in Figure 2.10 is as follows:

```
<h1>My Favorite Colors</h1>
<ul>
  <li>Blue</li>
  <li>Teal</li>
  <li>Red</li>
</ul>
```

? FAQ **Can I change the "bullet" in an unordered list?**

Back in the day before HTML5, the type attribute could be included with a <ul> tag to change the default list marker to a square (type="square") or open circle (type="circle"). However, be aware that using the type attribute on an unordered list is considered obsolete in HTML5 because it is decorative and does not convey meaning. No worries, though—there are CSS techniques to configure list markers (bullets) to display images and shapes.

Hands-On Practice 2.7 ──────────────────────────────

In this Hands-On Practice, you will use a heading and an unordered list on the same page. To create the web page shown in Figure 2.11, launch a text editor and open the template.html file from the chapter2 folder in the student files. Modify the title element and add h1, ul, and li tags to the body section as indicated by the following code:

FIGURE 2.11 *An unordered list.*

```
<!DOCTYPE html>
<html lang="en">
<head>
<title>Heading and List</title>
<meta charset="utf-8">
</head>
<body>
<h1>My Favorite Colors</h1>
<ul>
  <li>Blue</li>
  <li>Teal</li>
  <li>Red</li>
</ul>
</body>
</html>
```

Save your file as ul.html. Launch a browser and test your page. It should look similar to the page shown in Figure 2.11. You can compare your work with the solution in the student files (chapter2/ul.html).

───

Description List

A **description list** can be used to organize terms and their descriptions. The terms stand out and their descriptions can be as long as needed to convey your message. Each term begins on its own line at the margin. Each description begins on its own line and is indented. Description lists are also handy for organizing Frequently Asked Questions (FAQs) and their answers. The questions and answers are offset with indentation. Any type of information that consists of a number of corresponding terms and associated descriptions is well suited to being organized in a description list. See Figure 2.12 for an example of a web page that uses a description list.

Description lists begin with the **<dl>** tag and end with the **</dl>** tag. Each term or name in the list begins with the **<dt>** tag and ends with the **</dt>** tag. Each term description begins with the **<dd>** tag and ends with the **</dd>** tag.

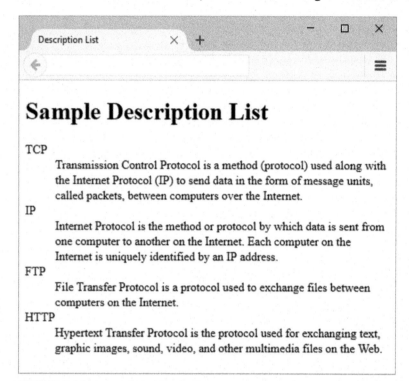

FIGURE 2.12 *A description list.*

 Hands-On Practice 2.8

In this Hands-On Practice, you will use a heading and a description list on the same page. To create the web page shown in Figure 2.12, launch a text editor and open the template.html file from the chapter2 folder in the student files. Modify the title element and add h1, dl, dd, and dt tags to the body section as indicated by the following code:

```
<!DOCTYPE html>
<html lang="en">
<head>
<title>Description List</title>
<meta charset="utf-8">
</head>
<body>
<h1>Sample Description List</h1>
<dl>
  <dt>TCP</dt>
    <dd>Transmission Control Protocol is a method (protocol) used along
with the Internet Protocol (IP) to send data in the form of message
units, called packets, between computers over the Internet.</dd>
  <dt>IP</dt>
    <dd>Internet Protocol is the method or protocol by which data is
sent from one computer to another on the Internet. Each computer on
the Internet is uniquely identified by an IP address.</dd>
  <dt>FTP</dt>
    <dd>File Transfer Protocol is a protocol used to exchange files
between computers on the Internet.</dd>
  <dt>HTTP</dt>
    <dd>Hypertext Transfer Protocol is the protocol used for exchanging
text, graphic images, sound, video, and other multimedia files on the
Web.</dd>
</dl>
</body>
</html>
```

Save your file as description.html. Launch a browser and test your page. It should look similar to the page shown in Figure 2.12. Don't worry if the word wrap is a little different—the important formatting is that each <dt> term should be on its own line and the corresponding <dd> description should be indented under it. Try to resize your browser window and notice how the word wrap on the description text changes. You can compare your work with the solution in the student files (chapter2/description.html).

FAQ Why does the text in my web page wrap differently than the examples?

The text may wrap a little differently because your screen resolution or browser viewport size may not be the same as those on the computer used for the screen captures. That's part of the nature of working with the Web—expect your web pages to look slightly different in the multitude of screen resolutions, browser viewport sizes, and devices that people will use to view your designs.

Special Entity Characters

In order to use special characters such as quotation marks, the greater than sign (>), the less than sign (<), and the copyright symbol (©) in your web page document, you need to use special characters, sometimes called entity characters.[1] For example, if you want to include a copyright line on your page as follows:

© Copyright 2022 My Company. All rights reserved.

you need to use the special character **©** to display the copyright symbol, as shown below:

```
&copy; Copyright 2022 My Company. All rights reserved.
```

Another useful special character is ** **, which stands for nonbreaking space. You may have noticed that web browsers treat multiple spaces as a single space. If you want multiple spaces to be displayed in your text, you can use   multiple times to indicate multiple blank spaces. This is acceptable if you simply need to tweak the position of an element a little. If you find that your web pages contain many   special characters in a row, you should use a different method to align elements, such as configuring the margin or padding with Cascading Style Sheets (see Chapter 6).

See Table 2.3 for a description of commonly used special characters and their codes.

TABLE 2.3 *Common Special Characters*

Character	Entity Name	Code
"	Quotation mark	&quot;
©	Copyright symbol	&copy;
&	Ampersand	&amp;
Empty space	Nonbreaking space	
'	Apostrophe	&apos;
—	Long dash	&mdash;
|	Vertical bar	|
<	Less than sign	&lt;
>	Greater than sign	&gt;

Hands-On Practice 2.9 ─────────────────────────────

Figure 2.13 shows the web page you will create in this Hands-On Practice. Launch a text editor and open the template.html file from the chapter2 folder in the student files.

Change the title of the web page to "Web Design Steps" by modifying the text between the `<title>` and `</title>` tags.

The sample page shown in Figure 2.13 contains a heading, an unordered list, and the copyright information. You will add these elements to your file next.

Configure the phrase, "Web Design Steps", as a level 1 heading (`<h1>`) as follows:

```
<h1>Web Design Steps</h1>
```

Now create the unordered list. The first line of each bulleted item is the title of the web design step. In the sample, each step title should be strong or stand out from the rest of the text. The code for the beginning of the unordered list follows:

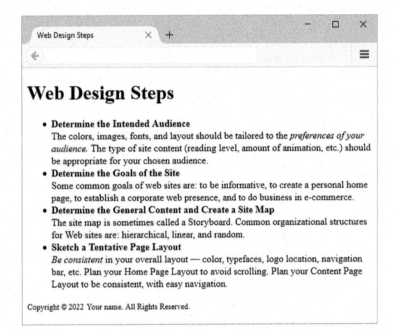

FIGURE 2.13 *Sample design.html.*

```
<ul>
  <li><strong>Determine the
  Intended Audience</strong>
  <br> The colors, images, fonts, and layout should be tailored to
  the <em> preferences of your audience.</em> The type of site content
  (reading level, amount of animation, etc.) should be appropriate for
  your chosen audience.</li>
```

Edit your design.html file and code the entire unordered list shown in Figure 2.13. Remember to code the closing `</ul>` tag at the end of the list. Finally, configure the copyright information in a paragraph and apply the small element. Use the special character `&copy;` for the copyright symbol. The code for the copyright line follows:

```
<p><small>Copyright &copy; 2022 Your name. All Rights
Reserved.</small></p>
```

Save your file as design.html. Launch a browser and test your page. How did you do? Compare your work to the sample in the student files (chapter2/design.html).

───

HTML Syntax Validation

VideoNote
HTML Validation

The W3C has a free Markup Validation Service available at http://validator.w3.org that will check your code for syntax errors and validate your web pages. **HTML validation** provides you with quick self-assessment—you can prove that your code uses correct syntax. In the working world, HTML validation serves as a quality assurance tool. Invalid code may cause browsers to render the pages slower than otherwise.

 Hands-On Practice 2.10

In this Hands-On Practice, you will use the W3C Markup Validation Service to validate a web page file. Launch a text editor and open the design.html file from the chapter2 folder in the student files.

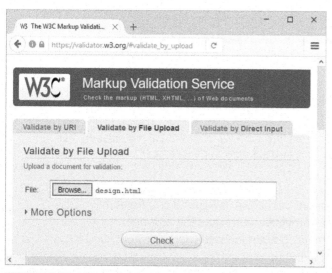

FIGURE 2.14 *Validate your page.*
Screenshots of W3C. Courtesy of W3C (World Wide Web Consortium)

1. We will add an error to the design.html page. Delete the first closing `</strong>` tag. This modification should generate several error messages.

2. Next, attempt to validate the design.html file. Launch a browser and visit the W3C Markup Validation Service file upload page at http://validator.w3.org and select the "Validate by File Upload" tab. Click the Browse button and select the chapter2/design.html file from your computer. Click the Check button to upload the file to the W3C site (Figure 2.14).

3. A results page will be displayed. Scroll down the page to view the errors, as shown in Figure 2.15.

4. Notice that the message indicates line 12, which is the first line after the missing closing `</strong>` tag. HTML error messages often point to a line that follows the error. The text of the message "End tag for li seen, but there were open elements" lets you know that something is wrong. It's up to you to figure out what it is. A good place to start is to check your container tags and make sure they are in pairs. In this case, that is the problem. You can scroll down to view the other errors. However, since multiple error messages are often displayed after a single error occurs, it's a good idea to fix one item at a time and then revalidate.

5. Edit the design.html file in a text editor and add the missing `</strong>` tag. Save the file. Launch a browser and visit http://validator.w3.org and select the "Validate by File Upload" tab. Click the Browse button and select your file. Click the Check button.

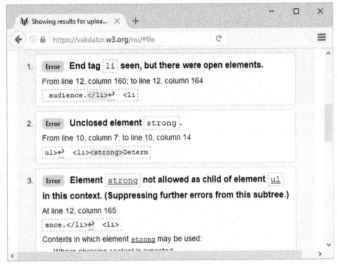

FIGURE 2.15 *The error indicates line 12. Screenshots of W3C. Courtesy of W3C (World Wide Web Consortium)*

6. Your display should be similar to that shown in Figure 2.16. Notice the message, "Document checking completed. No errors or warnings to show." This means that your page passed the validation test. Congratulations, your web page is valid!

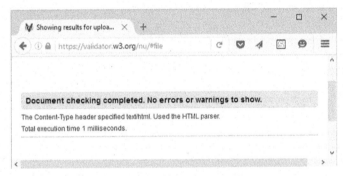

FIGURE 2.16 *The page has passed the validation test. Screenshots of W3C. Courtesy of W3C (World Wide Web Consortium)*

It's a good practice to validate your web pages. However, when validating code, use common sense. Since web browsers still do not completely follow W3C recommendations, there will be situations, such as when adding multimedia to a web page, when HTML code configured to work reliably across a variety of browsers and platforms will not pass validation.

In addition to the W3C validation service, there are other tools that you can use to check the syntax of your code. Try out the validators at the following websites:

* Validator.nu: https://html5.validator.nu

* Freeformatter.com: https://www.freeformatter.com/html-validator.html

Structural
Elements

HTML5 introduces a number of semantic structural elements that can be used along with the generic div element to configure specific areas on a web page. These new HTML5 header, nav, main, and footer elements are intended to be used in conjunction with div and other elements to structure web page documents in a more meaningful manner that indicates the purpose of each structural area. Figure 2.17 shows a diagram of a page (called a **wireframe**) that indicates how the structure of a web page could be configured with the header, nav, main, div, and footer elements.

header
nav

main		
div	div	div

footer

FIGURE 2.17 *Structural elements.*

The Div Element

The **div element** has been used for many years to configure a generic structural area or "division" on a web page as a block display with empty space above and below. A div element begins with a **`<div>`** tag and ends with a **`</div>`** tag. Use a div element when you need to format an area of a web page that may contain other block display elements such as headings, paragraphs, unordered lists, and even other div elements. You'll use Cascading Style Sheets (CSS) later in this book to style and configure the color, font, and layout of HTML elements.

The Header Element

The purpose of the HTML5 **header element** is to contain the headings of either a web page document or an area within the document such as a section or an article (more on the section element and article element in Chapter 8). The header element begins with the **`<header>`** tag and ends with the **`</header>`** tag. The header element is block display and typically contains one or more heading level elements (h1 through h6).

The Nav Element

The purpose of the HTML5 **nav element** is to contain a section of navigation links. The block display nav element begins with the **`<nav>`** tag and ends with the **`</nav>`** tag.

The Main Element

The purpose of the HTML5 **main** element is to contain the main content of a web page document. The block display main element begins with the **`<main>`** tag and ends with the **`</main>`** tag.

The Footer Element

The purpose of the HTML5 **footer element** is to contain the footer content of a web page or section of a web page. The block display footer element begins with the **<footer>** tag and ends with the **</footer>** tag.

 Hands-On Practice 2.11 ——————————————

In this Hands-On Practice, you will use structural elements as you create the Trillium Media Design home page, shown in Figure 2.18. Launch a text editor, and open the template.html file from the chapter2 folder in the student files. Edit the code as follows:

FIGURE 2.18 *Trillium home page.*

1. Modify the title of the web page by changing the text between the <title> and </title> tags to Trillium Media Design.

2. Position your cursor in the body section and code the header element with the text, "Trillium Media Design" in an h1 element:

```
<header>
  <h1>Trillium Media Design</h1>
</header>
```

3. Code a nav element to contain text that will indicate the main navigation for the website. Configure bold text (use the b element) and use the   special character to add extra blank space:

```
<nav>
  <b>Home   Services   Contact</b>
</nav>
```

4. Code the content within a main element that contains the h2 and paragraph elements:

```
<main>
  <h2>New Media and Web Design</h2>
  <p>Trillium Media Design offers a comprehensive range of
services to take your company&apos;s Web presence to the next
level.</p>
  <h2>Meeting Your Business Needs</h2>
  <p>Our expert designers will listen to you as they create a website
that helps to promote and grow your business.</p>
</main>
```

5. Configure the footer element to contain a copyright notice displayed in small font size (use the small element) and italic font (use the i element). Be careful to properly nest the elements as shown here:

```
<footer>
  <small><i>Copyright &copy; 2022 Your Name Here</i></small>
</footer>
```

Save your page as structure.html. Test your page in a browser. It should look similar to Figure 2.18. You can compare your work to the sample in the student files (chapter2/structure.html).

Practice with Structural Elements

Coding HTML is a skill and skills are best learned by practice. You'll get more practice coding a web page using structural elements in this section.

 Hands-On Practice 2.12

| header |
| nav |
| main |
| div |
| footer |

FIGURE 2.19 *Wireframe for Casita Sedona.*

In this Hands-On Practice, you will use the wireframe shown in Figure 2.19 as a guide as you create the Casita Sedona Bed & Breakfast web page, shown in Figure 2.20.

Launch a text editor and open the template.html file from the chapter2 folder in the student files. Edit the code as follows:

1. Modify the title of the web page by changing the text between the `<title>` and `</title>` tags to Casita Sedona.

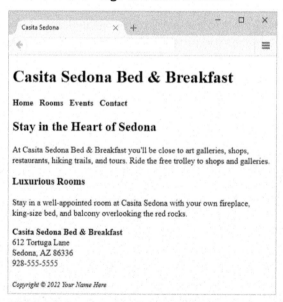

FIGURE 2.20 *Casita Sedona web page.*

2. Position your cursor in the body section and code the header element with the text, "Casita Sedona Bed & Breakfast" in an h1 element. Be sure to use the special character `&amp;` for the ampersand:

```
<header>
  <h1>
  Casita Sedona Bed &amp; Breakfast
  </h1>
</header>
```

3. Code a nav element to contain text that will indicate the main navigation for the website. Configure bold text (use the b element) and use the   special character to add extra blank space:

```
<nav>
  <b>
    Home  
    Rooms  
    Events  
    Contact
  </b>
</nav>
```

4. Code the content within a main element. Start with the h2 and paragraph elements:

```
<main>
<h2>Stay in the Heart of Sedona</h2>
  <p>At Casita Sedona Bed &amp; Breakfast you&apos;ll be close to
art galleries, shops, restaurants, hiking trails, and tours. Ride
the free trolley to shops and galleries.</p>
  <h3>Luxurious Rooms</h3>
  <p>Stay in a well-appointed room at Casita Sedona with your own
fireplace, king-size bed, and balcony overlooking the red rocks.</p>
</main>
```

5. Configure the company name, address, and phone number within a div element. Code the div element *within* the main element before the closing main tag. Use line break tags to display the name, address, and phone information on separate lines and to create extra empty space before the footer:

```
<div>
  <strong>Casita Sedona Bed &amp; Breakfast</strong><br>
  612 Tortuga Lane<br>
  Sedona, AZ 86336<br>
  928-555-5555<br>
</div>
```

6. Configure the footer element to contain a copyright notice displayed in small font size (use the small element) and italic font (use the i element). Be careful to properly nest the elements as shown here:

```
<footer>
  <small><i>Copyright &copy; 2022 Your Name Here</i></small>
</footer>
```

Save your page as casita.html. Test your page in a browser. It should look similar to Figure 2.20. You can compare your work to the sample in the student files (chapter2/casita.html).

More Structural Elements

You've just worked with the HTML5 header, nav, main, and footer elements. These HTML5 elements are used along with div and other elements to structure web page documents in a meaningful manner that defines the purpose of the structural areas. In this section, you'll explore more HTML5 structural elements.

The Section Element

The purpose of a **section element** is to indicate a "section" of a document, such as a chapter or topic. This block display element could contain header, footer, section, article, aside, figure, div, and other elements needed to configure the content.

The Article Element

The **article element** is intended to present an independent entry, such as a blog posting, comment, or e-zine article that could stand on its own. This block display element could contain header, footer, section, aside, figure, div, and other elements needed to configure the content.

The Aside Element

The **aside element** indicates a sidebar or other tangential content. This block display element could contain header, footer, section, aside, figure, div, and other elements needed to configure the content.

The Time Element

The **time element** represents a date or a time. The time element is not a structural element, but it is included here because it is useful to identity the date of content, such as an article on a web page or a blog entry. An optional `datetime` attribute can be used to specify a calendar date and/or time in machine-readable format.[2] Use YYYY-MM-DD for a date. Use a 24-hour clock and HH:MM for time.

Hands-On Practice 2.13

In this Hands-On Practice, you'll edit a web page document and apply the section, article, aside, and time elements to create the page with blog postings shown in Figure 2.21.

Launch a text editor and open the starter.html file from the chapter2 folder in the student files. Save the file as blog.html. Examine the source code.

FIGURE 2.21 *The blog page.*

1. Locate the title tag in the head section. Change the text within the title tags to "Lighthouse Bistro Blog."

2. Locate the opening main tag. Delete the HTML elements and text between the opening and closing main tags.

3. Code an aside element with the following content below the opening main tag. The HTML follows:

```
<aside>
  <p><i>Watch for the March
  Madness Wrap next month!</i></p>
</aside>
```

4. Code an opening section tag followed by an h2 element. The HTML follows:

```
<section>
<h2>Bistro Blog</h2>
```

5. Code two blog articles as shown below. Note the use of the header, h3, time, and paragraph elements. Also, code a closing section tag. The HTML follows:

```
<article>
  <header><h3>Valentine Wrap</h3></header>
  <time datetime="2022-02-01">February 1, 2022</time>
  <p>The February special sandwich is the Valentine Wrap &mdash;
  heart-healthy organic chicken with roasted red peppers on a
  whole wheat wrap.</p>
</article>
<article>
  <header><h3>New Coffee of the Day Promotion</h3></header>
  <time datetime="2022-01-12">January 12, 2022</time>
  <p>Enjoy the best coffee on the coast in the comfort of your
  home. We will feature a different flavor of our gourmet,
  locally roasted coffee each day with free bistro tastings and a
  discount on one-pound bags.</p>
</article>
</section>
```

Save your file. Display your blog.html page in a browser. It should look similar to the page shown in Figure 2.21. A sample solution is in the student files (chapter2/blog.html).

Anchor Element

Use the **a element** (commonly called the **anchor element**) to specify a **hyperlink,** often referred to as a *link*, to another web page or file that you want to display. Each anchor element begins with an **<a>** tag and ends with an **</a>** tag. The opening and closing anchor tags surround the text that the user can click to perform the hyperlink. Use the **href attribute** to configure the hyperlink reference, which identifies the name and location of the file to access.

Figure 2.22 shows a web page document with an anchor tag that configures a hyperlink to this book's website, http://webdevbasics.net.

The code for the anchor tag in Figure 2.22 is as follows:

```
<a href="http://webdevbasics.net">Basics of Web Design Textbook
Companion</a>
```

Notice that the href value is the URL for the website and will display the home page. The text that is typed between the two anchor tags displays on the web page as a hyperlink and is underlined by most browsers. When you move the mouse over a hyperlink, the cursor changes to a pointing hand, as shown in Figure 2.22.

 Hands-On Practice 2.14

To create the web page shown in Figure 2.22, launch a text editor and open the template.html file from the chapter2 folder in the student files. Modify the title element and add an anchor tag to the body section as indicated by the following code:

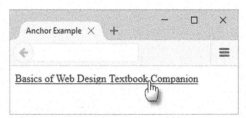

FIGURE 2.22 *Sample hyperlink.*

```
<!DOCTYPE html>
<html lang="en">
<head>
<title>Anchor Example</title>
<meta charset="utf-8">
</head>
<body>
<a href="http://webdevbasics.net">Basics of Web
Design Textbook Companion</a>
</body>
</html>
```

Save the document as anchor.html on your hard drive or flash drive. Launch a browser and test your page. It should look similar to the page shown in Figure 2.22. You can compare your work with the solution found in the student files (chapter2/anchor.html).

Can images be hyperlinks?

Yes. Although we'll concentrate on text hyperlinks in this chapter, it is also possible to configure an image as a hyperlink. You'll get practice with image links in Chapter 5.

Targeting Hyperlinks

You may have noticed in Hands-On Practice 2.14 that when a visitor clicks a hyperlink, the new web page automatically opens in the same browser window. You can configure the **target attribute** on an anchor tag with `target="_blank"` to open a hyperlink in a *new* browser window or browser tab. Note that you cannot control whether the web page opens in a new window or opens in a new tab—that is dependent on your visitor's browser configuration. To see the target attribute in action, try the example in the student files at chapter2/target.html.

Absolute Hyperlink

An **absolute hyperlink** indicates the absolute location of a resource on the Web. The hyperlink in Hands-On Practice 2.14 is an absolute hyperlink. Use an absolute hyperlink when you need to link to resources on other websites. The href value for an absolute hyperlink to the home page of a website includes the http://protocol and the domain name. The following hyperlink is an absolute hyperlink to the home page of this book's website:

```
<a href="http://webdevbasics.net">Basics of Web Design</a>
```

Note that if we want to access a web page other than the home page on the book's website, we can also include a specific folder name and file name. For example, the following anchor tag configures an absolute hyperlink for a file named chapter1.html located in a folder named 6e on this book's website:

```
<a href="http://webdevbasics.net/6e/chapter1.html">Chapter 1</a>
```

Relative Hyperlink

When you need to link to web pages within your site, use a **relative hyperlink**. The href value for a relative hyperlink does not begin with http:// and does not include a domain name. For a relative hyperlink, the href value will contain only the file name (or folder and file name) of the web page you want to display. The hyperlink location is relative to the page currently being displayed. For example, if you are coding a home page (index.html) for a website and want to link to a page named contact.html located in the same folder as index.html, you can configure a relative hyperlink as shown in the following code sample:

```
<a href="contact.html">Contact Us</a>
```

Block Anchor

It's typical to use anchor tags to configure phrases or even just a single word as a hyperlink. HTML5 provides a new function for the anchor tag—the block anchor. A block anchor can configure one or more entire elements (even those that display as a block, such as a div, h1, or paragraph) as a hyperlink. See an example in the student files (chapter2/block.html).

Accessibility and Hyperlinks

Visually challenged visitors who are using a screen reader can configure the software to display a list of the hyperlinks in the document. However, a list of hyperlinks is only useful if the text describing each hyperlink is actually helpful and descriptive. For example, on your college website, a "Search the course schedule" link would be more useful than a hyperlink that simply says "More information" or "click here." Keep this in mind as you are coding hyperlinks in your web pages.

Practice with Hyperlinks

The best way to learn how to code web pages is by actually doing it! In this section, you'll create three pages in a small website so that you can practice using the anchor tag to configure hyperlinks.

Site Map

Figure 2.23 displays the site map for your new website—a Home page with two content pages: a Services page and a Contact page.

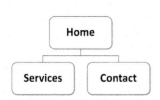

FIGURE 2.23 *Site map.*

A **site map** represents the structure, or organization, of pages in a website in a visual manner. Each page in the website is represented by a box on the site map. Review Figure 2.23 and notice that the Home page is at the top of the site map. The second level in a site map shows the other main pages of the website. In this very small three-page website, the other two pages (Services and Contact) are included on the second level. The main navigation of a website usually includes hyperlinks to the pages shown on the first two levels of the site map.

 Hands-On Practice 2.15

Figure 2.23 displays the site map for your new website—a home page (index.html) with two content pages: services page (services.html) and contact page (contact.html).

1. **Create a Folder.**   If you had printed papers to organize, you would probably store them in a paper folder. Web designers store and organize their computer files by creating a folder on a hard drive (or portable storage such as an SD card or a Flash drive) for each website. This helps them to be efficient as they work with many different

?FAQ   How do I create a new folder?
Before you begin to learn how to code web pages, it's a good idea to be comfortable using your computer for basic tasks such as creating a new folder. If you don't remember how to create a folder, review Hands-On Practice 1.1 for instructions.

websites. You will organize your own web design work by creating a new folder for each website and storing your files for that website in the new folder. Use your operating system to create a new folder named mypractice for your new website.

2. **Create the Home Page.** Use the Trillium Media Design web page (Figure 2.18) from Hands-On Practice 2.11 as a starting point for your new home page (shown in Figure 2.24). Copy the file from Hands-On Practice 2.11 (chapter2/structure.html) into your mypractice folder. Change the file name of structure.html to index.html. It's common practice to use the file name index.html for the home page of a website.

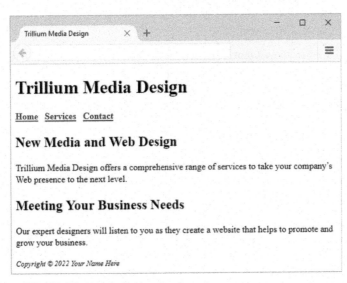

FIGURE 2.24 *New index.html web page.*

Launch a text editor and open the index.html file.

a. The navigation hyperlinks will be located within the nav element. You will edit the code within the nav element to configure three hyperlinks:

▶ The text "Home" will hyperlink to index.html

▶ The text "Services" will hyperlink to services.html

▶ The text "Contact" will hyperlink to contact.html

Modify the code within the nav element as given below:

```
<nav>
  <b>
    <a href="index.html">Home</a>  
    <a href="services.html">Services</a>  
    <a href="contact.html">Contact</a>
  </b>
</nav>
```

b. Save the index.html file in your mypractice folder. Test your page in a browser. It should look similar to Figure 2.24. You can compare your work to the sample in the student files (chapter2/2.15/index.html).

3. **Create the Services Page.** It is common practice to create a new web page based on an existing page. You will use the index.html file as a starting point for the new Services page, as shown in Figure 2.25.

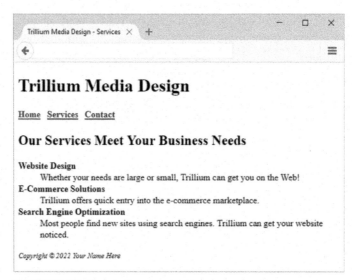

FIGURE 2.25 *The services.html web page.*

Open your index.html file in a text editor and save the file as services.html. Edit the code as indicated below:

a. Modify the title of the web page by changing the text between the `<title>` and `</title>` tags to "Trillium Media Design - Services". In order to create a consistent header, navigation, and footer for the web pages in this website, do not change the code within the header, nav, or footer elements.

b. Position your cursor in the body section and delete the code and text between the opening and closing main tags. Code the main page content (heading 2 and description list) for the Services page between the main tags as follows:

```
<h2> Our Services Meet Your Business Needs</h2>
  <dl>
    <dt><strong>Website Design</strong></dt>
      <dd>Whether your needs are large or small, Trillium can get
      you on the Web!</dd>
    <dt><strong>E-Commerce Solutions</strong></dt>
      <dd>Trillium offers quick entry into the e-commerce
      marketplace.</dd>
    <dt><strong>Search Engine Optimization</strong></dt>
      <dd>Most people find new sites using search engines.
      Trillium can get your website noticed.</dd>
  </dl>
```

c. Save the services.html file in your mypractice folder. Test your page in a browser. It should look similar to Figure 2.25. You can compare your work to the sample in the student files (chapter2/2.15/services.html).

4. **Create the Contact Page.** Use the index.html file as a starting point for the Contact page, as shown in Figure 2.26. Launch a text editor, open index.html, and save the file as contact.html. Edit the code as indicated below:

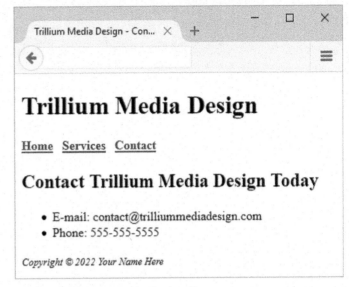

FIGURE 2.26 *The contact.html web page.*

   a. Modify the title of the web page by changing the text between the `<title>` and `</title>` tags to "Trillium Media Design - Contact". In order to create a consistent header, navigation, and footer for the web pages in this website, do not change the code within the header, nav, or footer elements.

   b. Position your cursor in the body section and delete the code and text between the opening main tag and the closing main tag. Code the main page content for the Contact page between the main tags:

```
<h2>Contact Trillium Media Design Today</h2>
   <ul>
     <li>E-mail: contact@trilliummediadesign.com</li>
     <li>Phone: 555-555-5555</li>
   </ul>
```

   c. Save the contact.html file in your mypractice folder. Test your page in a browser. It should look similar to Figure 2.26. Test your page by clicking each link. When you click the "Home" hyperlink, the index.html page should be displayed. When you click the "Services" hyperlink, the services.html page should be displayed. When you click the "Contact" hyperlink, the contact.html page should be displayed. You can compare your work to the sample in the student files (chapter2/2.15/contact.html).

 FAQ What if my relative hyperlink doesn't work?

Check the following:

▷ Did you save the files in the specified folder?

▷ Did you save the files with the names as requested? Use Windows Explorer or Finder (Mac users) to verify the actual names of the files you saved.

▷ Did you type the file names correctly in the anchor tag's href property? Check for typographical errors.

▷ When you place your mouse over a link, the file name of a relative link will be displayed in the status bar in the lower edge of the browser window. Verify that this is the correct file name. On many operating systems, such as UNIX or Linux, the use of uppercase and lowercase letters in file names matters—make sure that the file name and the reference to it are in the same case.

E-Mail Hyperlinks

The anchor tag can also be used to create e-mail hyperlinks. An **e-mail hyperlink** will automatically launch the default mail program configured for the browser. It is similar to an external hyperlink with the following two exceptions:

▶ It uses `mailto:` instead of `http://`.
▶ It launches the default e-mail application for the visitor's browser with your e-mail address as the recipient.

For example, to create an e-mail hyperlink to the e-mail address help@webdevbasics.net, code the following:

```
<a href="mailto:help@webdevbasics.net">help@webdevbasics.net</a>
```

It is good practice to place the e-mail address both on the web page and within the anchor tag. Not everyone has an e-mail program configured with his or her browser. By placing the e-mail address in both places, you increase usability for all of your visitors.

 Hands-On Practice 2.16 ——————————————————————————

In this Hands-On Practice, you will modify the Contact page (contact.html) of the website you created in Hands-On Practice 2.15 and configure an e-mail link in the page content area. Launch a text editor and open the contact.html file from your mypractice folder. This example uses the contact.html file found in the student files in the chapter2/2.15 folder.

FIGURE 2.27 *An e-mail hyperlink has been configured on the Contact page.*

Configure the e-mail address in the content area as an e-mail hyperlink as given below:

```
<li>E-mail:
<a href="mailto:contact@trilliummediadesign.com">contact@trilliummediadesign.com</a>
</li>
```

Save and test the page in a browser. The browser display should look similar to the page shown in Figure 2.27. Compare your work with the sample in the student files (chapter2/2.16/contact.html).

Quick TIP

Free web-based e-mail is offered by many providers, such as Google, Outlook, and so on. You can create one or more free e-mail accounts to use when communicating with new websites or signing up for free services, such as newsletters. This will help to organize your e-mail into those you need to access and respond to right away (such as school, work, or personal messages) and those you can get to at your convenience.

FAQ

Won't displaying my actual e-mail address on a web page increase spam?

Yes and no. While it's possible that some unethical spammers may harvest web pages for e-mail addresses, the chances are that your e-mail application's built-in spam filter will prevent your inbox from being flooded with messages. When you configure an easily readable e-mail hyperlink, you increase the usability of your website for your visitors in the following situations:

▶ The visitor may be at a public computer with no e-mail application configured. In this case, when the e-mail hyperlink is clicked, an error message may display, and the visitor will have difficulty contacting you using the e-mail link.

▶ The visitor may be at a private computer but may prefer not to use the e-mail application (and address) that is configured by default to work with the browser. Perhaps he or she shares the computer with others, or perhaps he or she wishes to preserve the privacy of the default e-mail address.

If you prominently displayed your actual e-mail address, in both of these situations, the visitor can still access your e-mail address and use it to contact you (in either their e-mail application or via a web-based e-mail system such as Google's Gmail). The result is a more usable website for your visitors.

Review and Apply

Review Questions

1. Which tag is used to hyperlink web pages to each other?

   a. `<br>` **b.** `<hyperlink>`

   c. `<a>` **d.** `<link>`

2. Which tag pair configures the largest heading?

   a. `<h1> </h1>`

   b. `<h9> </h9>`

   c. `<h type="largest"> </h>`

   d. `<h6> </h6>`

3. Which tag configures the following text or element to display on a new line?

   a. `<new line>` **b.** `<nl>`

   c. `<br>` **d.** `<line>`

4. Which tag pair configures a paragraph?

   a. `<para> </para>`

   b. `<paragraph> </paragraph>`

   c. `<p> </p>`

   d. `<body> </body>`

5. Which of the following is an HTML5 element used to indicate navigational content?

   a. nav **b.** header

   c. footer **d.** a

6. When should you code an absolute hyperlink?

   a. when linking to a web page that is internal to your website

   b. when linking to a web page that is external to your website

   c. always; the W3C prefers absolute hyperlinks

   d. never; absolute hyperlinks are obsolete

7. Which tag pair is the best choice to emphasize text with italic font on a web page?

   a. `<b> </b>`

   b. `<strong> </strong>`

   c. `<em> </em>`

   d. `<bold> </bold>`

8. Which tag configures a horizontal line on a web page?

   a. `<br>` **b.** `<hr>`

   c. `<line>` **d.** `<h1>`

9. Which type of HTML list will automatically number the items for you?

   a. numbered list **b.** ordered list

   c. unordered list **d.** description list

10. Which statement is true?

   a. The W3C Markup Validation Service describes how to fix the errors in your web page.

   b. The W3C Markup Validation Service lists syntax errors in a web page.

   c. The W3C Markup Validation Service is only available to W3C members.

   d. None of the above statements are true.

Review Answers

1. c 2. a 3. c 4. c 5. a 6. b 7. c 8. b 9. b 10. b

Hands-On Exercises

1. Write the markup language code to display your name in the largest-size heading element.
2. Write the markup language code for an unordered list to display the days of the week.

3. Write the markup language code for an ordered list that uses uppercase letters to order the items. This ordered list will display the following: Spring, Summer, Fall, and Winter.

4. Think of a favorite quote by someone you admire. Write the HTML code to display the person's name in a heading and the quote in a blockquote.

5. Modify the following code snippet to indicate that the bolded text has strong importance:

```
<p>A diagram of the organization of a website is called a <b>site
map</b> or <b>storyboard</b>. Creating the <b>site map</b> is one
of the initial steps in developing a website.</p>
```

6. Write the code to create an absolute hyperlink to your school's website.

7. Write the code to create a relative hyperlink to a web page named clients.html.

8. Create a web page about your favorite musical group. Include the name of the group, the members of the group, a hyperlink to the group's website, your favorite three (or fewer if the group is new) album releases, and a brief review of each album. Be sure to use the following elements: html, head, title, meta, body, header, footer, main, h1, h2, p, ul, li, and a. Configure your name in an e-mail link in the page footer area. Save the page as band.html. Open your file in a text editor and print the source code for the page. Display your page in a browser and print the page. Hand in both printouts to your instructor.

Focus on Web Design

Markup language code alone does not make a web page—design is very important. Access the Web and find two web pages—one that is appealing to you and one that is unappealing to you. Print each page. Create a web page that answers the following questions for each of your examples.

a. What is the URL of the website?

b. Is the page appealing or unappealing? List three reasons for your answer.

c. If the page is unappealing, what would you do to improve it?

d. Would you encourage others to visit this site? Why or why not?

Case Study

The following case studies continue throughout most of the text. This chapter introduces each website scenario, presents the site map, and directs you to create two pages for the site.

Pacific Trails Resort Case Study

Melanie Bowie is the owner of Pacific Trails Resort, located right on the California North Coast. The resort offers a quiet getaway with luxury camping in yurts along with an upscale lodge for dining and visiting with fellow guests. The target audience for Pacific Trails Resort is couples who enjoy nature and hiking. Melanie would like a website that emphasizes the uniqueness of the location and accommodations. She would like the website to include a

FIGURE 2.28 *Pacific Trails Resort site map.*

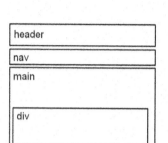

FIGURE 2.29 *Pacific Trails Resort wireframe page layout.*

home page, a page about the special yurt accommodations, a reservations page with a contact form, and a page to describe the activities available at the resort.

A site map for the Pacific Trails Resort website is shown in Figure 2.28. The site map describes the architecture of the website—a Home page with three main content pages: Yurts, Activities, and Reservations.

Figure 2.29 displays a wireframe sketch of the page layout for the Pacific Trails Resort website. The wireframe contains a header area, a navigation area, a main content area, and a footer area for copyright information.

You have three tasks in this case study:

1. Create a folder for the Pacific Trails Resort website.
2. Create the Home page: index.html.
3. Create the Yurts page: yurts.html.

Task 1: Create a folder on your hard drive or portable storage device (a thumb drive or an SD card) called pacific to contain your Pacific Trails Resort website files.

Task 2: The Home Page. You will use a text editor to create the Home page for the Pacific Trails Resort website. The Home page is shown in Figure 2.30.

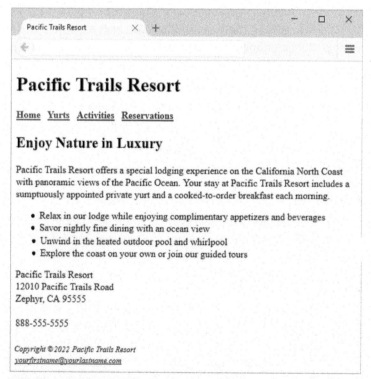

FIGURE 2.30 *Pacific Trails Resort index.html.*

Launch a text editor and create a web page document with the following specifications:

1. **Web Page Title:** Use a descriptive page title—the company name is a good choice for a business website.

2. **Wireframe Header Area:** Code the header element with the text, "Pacific Trails Resort" within an h1 heading element.

3. **Wireframe Navigation Area:** Place the following text within a nav element with bold text (use the `<b>` element):

   Home   Yurts   Activities   Reservations

   Code anchor tags so that "Home" links to index.html, "Yurts" links to yurts.html, "Activities" links to activities.html, and "Reservations" links to reservations.html. Add extra blank spaces between the hyperlinks with the ` ` special character as needed.

4. **Wireframe Main Content Area:** Code the page content within a main element. Use Hands-On Practices 2.11 and 2.12 as a guide.

   a. Place the following within an h2 element: Enjoy Nature in Luxury

   b. Place the following content in a paragraph:

      Pacific Trails Resort offers a special lodging experience on the California North Coast with panoramic views of the Pacific Ocean. Your stay at Pacific Trails Resort includes a sumptuously appointed private yurt and a cooked-to-order breakfast each morning.

   c. Place the following content in an unordered list:

      Relax in our lodge while enjoying complimentary appetizers and beverages
      Savor nightly fine dining with an ocean view
      Unwind in the heated outdoor pool and whirlpool
      Explore the coast on your own or join our guided tours

   d. Contact information:

      Place the address and phone number information within a div below the unordered list. Use line break tags to help you configure this area and add extra space between the phone number and the footer area.

      Pacific Trails Resort
      12010 Pacific Trails Road
      Zephyr, CA 95555
      888-555-5555

5. **Wireframe Footer Area:** Configure the copyright and e-mail address information within a footer element. Also configure small text size (use the `<small>` element) and italics font style (use the `<i>` phrase element). The copyright information is

   Copyright © 2022 Pacific Trails Resort

   Place your name in an e-mail link on the line under the copyright.

   The web page in Figure 2.30 may seem a little sparse, but don't worry. As you gain experience and learn to use more advanced techniques, your pages will look more professional. White space (blank space) on the page can be added with `<br>` tags where needed. Your page does not need to look exactly the same as the sample. Your goal at this point should be to practice and get comfortable using HTML. Save your file in the pacific folder and name it index.html.

Pacific Trails Resort

Home Yurts Activities Reservations

The Yurts at Pacific Trails

What is a yurt?
Our luxury yurts are permanent structures four feet off the ground. Each yurt is fully enclosed with canvas walls, a wooden floor, and a roof dome that can be opened.

How are the yurts furnished?
Each yurt is furnished with a queen-size bed with down quilt and gas-fired stove. Your luxury camping experience includes electricity and a sink with hot and cold running water. Shower and restroom facilities are located in the lodge.

What should I bring?
Most guests pack comfortable walking shoes and plan to dress for changing weather with light layers of clothing. It's also helpful to bring a flashlight and a sense of adventure!

Copyright © 2022 Pacific Trails Resort
yourfirstname@yourlastname.com

FIGURE 2.31 *The new Yurts page.*

Task 3: The Yurts Page. Create the Yurts page shown in Figure 2.31. A productivity technique is to create new pages based on existing pages so you can benefit from your previous work. Your new Yurts page will use the index.html page as a starting point. Open the index.html page for the Pacific Trails Resort website in a text editor. Select File > Save As and save the file with the new name yurts.html in the pacific folder.

Now you are ready to edit the page.

1. **Web Page Title:** Modify the page title. Change the text between the `<title>` and `</title>` tags to Pacific Trails Resort :: Yurts.

2. **Wireframe Main Content Area:**
   a. Replace the text within the `<h2>` tags with The Yurts at Pacific Trails.
   b. Delete the Home page paragraph, unordered list, and the contact information.

   c. The Yurts page contains a list with questions and answers. Add this content to the page using a description list. Use the `<dt>` element to contain each question. Configure the question to display in bold text (use the `<strong>` element). Use the `<dd>` element to contain the answer to the question. The questions and answers are as follows:

   What is a yurt?

   Our luxury yurts are permanent structures four feet off the ground. Each yurt is fully enclosed with canvas walls, a wooden floor, and a roof dome that can be opened.

   How are the yurts furnished?

   Each yurt is furnished with a queen-size bed with down quilt and gas-fired stove. Your luxury camping experience includes electricity and a sink with hot and cold running water. Shower and restroom facilities are located in the lodge.

   What should I bring?

   Most guests pack comfortable walking shoes and plan to dress for changing weather with light layers of clothing. It's also helpful to bring a flashlight and a sense of adventure!

Save your page and test it in a browser. Test the hyperlink from the yurts.html page to index.html. Test the hyperlink from the index.html page to yurts.html. If your links do not work, review your work with close attention to these details:

- Verify that you have saved the pages with the correct names in the correct folder.
- Verify your spelling of the page names in the anchor tags.
- After you make changes, test again.

Path of Light Yoga Studio Case Study

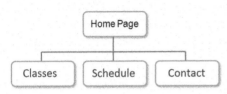

FIGURE 2.32 *Path of Light site map.*

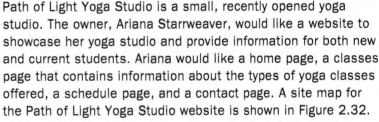

FIGURE 2.33 *Path of Light wireframe.*

Path of Light Yoga Studio is a small, recently opened yoga studio. The owner, Ariana Starrweaver, would like a website to showcase her yoga studio and provide information for both new and current students. Ariana would like a home page, a classes page that contains information about the types of yoga classes offered, a schedule page, and a contact page. A site map for the Path of Light Yoga Studio website is shown in Figure 2.32. The site map describes the architecture of the website, which consists of Home page with three main content pages: Classes, Schedule, and Contact.

Figure 2.33 displays a wireframe sketch of the page layout for the website. It contains a site logo, a navigation area, a main content area, and a footer area for copyright information. You have three tasks in this case study:

1. Create a folder for the Path of Light Yoga Studio website.

2. Create the Home page: index.html.

3. Create the Classes page: classes.html.

Task 1: Create a folder on your hard drive or portable storage device (a thumb drive or an SD card) called yoga to contain your Path of Light Yoga Studio website files.

Task 2: The Home Page. You will use a text editor to create the Home page for the Path of Light Yoga Studio website. The Home page is shown in Figure 2.34.

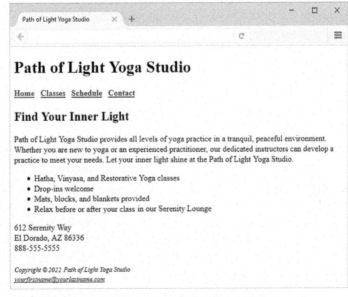

FIGURE 2.34 *Path of Light Yoga Studio index.html.*

Launch a text editor and create a web page with the following specifications:

1. **Web Page Title:** Use a descriptive page title. The company name is a good choice for a business website.

2. **Wireframe Header Area:** Code the header element with the text, "Path of Light Yoga Studio" within an h1 heading element.

3. **Wireframe Navigation Area:** Place the following text within a nav element with bold text (use the `<b>` element):

   Home   Classes   Schedule   Contact

   Code anchor tags so that "Home" links to index.html, "Classes" links to classes.html, "Schedule" links to schedule.html, and "Contact" links to contact.html. Add extra blank spaces between the hyperlinks with the ` ` special character as needed.

4. **Wireframe Main Content Area:** Code the page content within a main element. Use Hands-On Practices 2.11 and 2.12 as a guide.

   a. Code the following text within an h2 element:

   Find Your Inner Light

   b. Configure the following sentences in a paragraph:

   Path of Light Yoga Studio provides all levels of yoga practice in a tranquil, peaceful environment. Whether you are new to yoga or an experienced practitioner, our dedicated instructors can develop a practice to meet your needs. Let your inner light shine at the Path of Light Yoga Studio.

   c. Configure the following content in an unordered list:

   Hatha, Vinyasa, and Restorative Yoga classes
   Drop-ins welcome
   Mats, blocks, and blankets provided
   Relax before or after your class in our Serenity Lounge

   d. Code the following address and phone number contact information within a div element. Use line break tags to help you configure this area and add extra space between the phone number and the footer area.

   612 Serenity Way
   El Dorado, AZ 86336
   888-555-5555

5. **Wireframe Footer Area:** Configure the following copyright and e-mail link information within a footer element. Format it with small text size (use the `<small>` tag) and italics font style (use the `<i>` tag).

   Copyright © 2022 Path of Light Yoga Studio
   Place your name in an e-mail link on the line under the copyright.

   The page in Figure 2.34 may seem a little sparse, but don't worry; as you gain experience and learn to use more advanced techniques, your pages will look more professional. White space (blank space) on the page can be added with `<br>` tags where needed. Your page does not need to look exactly the same as the sample. Your goal at this point should be to practice and get comfortable using HTML. Save your page in the yoga folder, and name it index.html.

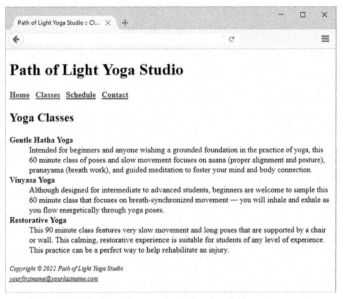

FIGURE 2.35 *Path of Light Yoga Studio classes.html.*

Task 3: The Classes Page. Create the Classes page as shown in Figure 2.35. A technique that improves productivity is to create new pages based on existing pages so that you can benefit from your previous work. Your new Classes page will use the index.html page as a starting point.

Open the index.html page for the Path of Light Yoga Studio website in a text editor. Select File > Save As and save the file with the new name classes.html in the yoga folder. Now you are ready to edit the page.

1. **Web Page Title:** Modify the page title. Change the text between the `<title>` and `</title>` tags to the following:

   Path of Light Yoga Studio :: Classes

2. **Wireframe Main Content Area:**

   a. Delete the Home page content paragraph, unordered list, and contact information.

   b. Configure the following text in the heading 2 element:
   Yoga Classes

   c. Use a description list to configure information about the yoga classes. Configure the name of each class to have strong importance and bold font weight (use the `<strong>` tag) within a `<dt>` tag. Configure `<dd>` tags to contain the class descriptions. The information follows:

   Gentle Hatha Yoga
   Intended for beginners and anyone wishing a grounded foundation in the practice of yoga, this 60 minute class of poses and slow movement focuses on asana (proper alignment and posture), pranayama (breath work), and guided meditation to foster your mind and body connection.

Vinyasa Yoga

Although designed for intermediate to advanced students, beginners are welcome to sample this 60 minute class that focuses on breath-synchronized movement—you will inhale and exhale as you flow energetically through yoga poses.

Restorative Yoga

This 90 minute class features very slow movement and long poses that are supported by a chair or wall. This calming, restorative experience is suitable for students of any level of experience. This practice can be a perfect way to help rehabilitate an injury.

Save your page and test it in a browser. Test the hyperlink from the classes.html page to index.html. Test the hyperlink from the index.html page to classes.html. If your links do not work, review your work with close attention to these details:

- Verify that you have saved the pages with the correct names in the correct folder.
- Verify your spelling of the page names in the anchor tags.

Test again after you make changes.

Endnotes

1. "Character Entity Reference Chart." *W3C Public CVS Repository*, W3C, dev.w3.org/html5/html-author/charref.

2. "HTML Living Standard 4.5.14 The Time Element." *HTML Standard*, W3C, html.spec.whatwg.org/#the-time-element.

CHAPTER 3

Web Design Basics

As a website visitor, you have probably found that some websites are appealing and easy to use, while others seem awkward or just plain annoying. What separates the good from the bad? This chapter discusses recommended website design practices. The topics include site organization, site navigation, page design, choosing a color scheme, text design, graphic design, and accessibility considerations.

You'll learn how to...

- Describe the most common types of website organization
- Describe principles of visual design
- Design for your target audience
- Create clear, easy-to-use navigation
- Improve the readability of the text on your web pages
- Use graphics appropriately on web pages

- Choose a color scheme for your website
- Apply the concept of universal design to web pages
- Describe web page layout design techniques
- Describe the concept of responsive web design
- Apply best practices of web design

Your Target Audience

Whatever your personal preferences, your website should appeal to your **target audience**—the people who will use your site. Your intended target audience may be specific, such as kids, college students, young couples, or seniors, or you may intend your site to appeal to everyone. The purpose and goals of your visitors will vary—they may be casually seeking information, performing research for school or work, comparison shopping, job hunting, and so on. The design of a website should appeal to and meet the needs of the target audience.

FIGURE 3.1 *The compelling graphic draws you in.*

For example, the web page shown in Figure 3.1 features compelling graphics and has a different look and feel from the text- and link-intensive web page displayed in Figure 3.2.

The first site engages you, draws you in, and invites exploration. The second site provides you with text-based information so that you can quickly get down to work. Keep your target audience in mind as you explore the web design practices in this chapter.

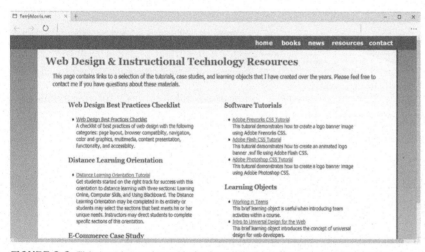

FIGURE 3.2 *This text-intensive website immediately offers numerous choices.*

Browsers

Just because your web page looks great in your favorite browser doesn't automatically mean that all browsers will render it well. Expect your website to be visited by people using a variety of browsers. StatCounter[1] reported the market share of the top five desktop browsers in a recent month as Chrome (68%), Firefox (9%), Safari (9%), Edge (5%), and Internet Explorer (4%). The market share of the top five mobile browsers was reported by StatCounter GlobalStats[2]: Chrome (61%), Safari (25%), Samsung Internet (6%), UC Browser (4%), and Opera (1%).

Apply the principle of **progressive enhancement**: Design a website so that it looks good in commonly used browsers and then add enhancements with CSS and/or HTML5 for display in the most recent versions of browsers.

Try to test your pages with the most popular versions of browsers on both PC and Mac operating systems. Many web page components, including default text size and default margin size, are different among browsers, browser versions, and operating systems. Also try to test your website on other types of devices, such as tablets and smartphones.

Screen Resolution

Your website visitors will use a variety of devices and use many different screen resolutions. A recent survey by StatCounter[3] reported the use of more than 20 screen resolutions in a recent month, with the top four being 360×640 (with 23%), 1366×768 (9%), 1920×1080 (89%), and 375×667 (4%). Observe that this report lists a mobile screen resolution 360×640 as the most popular. It is becoming more important than ever to design web pages that can be used on both desktops and mobile devices. In Chapter 8, you'll explore CSS media queries, which is a technique for configuring a web page to display well on various screen resolutions.

FAQ How can I create web pages that look exactly the same on all browsers?

You can't. Design with the most popular browsers and screen resolutions in mind, but expect your web pages to look slightly different when displayed by different browsers and on monitors with different screen resolutions. Expect web pages to look even more different when displayed on mobile devices. You'll learn about responsive web design techniques later in this chapter.

Website Organization

How will visitors move around your site? How will they find what they need? This is largely determined by the website's organization or architecture. There are three common types of website organization:

- ▶ Hierarchical
- ▶ Linear
- ▶ Random (sometimes called web organization)

A diagram of the organization of a website is called a **site map**. Creating the site map is one of the initial steps in developing a website.

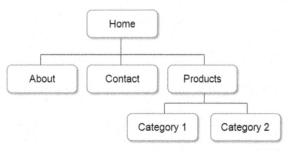

FIGURE 3.3 *Hierarchical site organization.*

Hierarchical Organization

Most websites use **hierarchical organization**. A site map for hierarchical organization, such as the one shown in Figure 3.3, is characterized by a clearly defined home page with links to major site sections. Web pages within sections are placed as needed. The home page plus the first level of pages in a hierarchical site map typically indicate the hyperlinks that will be displayed on the main navigation bar of each web page within the website.

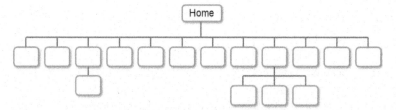

FIGURE 3.4 *This site design uses a shallow hierarchy.*

It is important to be aware of pitfalls of hierarchical organization. Figure 3.4 shows a site design that is too shallow—there could be too many major site sections. This site design needs to be organized into fewer, easily managed topics or units of information, a process called **chunking**. In the case of web page design, each unit of information is a page. Nelson Cowan, a research psychologist at the University of Missouri, found that adults are typically able to keep about four items or chunks of items (such as the three parts of a phone number 888-555-5555) in their short-term memory.[4] Following this principle, be aware of the number of major navigation links and try to group them into visually separate sections on the page, with each group having no more than about four links.

Another potential design pitfall of hierarchical website design is creating a site whose structure is too deep. Figure 3.5 shows an example of this. A visitor who cannot get what he or she wants in a few mouse clicks will begin to feel frustrated and may leave your site. This rule may be very difficult to satisfy on a large site, but in general, the goal is to organize your site so that your visitors can easily navigate from page to page within the site structure.

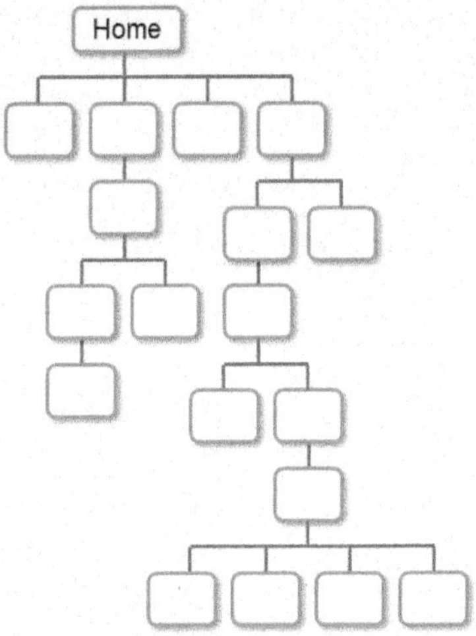

Linear Organization

Linear organization, shown in Figure 3.6, is useful when the purpose of a website or series of pages within a site is to provide a tutorial, tour, or presentation that needs to be viewed sequentially.

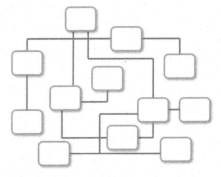

FIGURE 3.5 *This site design uses a deep hierarchy.*

FIGURE 3.6 *Linear site organization.*

In linear organization, the pages are viewed one after another. Some websites use hierarchical organization in general but with linear organization in a few small areas.

Random Organization

Random organization (sometimes called web organization) offers no clear path through the site, as shown in Figure 3.7. There is often no clear home page and no discernible structure. Random organization is not as common as hierarchical or linear organization and is usually found only on artistic sites or sites that strive to be especially different and original. This type of organization is typically not used for commercial websites.

FIGURE 3.7 *Random site organization.*

FAQ

What's a good way to organize my site map?

Sometimes it is difficult to begin creating a site map for a website. Some design teams meet in a room with a blank wall and a package of large Post-it Notes. They write the titles of topics and subtopics needed on the site on the Post-it Notes. They arrange the notes on the wall and discuss until a site structure evolves and there is consensus within the group. If you are not working in a group, you can try this on your own and then discuss the way you have chosen to organize the website with a friend or fellow student.

Principles of
Visual Design

VideoNote
*Principles of
Visual Design*

There are four visual design principles that you can apply to the design of just about anything: repetition, contrast, proximity, and alignment. Whether you are designing a web page, a button, a logo, a DVD cover, a brochure, or a software interface, the design principles of repetition, contrast, proximity, and alignment will help to create the "look" (visual aesthetic), of your project and will determine whether your message is effectively communicated.

Repetition: Repeat Visual Components Throughout the Design

When applying the principle of **repetition**, the web designer repeats one or more components throughout the page. The repeating aspect ties the work together. Figure 3.8 displays the home page for a bed and breakfast business. The page design demonstrates the use of repetition in a variety of design components, including color, shape, font, and images.

▶ The large photograph on the web page draws the viewer's attention. The smaller image repeats some of the same colors: terracotta, green, off-white, and gray. Colors that either occur in the photographs (off-white and gray) or that coordinate with the photographs (a dark red/rust) are repeated in other elements on the page. The rust color is used in the headings, navigation, e-mail address box, call to action button backgrounds, and horizontal line. The page background is off-white.

▶ The call-to-action "Book Now" and "Sign Up" buttons both have a rectangular shape and format with heading, content, and button.

▶ The use of only two font typefaces on the page also demonstrates repetition and helps to create a cohesive look.

Whether it is color, shape, font, or image, repetition helps to unify a design.

Contrast: Add Visual Excitement and Draw Attention

To apply the principle of **contrast**, emphasize the differences between page elements in order to make the design interesting and direct attention. There should be good contrast between the background color and the text color on a web page. If there is too little contrast, the text will be difficult to read. Notice how the upper right navigation area in Figure 3.8 uses a dark text color that has good contrast with the light background color. The BOOK NOW call to action button has a dark background that both contrasts well with the light text and serves to make it stand out from the navigation. The area under the large photograph features a light background that has good contrast with the dark text. The dark text in the footer area contrasts well with the light background color.

FIGURE 3.8 *The design principles of repetition, contrast, proximity, and alignment are applied on this web page.*

Proximity: Group Related Items

When designers apply the principle of **proximity**, related items are placed physically close together. Unrelated items should have space separating them. In Figure 3.8, the horizontal navigation links are all placed in close proximity to each other. This creates a visual group on the page and makes the navigation easier to use. Proximity is used well on this page to group related elements such as each heading and its related paragraph and call to action button.

Alignment: Align Elements to Create Visual Unity

Another principle that helps to create a cohesive web page is **alignment**. When applying this principle, the designer organizes the page so that each element placed has some alignment (vertical or horizontal) with another element on the page. The page shown in Figure 3.8 also applies this principle. Notice how the page components under the large photograph are vertically aligned in columns of equal height.

Repetition, contrast, proximity, and alignment are four visual design principles that can greatly improve your web page designs. If you apply these principles effectively, your web pages will look more professional and you will communicate your message more clearly. Keep these principles in mind as you design and build web pages.

Design to Provide for Accessibility

In Chapter 1, you were introduced to the concept of universal design. Let's take a closer look in this section at how the concept of universal design can apply to web design.

Who Benefits from Universal Design and Increased Accessibility?

Consider the following scenarios:

▶ Maria is a young woman in her twenties with physical challenges who cannot manipulate a mouse and who uses a keyboard with much effort. Accessible web pages designed to function without a mouse will help Maria access content.

▶ Leotis is a college student who is deaf and wants to be a web developer. Captions for audio/video content and transcripts will provide Leotis access to content.

▶ Jim is a middle-aged man who has a dial-up Internet connection and is using the Web for personal enjoyment. Alternate text for images and transcripts for multimedia will provide Jim improved access to content.

▶ Nadine is a mature woman with age-related macular degeneration who has difficulty reading small print. Web pages designed so that text can be enlarged in the browser will make it easier for Nadine to read.

▶ Karen is a college student using a smartphone to access the Web. Accessible content organized with headings and lists will make it easier for Karen to surf the Web on a mobile device.

▶ Prakesh is a man in his thirties who is legally blind and needs access to the Web to do his job. Web pages designed to be accessible (which are organized with headings and lists, display descriptive text for hyperlinks, provide alternate text descriptions for images, and are usable without a mouse) will help Prakesh access content when using a screen reader application such as JAWS or Window-Eyes.

All of these individuals benefit from web pages designed with accessibility in mind. A web page that is designed to be accessible is typically more usable for all—even a person who has no physical challenges and is using a broadband connection benefits from the improved presentation and organization of a well-designed web page.

Accessible Design Can Benefit Search Engine Listing

Search engine programs (commonly referred to as bots or spiders) walk the Web and follow hyperlinks on websites. An accessible website with descriptive page titles that is well organized with headings, lists, descriptive text for hyperlinks, and alternate text for images is more visible to search engine robots and may result in better ranking.

Legal Requirements

The Internet and World Wide Web are such a pervasive part of our culture that accessibility is mandated by laws in the United States. Section 508 of the Rehabilitation Act requires electronic and information technology, including web pages, used by federal agencies to be accessible to people with disabilities. The accessibility recommendations presented in this text are intended to satisfy the Section 508 standards and the W3C Web Accessibility Initiative guidelines. In 2017, an update to Section 508 Standards became official which requires meeting the requirements of WCAG 2.0 Level A & AA Success Criteria.[5]

Accessibility Is the Right Thing to Do

The federal government is promoting accessibility by law, and the private sector is following its lead. The W3C is quite active in this cause and has created the Web Accessibility Initiative (WAI) to create guidelines and standards applicable to web content developers, authoring-tool developers, and browser developers. The following four content accessibility principles are essential to conformance with WCAG 2.0—**P**erceivable, **O**perable, **U**nderstandable, and **R**obust—referred to by the acronym **POUR**.

1. Content must be **Perceivable**. Perceivable content is easy to see or hear. Any graphic or multimedia content should be available in a text format, such as text descriptions for images, closed captions for videos, and transcripts for audio.

2. Interface components in the content must be **Operable**. Operable content has interactive features, such as navigation forms, that can be used or operated with either a mouse or keyboard. Multimedia content should be designed to avoid flashing, which may cause a seizure.

3. Content and controls must be **Understandable**. Understandable content is easy to read, organized in a consistent manner, and provides helpful error messages when appropriate.

4. Content should be **Robust** enough to work with current and future user agents, including assistive technologies. Robust content is written to follow W3C Recommendations and should be compatible with multiple operating systems, browsers, and assistive technologies such as screen reader applications.

The W3C has approved a new version of WCAG, called WCAG 2.1, which extends the guidelines in WCAG 2.0.[6] The WCAG 2.1 Quick Reference in the Appendix contains a brief list of guidelines for designing accessible web pages.

As you work through this book, you'll learn to include accessibility features as you create practice pages. You've already discovered the importance of configuring the title tag, heading tags, and descriptive text for hyperlinks in Chapters 1 and 2. You're already well on your way to creating accessible web pages!

Use of Text

Writing for the Web

Long-winded sentences and explanations are often found in academic textbooks and romance novels, but they really are not appropriate on a web page. Long blocks of text and long paragraphs are difficult to read on the Web. The following suggestions will help to increase the readability of your web pages.

▶ Be concise. Use the text equivalent of sound bites—short sentences and phrases.

▶ Organize the page content with headings and subheadings.

▶ Use lists to help text stand out and make content easier to read.

The web page shown in Figure 3.9 provides an example of using headings and brief paragraphs to organize web page content so that it is easy to read and visitors can quickly find what they need.

FIGURE 3.9 *The web page content is well organized with headings.*

Text Design Considerations

You may be wondering how to know whether a web page is easy to read. Readable text is crucial to providing content of value for your web page visitors. Carefully consider the typeface, size, weight, and color when you select fonts for your web pages. The following are some suggestions that will help increase the readability of your pages:

▶ **Use Common Fonts**

Use common fonts such as Arial, Verdana, or Times New Roman. Remember that the web page visitor must have the font installed on his or her computer in order for that particular font to appear. Your page may look great with Gill Sans Ultra Bold Condensed, but if your visitor doesn't have this typeface, the browser's default font will be displayed.

⏵ Carefully Choose Fonts

Serif fonts, such as Times New Roman, were originally developed for printing text on paper, not for displaying text on a computer monitor. Sans-serif fonts such as Verdana and Tahoma were specifically designed for display on screens.

⏵ Check Font Size

Be aware that fonts display smaller on a Mac than on a PC. Even within the PC platform, the default font size displayed by browsers may not be the same. Consider creating prototype pages of your font size settings to test on a variety of browsers and screen resolution settings.

⏵ Check Font Weight

Bold or *emphasize* important text (use the `<strong>` element for bold and the `<em>` element to configure italics). However, be careful not to bold everything—that has the same effect as bolding nothing.

⏵ Check Font Color for Contrast

Use appropriate color combinations. Newbie web designers sometimes choose color combinations for web pages that they would never dream of using in their wardrobe. An easy way to choose colors that contrast well and look good together is to select colors from an image or logo that you will use for your site. Online tools exist that can help you verify that your page background color properly contrasts with your text and hyperlink colors.[7,8]

⏵ Check Line Length

Be aware of line length—use empty space and multiple columns if possible. Christian Holst at the Baymard Institute recommends using between 50 and 60 characters per line for readability.[9] Look ahead to Figure 3.37 for examples of text placement on a web page.

⏵ Check Alignment

A paragraph of centered text is more difficult to read than left-aligned text.

⏵ Carefully Choose Text in Hyperlinks

Use hyperlinks for keywords and descriptive phrases. Do not hyperlink entire sentences. Also avoid the use of the words "click here" in hyperlinks—users know what to do by now.

⏵ Check Spelling and Grammar

Unfortunately, many websites contain misspelled words. Most web-authoring tools have built-in spell checkers; consider using this feature.

Finally, be sure that you proofread and test your site thoroughly. It's very helpful if you can find web developer buddies—you check their sites, and they check yours. It's always easier to see someone else's mistake than your own.

Web Color Palette

Red: #FF0000

Green: #00FF00

Blue: #0000FF

Black: #000000

White: #FFFFFF

Grey: #CCCCCC

FIGURE 3.10
Sample colors.

Computer monitors display color as a combination of different intensities of red, green, and blue, also known as **RGB color**. RGB intensity values are numeric from 0 to 255.

Each RGB color has three values, one each for red, green, and blue. These are always listed in the same order (red, green, and blue) and specify the numerical value of each color (see examples in Figure 3.10). You will usually use hexadecimal color values to specify RGB color on web pages.

Hexadecimal Color Values

Hexadecimal is the name for the base 16 numbering system, which uses the characters 0, 1, 2, 3, 4, 5, 6, 7, 8, 9, A, B, C, D, E, and F to specify numeric values.

Hexadecimal color values specify RGB color with numeric value pairs ranging from 00 to FF (0 to 255 in base 10). Each pair is associated with the amount of red, green, and blue displayed. Using this notation, one would specify the color red as #FF0000 and the color blue as #0000FF. The # symbol signifies that the value is hexadecimal. You can use either uppercase or lowercase letters in hexadecimal color values; #FF0000 and #ff0000 both configure the color red.

Don't worry—you won't need to do calculations to work with web colors. Just become familiar with the numbering scheme. See Figure 3.11 for example color values on a partial color chart.

#FFFFFF	#FFFFCC	#FFFF99	#FFFF66	#FFFF33	#FFFF00
#FFCCFF	#FFCCCC	#FFCC99	#FFCC66	#FFCC33	#FFCC00
#FF99FF	#FF99CC	#FF9999	#FF9966	#FF9933	#FF9900
#FF66FF	#FF66CC	#FF6699	#FF6666	#FF6633	#FF6600
#FF33FF	#FF33CC	#FF3399	#FF3366	#FF3333	#FF3300
#FF00FF	#FF00CC	#FF0099	#FF0066	#FF0033	#FF0000

FIGURE 3.11 *Partial color chart.*

Web-Safe Colors

It is easy to tell whether a color is a web-safe color—check the hexadecimal color values.

Web-Safe Hexadecimal Values

00, 33, 66, 99, CC, FF

The color chart in this book's Web Safe Color Palette Appendix displays colors that follow this numbering scheme—they comprise the Web-Safe Color Palette.

Accessibility and Color

While color can help you create a compelling web page, keep in mind that not all of your
visitors will see or be able to distinguish between colors. Some visitors will use a screen
reader and will not experience your colors, so your information must be clearly conveyed
even if the colors cannot be viewed. According to Color Blindness Awareness, 1 in 12 men
and 1 in 200 women experience some type of color perception deficiency.[10]

Color choices can be crucial. For example, red text on a blue background, as shown in
Figure 3.12, is usually difficult for everyone to read. Also avoid using color combinations
that can be difficult for individuals with color perception deficiency such as green and red,
blue and purple, and green and brown.[11] White, black, and shades of blue and yellow are
easier for most people to discern.

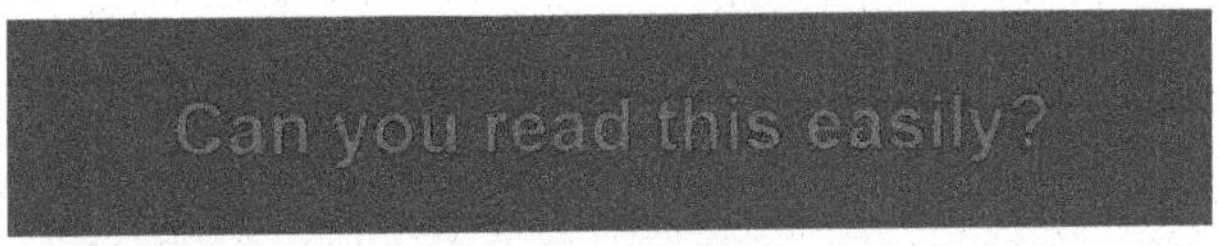

Can you read this easily?

FIGURE 3.12 *Some color combinations are difficult
to read.*

Choose text and background colors with enough contrast so the text can be easily read.
The WCAG guidelines recommend a contrast ratio of 4.5:1 for standard text. If the text has
a large font, the contrast ratio can be as low as 3:1. The following online tools verify the
contrast level of your text and background colors.

▶ **WebAIM Contrast Checker:**

https://webaim.org/resources/contrastchecker/

▶ **Deque Color Contrast Analyzer:**

https://dequeuniversity.com/color-contrast

▶ **Snook.ca Colour Contrast Check:**

https://snook.ca/technical/colour_contrast/colour.html

Design for Your Target Audience

The first section in this chapter introduced the importance of designing for your target audience. In this section, we consider how to use color, graphics, and text to appeal to a target audience.

FIGURE 3.13 *A web page intended to appeal to children.*

Appealing to Children and Preteens

Younger audiences, such as children and preteens, prefer bright, lively colors. The web page shown in Figure 3.13 features bright graphics, lots of color, and interactivity.

FIGURE 3.14 *Many teens and young adults find dark sites appealing.*

Appealing to Young Adults

Individuals in their late teens and early twenties generally prefer dark background colors with occasional use of bright contrast, music, and dynamic navigation. Figure 3.14 shows a web page designed for this age group. Note how it has a completely different look and feel from the site designed for young children.

Appealing to Everybody

If your goal is to appeal to everyone, follow the example of the popular Amazon.com and eBay.com websites in their use of color. These sites display a neutral white background with splashes of color to add interest and highlight page areas. Use of white as a background color was found to be quite popular by Jakob Nielsen and Marie Tahir in *Homepage Usability: 50 Websites Deconstructed*, a book that analyzed 50 top websites. According to this study, 84% of the sites used white as the background color and 72% used black as the text color. This maximized the contrast between text and background—providing maximum ease of reading.

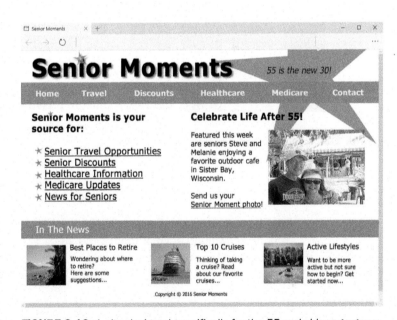

FIGURE 3.15 *A compelling graphic along with white background for the content area.*

You'll also notice that websites targeting "everyone" often include compelling visual graphics. The web page shown in Figure 3.15, provides the text content on a white background for maximum contrast while engaging the visitor with a large graphic, called a **hero**, intended to grab attention and entice the visitor to want to explore the website.

Appealing to Older Adults

For an older target audience, light backgrounds, well-defined images, and large text are appropriate. The web page shown in Figure 3.16 is an example of a web page intended for the 55-and-older age group.

FIGURE 3.16 *A site designed specifically for the 55-and-older age group.*

Choosing a Color Scheme

A compelling color scheme can attract and engage your website visitors while a garish color scheme can drive them away. This section introduces several methods for choosing a color scheme.

Color Scheme A:

Color Scheme B:

FIGURE 3.17 *A color scheme selected from a photo.*

Color Scheme Based on an Image

One of the easiest ways to select a color scheme for your website is to start with an existing graphic image, such as a logo or a photograph of nature. If the organization already has a logo, select colors from the logo for use as the basis of your color scheme.

Another option is to use a photograph that captures the mood of the website—you can create a color scheme using colors found in the image. Figure 3.17 shows a photograph along with two potential color schemes created by selecting colors from the image.

If you are comfortable using a graphic application, you can use the color picker tool within the application to determine the colors used in an image.

Another option is to use an online tool to generate a color palette based on an image:

▶ **Degraeve.com:**
https://www.degraeve.com/color-palette/index.php

▶ **Canva.com:**
https://www.canva.com/colors/color-palette-generator/

Even if you use an existing graphic as the basis for a color scheme, it's helpful to have a working knowledge of **color theory,** the study of color and its use in design. A starting point is to explore the color wheel.

Color Wheel

A **color wheel** (see Figure 3.18) is a circle of colors depicting the primary colors (red, yellow, and blue), the secondary colors (orange, violet, and green), and the tertiary colors (yellow-orange, red-orange, red-violet, violet-blue, blue-green, and yellow-green).

FIGURE 3.18 *Color wheel.*

Shades, Tints, and Tones

There is no need to restrict your choices to the web-safe color palette. Modern monitors can display millions of colors. You are free to choose a shade, tint, or tone of a color. Figure 3.19 shows four swatches: yellow, a shade of yellow, a tint of yellow, and a tone of yellow.

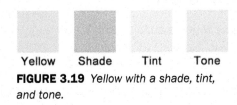

Yellow   Shade   Tint   Tone

FIGURE 3.19 *Yellow with a shade, tint, and tone.*

▶ A **shade** of a color is darker than the original color and is created by mixing the color with black.

▶ A **tint** of a color is lighter than the original color and is created by mixing the color with white.

▶ A **tone** of a color has less saturation than the original color and is created by mixing the color with gray.

Next, let's explore the six commonly used types of color schemes: monochromatic, analogous, complementary, split complementary, triadic, and tetradic.

Monochromatic Color Scheme

FIGURE 3.20 *Monochromatic color scheme.*

Figure 3.20 shows a **monochromatic color scheme** which consists of shades, tints, or tones of the same color. You can determine these values yourself, or use an online tool provided by one of the following resources:

▶ Color Blender: https://meyerweb.com/eric/tools/color-blend
▶ Pine Tools: https://pinetools.com/monochromatic-colors-generator
▶ HexColorCodes: https://hexcolorcodes.org/monochromatic-colors-generator

Analogous Color Scheme

FIGURE 3.21 *Analogous color scheme.*

To create an **analogous color scheme**, select a main color and the two colors that are adjacent to it on the color wheel. Figure 3.21 displays an analogous color scheme with orange, red-orange, and yellow-orange.

When you design a web page with an analogous color scheme, the main color is the most dominant on the web page. The adjacent colors are typically configured as accents. Be sure that the main content of the page is easy to read and use neutrals white, off-white, gray, black, or brown along with the analogous color scheme.

Complementary Color Scheme

FIGURE 3.22 *Complementary color scheme.*

A **complementary color scheme** consists of two colors that are opposite each other on the color wheel. Figure 3.22 displays a complementary color scheme with yellow and violet.

When you design a web page with a complementary color scheme, choose one color to be the main or dominant color. The other color is considered to be the **complement.** Configure the complement along with colors adjacent to the dominant color as accents. Use neutrals white, off-white, gray, black, or brown as part of a complementary color scheme.

Split Complementary Color Scheme

A **split complementary color scheme** is comprised of a main color, the opposite color on the color wheel (the complement) and two colors adjacent to the complement. Figure 3.23 shows a split complementary color scheme with yellow (main), violet (complementary), red-violet, and blue violet.

FIGURE 3.23 *Split complementary color scheme.*

Triadic Color Scheme

Choose three colors that are equidistant on the color wheel to create a **triadic color scheme**. Figure 3.24 displays a triadic color scheme with blue-green (teal), yellow-orange, and red-violet.

FIGURE 3.24 *Triadic color scheme.*

Tetradic Color Scheme

Figure 3.25 shows a **tetradic color scheme** which consists of four colors which are two complementary pairs. For example, the complementary pair yellow and violet along with the complementary pair yellow-green and red-violet make up a tetradic color scheme.

FIGURE 3.25 *Tetradic color scheme.*

Implementing a Color Scheme

When designing a web page with a color scheme, one color is typically dominant. The other colors are configured as accents such as colors for headings, subheadings, borders, list markers, and backgrounds.

No matter what your color scheme is, you will typically also use neutral colors such as white, off-white, gray, black, or brown. Selecting the best color scheme for your website often takes some trial and error.

Feel free to use tints, shades, or tones of the primary, secondary, and tertiary colors.

There are so many colors to choose from! The following resources can help you choose a color scheme for your website:

▶ https://www.colorbook.io/pages/colorschemegenerator
▶ https://color.adobe.com
▶ https://www.colorspire.com

Use of Graphics and Multimedia

As shown in Figure 3.15, a compelling graphic can be an engaging element on a web page. However, be aware that you should avoid relying on images to convey meaning. Some individuals may not be able to see your images and multimedia—they may be accessing your site with a mobile device or using an assistive technology such as a screen reader to visit your page. You may need to include text descriptions of important concepts or key points that a graphic image or multimedia file conveys. In this section, you'll explore recommended techniques for use of graphics and multimedia on web pages.

File Size and Dimensions Matter

Image optimization is the process of creating an image with the lowest file size that still renders a good-quality image—balancing image quality and file size. The dimensions of the image should be as close to possible as the actual display size to enable speedy browser rendering of the image. Other approaches to image optimization are to crop an image or create a thumbnail image that links to a larger version of the image. Adobe Photoshop and GIMP (GNU Image Manipulation Program) are often used by web professionals to optimize images for the Web. A free online tool for image editing and optimization is Pixlr Editor at https://www.pixlr.com/editor.

Antialiased

FIGURE 3.26 *Antialiased text.*

FIGURE 3.27 *This graphic has a jagged look and was not saved using antialiasing.*

Antialiased/Aliased Text Considerations

Refer back to Figure 3.13 and notice how easy it is to read the text in the navigation buttons—the text in each button is antialiased text. **Antialiasing** introduces intermediate colors to smooth jagged edges in digital images. Graphic applications can be used to create antialiased text images. The graphic shown in Figure 3.26 was created using antialiasing. Figure 3.27 displays an image created without antialiasing; note the jagged edges.

Use Only Necessary Multimedia

Use animation and multimedia only if it will add value to your site. Limit the use of animated items. Only use animation if it makes the page more effective. Consider limiting how long an animation plays.

In general, younger audiences find animation more appealing than older audiences. The web page shown in Figure 3.13 is geared to children and uses lots of animation. This would be too much animation for a website targeted to adult shoppers. However, a well-done navigation animation or an animation that describes a product or service could be appealing to almost any target group, as shown in Figure 3.28. You'll work with new CSS properties to add animation and interactivity to web pages in Chapters 7 and 11.

FIGURE 3.28 *The slideshow adds visual interest and interactivity.*

Provide Alternate Text

Each image on your web page should be configured with alternate text. (See Chapter 5 for an introduction to configuring images on web pages.) Alternate text may be displayed instead of the image when the page is displayed by mobile devices, while the image is loading (when the image is slow to load), and when a browser is configured to not show images. Alternate text is also read aloud when a person with a disability uses a screen reader to access your website.

To satisfy accessibility requirements, also provide alternate text equivalents for multimedia, such as video and audio. A text transcript of an audio recording can be useful not only to those with hearing challenges but also to individuals who prefer to read when accessing new information. In addition, the text transcript may be accessed by a search engine and used when your site is categorized and indexed. Captions help to provide accessibility for video files. See Chapter 11 for more on accessibility and multimedia.

More Design Considerations

Load Time

The last thing you want to happen is for your visitors to leave your page before it has even finished loading! **Perceived load time** is the amount of time a web page visitor is aware of waiting while your page is loading. Since visitors often leave a website if a page takes too long to load, it is important to shorten their perception of waiting. Web usability expert Jakob Nielsen reports that visitors will often leave a page after waiting more than 10 seconds.[12]

According to a recent study by the PEW Internet and American Life Project, 73% of adult Americans have access to broadband at home.[13] Even with high bandwidth available to many of your visitors, keep in mind that 27% of households do not have broadband Internet access. PEW also reports that 17% of American adults depend on their smartphones for Internet access.[14] Reduce the file size of images and apply new lazy loading techniques for images (introduced in Chapter 8) to help decrease load time on all devices.

Mobile Devices

A recent StatCounter survey showed that mobile devices (52%) have more market share than desktops (45%) and tablets (3%).[15] Apply the principle of progressive enhancement. Design your website so that it looks good and loads quickly when displayed in mobile devices and the browsers commonly used by your target audience; then add enhancements with CSS and/or HTML5 to take advantage of the capabilities of modern browsers. Always try to test your pages with the most popular versions of desktop browsers (on both PC and Mac operating systems) and popular mobile devices such as smartphones and tablets. Many web page components, including default text size and default margin size, are different among devices, operating systems, and browsers. Chapter 8 will introduce techniques for designing web pages that display well on both mobile devices and standard desktop computer.

Adequate Empty Space

Placing empty space in areas around blocks of text increases the readability of the page. Placing empty space around graphics helps them to stand out. Allow for some blank space between blocks of text and images. How much is adequate? It depends—experiment until the page is likely to look appealing to your target audience. Chapter 6 will introduce using Cascading Style Sheets (CSS) to configure empty space on web pages.

Single Page Website

A **single page website** (sometimes called a one page website) is a website that contains one very long page (a single HTML file) with a clearly defined navigation area, usually at the top of the page. This navigation takes you to specific areas on the page. You can also scroll up and down to see the entire page. The first view of the page often displays a large attention-grabbing header element. Since the website is on one long page, there should be clearly separated sections for the information topics. Single page websites often feature large compelling images, so care must be taken to optimize them to prevent a slow-loading page. A single page website can be quick to create and may be a good choice for a portfolio website, a brochure website, or an initial website for a small business. You will explore creating a single page website in Chapter 7.

Parallax Scrolling

You may have noticed single page websites in which the background images scrolled at a different speed than the text content—this is a technique called **parallax scrolling**. There are many ways to accomplish this effect, some require the use of JavaScript or advanced CSS techniques. In Chapter 7, you will use CSS to create the effect of parallax scrolling.

Flat Web Design Trend

Flat web design is a minimalistic design style with a focus on simplicity, blocks of color, empty space between design elements, hero images, and use of typography. Because the design is so minimalistic and uncluttered, web pages with flat design often feature vertical scrolling.

Flat web design initially avoided the use of 3D effects such as drop shadows and gradients. However, as time went along, Flat Design 2.0 emerged which includes gradients, shadows, expanded color palettes, animation,

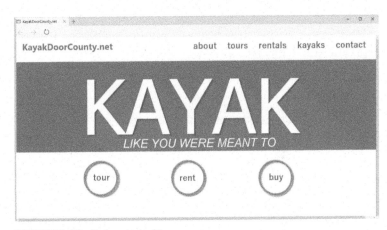

FIGURE 3.29 *Flat web design.*

and ghost buttons which are buttons with transparent backgrounds. The page shown in Figure 3.29 shows a flat minimalistic design that emphasizes typography, has large blocks of color, and has quite a bit of empty space.

Navigation Design

Ease of Navigation

Sometimes web developers are so close to their sites that they can't see the forest for the trees. A new visitor will wander onto the site and not know what to click or how to find the information he or she seeks. Clearly labeled navigation on each page is helpful and should be in the same location on each page for maximum usability.

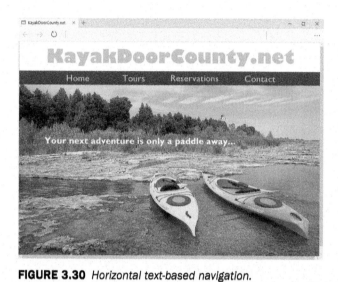

FIGURE 3.30 *Horizontal text-based navigation.*

Navigation Bars

Clear **navigation bars,** either graphic or text based, make it obvious to website users where they are and where they can go next. It's quite common for site-wide navigation to be located in either a horizontal navigation bar placed under the header (see Figure 3.30) or in a vertical navigation bar on the left side of the page (see Figure 3.31). Less common is a vertical navigation bar on the right side of the page—this area can be cut off at lower screen resolutions.

Breadcrumb Navigation

A **breadcrumb trail** indicates the path of web pages a visitor has viewed during the current session. Figure 3.31 shows a page with a vertical navigation area in addition to the breadcrumb trail navigation above the main content area that indicates the pages the visitor has viewed during this visit: Home > Tours > Half-Day Tours > Europe Lake Tour. Visitors can easily retrace their steps or jump back to a previously viewed page. This page demonstrates that a website may use more than one type of navigation.

FIGURE 3.31 *Visitors can follow the "breadcrumbs" to retrace their steps.*

Using Graphics for Navigation

Sometimes graphics are used to provide navigation, as in the pink navigation buttons on the web page shown in Figure 3.13. The "text" for the navigation is actually stored in image files. Be aware that using graphics for navigation is an outdated design technique. A website with text navigation is more accessible and more easily indexed by search engines.

Even when image hyperlinks instead of text hyperlinks provide the main navigation of the site, you can use two techniques that provide for accessibility:

▶ Configure each image with an alternate text description (see Chapter 5).

▶ Configure text hyperlinks in the footer area.

Dynamic Navigation

In your experiences visiting websites, you've probably encountered navigation menus that display additional options when your mouse cursor moves over an item. This is dynamic navigation, which provides a way to offer many choices to visitors while at the same time avoid overwhelming them. Instead of showing all the navigation links all the time, menu items are dynamically displayed (typically using a combination of HTML and CSS) as appropriate. The additional items are made available when a related top-level menu item is selected by the cursor. In Figure 3.32, "Tours" has been selected, causing the vertical menu to appear.

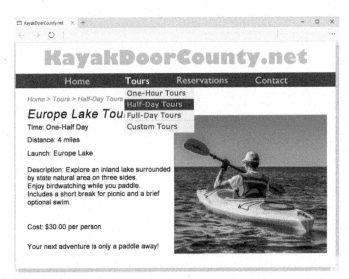

FIGURE 3.32 *Dynamic navigation with HTML, CSS, and JavaScript.*

Site Map

Even with clear and consistent navigation, visitors sometimes may lose their way on large websites. A site map, also referred to as a site index, provides an outline of the organization of the website with hyperlinks to each major page. This can help visitors find another route to get to the information they seek, as shown in Figure 3.33.

Site Search Feature

Note the search feature on the right side of the web page in Figure 3.33. The site search feature helps visitors find information that is not apparent from the navigation or the site map.

FIGURE 3.33 *This large site offers a site search and a site map to visitors.*

Wireframes and Page Layout

A **wireframe** is a sketch or diagram of a web page that shows the structure (but not the detailed design) of basic page elements such as the header, navigation, content area, and footer. Wireframes are used as part of the design process to experiment with various page layouts, develop the structure and navigation of the site, and provide a basis for communication among project members. Note that the exact content (text, images, logo, and navigation) does not need to be placed in the wireframe diagram—the wireframe depicts the overall structure of the page.

Figures 3.34, 3.35, and 3.36 show wireframe diagrams of three possible page designs with horizontal navigation. The wireframe in Figure 3.34 is adequate and may be appropriate for when the emphasis is on text information content, but it's not very engaging.

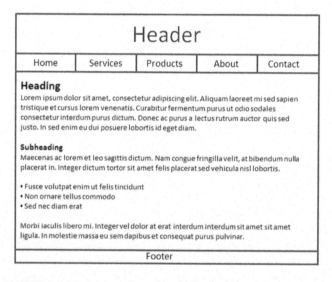

FIGURE 3.34 *An adequate page layout.*

Figure 3.35 shows a diagram of a web page containing similar content formatted in three columns of varying widths with a header area, navigation area, content area (with headings, subheadings, paragraphs, and unordered list), and a footer area.

Figure 3.36 shows a diagram of the same content but formatted in three columns with a header area, top navigation area, large hero image, content area (with headings, subheadings, paragraphs, and unordered lists), and a footer area. Notice how the use of columns and images in Figures 3.35 and 3.36 increase the appeal of the page.

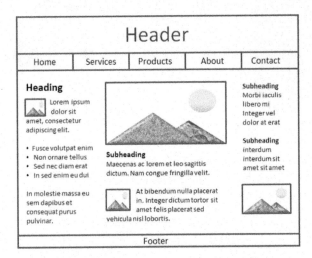

FIGURE 3.35 *The image and columns make this page layout more interesting.*

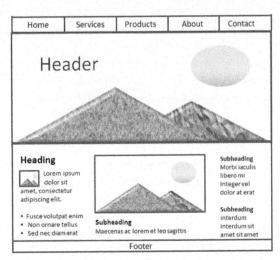

FIGURE 3.36 *This wireframe page layout uses a top navigation area and a hero image.*

The wireframe in Figure 3.37 displays a web page with a header, navigation area, hero image, content area (with heading and subheadings, image, paragraphs, and unordered list), and a footer area.

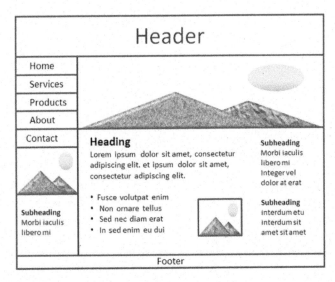

FIGURE 3.37 *Wireframe with vertical navigation.*

Often the page layout for the home page is different from the page layout used for the content pages. Even in this situation, a consistent logo header, navigation, and color scheme will produce a more cohesive website. You'll learn to use Cascading Style Sheets (CSS) along with HTML to configure color, text, and layout as you work through this book. In the next section, you will explore two commonly used layout design techniques: fixed layout and fluid layout.

Fixed and Fluid Layouts

Now that you have been introduced to wireframes as a way to sketch page layout, let's explore two commonly used design techniques to implement those wireframes: fixed layout and fluid layout.

FIGURE 3.38 *This page is configured with a fixed layout design.*

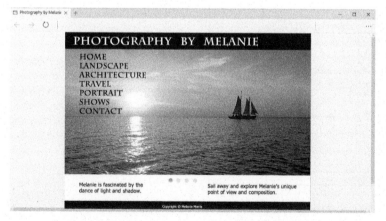

FIGURE 3.39 *This fixed-width, centered content is balanced on the page by left and right margins.*

Fixed Layout

The **fixed layout** technique is sometimes referred to as a solid or "ice" design. The web page content has a fixed width and may hug the left margin as shown in Figure 3.38.

Notice the empty space in the right side of the browser viewport in Figure 3.38. To avoid this unbalanced look, a popular method to create a fixed layout design is to configure the content with a specific width in pixels (such as 960px) and center it in the browser viewport as shown in Figure 3.39. As the browser is resized, it will expand or contract the left and right margin areas to center the content in the viewport. In Chapter 6, you'll learn how to use Cascading Style Sheets (CSS) to configure width and center content.

Fluid Layout

The **fluid layout** technique, sometimes referred to as a "liquid" layout, results in a fluid web page with content typically configured with percentage values for widths—often taking up 100% of the browser viewport. The content will flow to fill whatever size browser window is used to display it, as shown in Figure 3.40. One disadvantage of liquid layout is that when displayed in maximized browser viewports using high screen resolutions, the lines of text may be quite wide and become difficult to scan and read.

Figure 3.41 shows an adaptation of liquid layout that utilizes a 100% width for the header and navigation area along with an 80% width for the centered page content. Compare this to Figure 3.40, the centered content area grows and shrinks as the browser viewport is resized. Readability can be ensured by using CSS to configure a maximum width value for this area.

Websites designed using fixed and fluid layout techniques can be found throughout the Web. Fixed-width layouts provide the web developer with the most control over the page configuration but can result in pages with large empty areas when viewed at higher screen resolutions. Fluid designs may become less readable when viewed at high screen resolutions due to the page stretching to fill a wider area than originally intended by the developer. Configuring a maximum width on text content areas can alleviate the text readability issues. Even when using an overall fluid layout, portions of the design can be configured with a fixed width. Whether employing a fixed or fluid layout, web pages with centered content are typically pleasing to view on a variety of desktop screen resolutions.

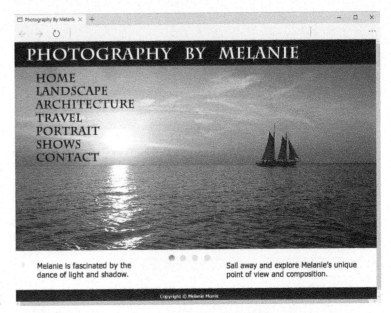

FIGURE 3.40 *This fluid layout expands to fill 100% of the browser viewport.*

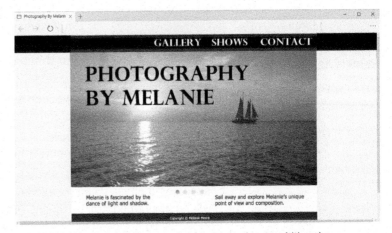

FIGURE 3.41 *This fluid layout also has a maximum width value configured for the centered content area.*

Design for the Mobile Web

Coding techniques to configure responsive web page layouts that display differently on desktop browsers and mobile devices will be introduced in Chapter 8. Figures 3.42 and 3.43 show the same website but look different. Figure 3.42 depicts the desktop browser display. Figure 3.43 shows the display in a small mobile device. Let's explore some design considerations for mobile display.

FIGURE 3.42 *Desktop browser display.*

FIGURE 3.43 *Mobile display.*

Mobile Web Design Considerations

Mobile web users are typically on-the-go, need information quickly, and may be easily distracted. A web page that is optimized for mobile access should try to serve these needs. Take a moment to review Figures 3.42 and 3.43 and observe how the design of the mobile website addresses the following design considerations:

▶ **Small screen size.** The size of the header area is reduced to accommodate a small screen display. It's also common to configure nonessential content, such as sidebar content to not display on a mobile device.

▶ **Low bandwidth (slow connection speed).** Note that a smaller image is displayed on the mobile version of the web page.

▶ **Font, color, and media issues.** Common font typefaces are utilized. There is also good contrast between text and background color.

▶ **Awkward controls, limited processor, and limited memory.** The mobile website uses a single-column page layout that facilitates keyboard tabbing and will be easy to control by touch. The page is mostly text, which will be quickly rendered by a mobile browser.

▶ **Functionality.** A single-column layout is utilized with navigation areas that can be easily selected with a fingertip. The W3C recommends a target size of at least 44 x 22 pixels for controls requiring tapping such a navigation hyperlink.

Let's build on this base of design considerations and expand them, while highlighting W3C recommended best practices for the mobile web.[16]

Optimize Layout for Mobile Use

A single-column page layout (Figure 3.44) with a small header, key navigation links, content, and page footer works well for a mobile device display. Mobile screen resolutions vary greatly (for example, 320×480, 360×640, 375×667, 640×690, and 720×1280). W3C recommendations include the following:

▶ Limit scrolling to one direction.

▶ Use heading elements.

- Use lists to organize information (such as unordered lists, ordered lists, and description lists).
- Avoid using tables (see Chapter 9) because they typically force both horizontal and vertical scrolling on mobile devices.
- Provide labels for form controls (see Chapter 10).
- Avoid using pixel units in style sheets.
- Avoid absolute positioning in style sheets.
- Hide content that is not essential for mobile use.

FIGURE 3.44

Wireframe for a typical single-column page layout.

Optimize Navigation for Mobile Use

Easy-to-use navigation is crucial on a mobile device. The W3C recommends the following:

- Provide minimal navigation near the top of the page.
- Provide consistent navigation.
- Avoid hyperlinks that open files in new windows or pop-up windows.
- Try to balance both the number of hyperlinks on a page and the number of levels of links needed to access information.

Optimize Graphics for Mobile Use

Graphics can help to engage visitors, but be aware of the following W3C recommendations for mobile use:

- Avoid displaying images that are wider than the screen width (assume a 320-pixel screen width on a smartphone display).
- Configure alternate small, optimized background images.
- Some mobile browsers will downsize all images, so images with text can be difficult to read.
- Avoid the use of large graphic images.
- Specify the size of images.
- Provide alternate text for graphics and other nontext elements.

Optimize Text for Mobile Use

It can be difficult to read text on a small mobile device. The following W3C recommendations will aid your mobile visitors:

- Configure good contrast between text and background colors.
- Use common font typefaces.
- Configure font size with em units or percentages.
- Use a short, descriptive page title.

Responsive Web Design

Responsive web design is a term coined by noted web developer Ethan Marcotte to describe progressively enhancing a web page for different viewing contexts (such as smartphones and tablets) through the use of coding techniques, including fluid, flexible layouts, flexible images, and media queries.[17] In Chapter 8, you'll learn to configure responsive layouts with CSS (flexbox and grid layout systems), configure flexible images, and code CSS media queries, which is a technique for configuring a web page to display well at various screen resolutions.

The Media Queries website (https://mediaqueri.es) showcases a gallery of sites that demonstrate this method for responsive web design.[18] The screen captures in the Media Queries gallery display web pages at the following screen widths: 320px (smartphone display), 768px (tablet portrait display), 1024px (netbook display and landscape tablet display), and 1600px (large desktop display).

You might be surprised to discover that Figures 3.45–3.47 are actually the same web page .html file that is configured with CSS to display differently, depending on the viewport size detected by media queries. Figure 3.45 shows the standard desktop browser display.

Display for tablets using portrait orientation is depicted in Figure 3.46. Figure 3.47 shows the web page displayed on a mobile device such as a smartphone—note the reduction of the logo area, removal of images, and prominent phone number.

You'll explore how to configure web pages with responsive coding techniques in Chapter 8.

FIGURE 3.45 *Desktop display of the web page.*

FIGURE 3.46 *Portrait orientation tablet display of the web page.*

FIGURE 3.47 *Smartphone display of the web page.*

Web Design Best Practices Checklist

Use Table 3.1 as a guide to help you create easy-to-read, usable, and accessible web pages.

TABLE 3.1 *Web Design Best Practices Checklist*

Page Layout Criteria

☐ 1. Consistent site header/logo

☐ 2. Consistent navigation area

☐ 3. Informative page title that includes the company/organization/site name

☐ 4. Page footer area—copyright, last update, contact e-mail address

☐ 5. Good use of basic design principles: repetition, contrast, proximity, and alignment

☐ 6. Balance of text/graphics/white space on page

☐ 7. Home page downloads within 10 seconds on a mobile device

☐ 8. Viewport meta tag is used to enhance display on smartphone

☐ 9. Responsive page layout is configured for smartphone and tablet display

Navigation Criteria

☐ 1. Main navigation links are clearly and consistently labeled

☐ 2. Navigation is structured within an unordered list

☐ 3. When the main navigation consists of images and/or multimedia, the page footer area contains plain text hyperlinks (accessibility)

☐ 4. Navigational aids, such as site map, skip to content link, or breadcrumbs, are used

Color and Graphics Criteria

☐ 1. Use of different colors is limited to a maximum of three or four plus neutrals

☐ 2. Color is used consistently

☐ 3. Background and text colors have good contrast

☐ 4. Color is not used alone to convey meaning (accessibility)

☐ 5. Use of color and graphics enhances rather than distracts from the site

☐ 6. Graphics are optimized and do not slow download significantly

☐ 7. Each graphic used serves a clear purpose

☐ 8. Image elements use the alt attribute to configure alternate text (accessibility)

☐ 9. Animated images do not distract from the site and do not loop endlessly

Multimedia Criteria

☐ 1. Each audio or video file used serves a clear purpose

☐ 2. The audio or video files used enhance rather than distract from the site

☐ 3. Captions or transcripts are provided for each audio or video file used (accessibility)

☐ 4. The file size is indicated for audio and video downloads

Content Presentation Criteria

☐ 1. Common fonts such as Arial or Times New Roman are used

☐ 2. Techniques of writing for the Web are applied: headings, subheadings, bulleted lists, short sentences in brief paragraphs, use of empty space

☐ 3. Fonts, font sizes, and font colors are consistently used

☐ 4. Content provides meaningful, useful information

☐ 5. Content is organized in a consistent manner

☐ 6. Information is easy to find (minimal clicks)

☐ 7. Timeliness: The date of the last revision and/or copyright date is accurate

☐ 8. Content is free of typographical and grammatical errors

☐ 9. Avoids the use of "Click here" when writing text for hyperlinks

☐ 10. Hyperlinks use a consistent set of colors to indicate visited/nonvisited status

☐ 11. Alternate text equivalent of content is provided for graphics and media (accessibility)

Functionality Criteria

☐ 1. All internal hyperlinks work

☐ 2. All external hyperlinks work

☐ 3. All forms function as expected

☐ 4. No error messages are generated by the pages

Additional Accessibility Criteria

☐ 1. Use attributes designed to improve accessibility such as alt and title where appropriate

☐ 2. The html element's lang attribute indicates the spoken language of the page

Browser Compatibility Criteria

☐ 1. Displays on current versions of Edge, Internet Explorer, Firefox, Safari, Chrome, and Opera

☐ 2. Displays on popular mobile devices (including tablets and smartphones)

CHAPTER 3

Review and Apply

Review Questions

1. Which of the following is a sketch or diagram of a web page that shows the structure (but not the detailed design) of basic page elements?
   a. drawing
   b. HTML code
   c. site map
   d. wireframe

2. Which of the following are the three most common methods of organizing websites?
   a. horizontal, vertical, and diagonal
   b. hierarchical, linear, and random
   c. accessible, readable, and maintainable
   d. none of the above

3. Which of the following is not a web design recommended practice?
   a. design your site to be easy to navigate
   b. colorful pages appeal to everyone
   c. design your pages to load quickly
   d. limit the use of animated items

4. Which are the four principles of the Web Content Accessibility Guidelines?
   a. contrast, repetition, alignment, and proximity
   b. perceivable, operable, understandable, and robust
   c. accessible, readable, maintainable, and reliable
   d. hierarchical, linear, random, and sequential

5. Which of the following would a consistent website design *not* have?
   a. a similar navigation area on each content page
   b. the same fonts on each content page
   c. a different background color on each page
   d. the same logo in the same location on each content page

6. Which of the following is the design technique used to create pages that stretch to fill the browser window?
   a. fixed
   b. fluid
   c. wireframe
   d. hero

7. Which of the following recommended design practices applies to a website that uses images for its main site navigation?
   a. provide alternative text for the images
   b. place text links at the bottom of the page
   c. both a and b
   d. no special considerations are needed

8. Which of the following is a mobile web design best practice?
   a. configure a single-column page layout
   b. configure a multiple-column page layout
   c. avoid using lists to organize information
   d. embed text in images wherever possible

9. Which of the following should you do when creating text hyperlinks?
   a. create the entire sentence as a hyperlink
   b. include the words "click here" in your text
   c. use a key phrase as a hyperlink
   d. none of the above

10. Which of the following is a color scheme that consists of two colors opposite each other on the color wheel?
    a. contrasting
    b. analogous
    c. split complementary
    d. complementary

Review Answers

1. d 2. b 3. b 4. b 5. c 6. b 7. c 8. a 9. c 10. d

Hands-On Exercise

1. **Website Design Evaluation.** In this chapter, you've explored web page design, including navigation design techniques and the design principles of contrast, repetition, alignment, and proximity. In this Hands-On Exercise, you'll review and evaluate the design of a website. Your instructor may provide you with the URL of a website to evaluate. If not, choose a website to evaluate from the following list of URLs:

   https://www.telework.gov

   https://www.dcmm.org

   https://www.sedonalibrary.org

   https://bostonglobe.com

   https://www.alistapart.com

   Visit the website you are evaluating. Write a paper that includes the following information:
   a. URL of the website
   b. Name of the website
   c. Target audience
   d. Screenshot of the home page
   e. Indicate the type(s) of navigation evident.
   f. Describe how the design principles of contrast, repetition, alignment, and proximity are applied. Be specific.
   g. Complete the Web Design Best Practices Checklist (see Table 3.1).
   h. Recommend three improvements for the website.

2. **Responsive Web Design.** Visit the Media Queries website at https://mediaqueri.es to view a gallery of sites that demonstrate responsive web design. Choose one of the example responsive websites to explore. Write a paper that includes the following:
   a. URL of the website
   b. Name of the website
   c. Target audience
   d. Three screenshots of the website (desktop display, tablet display, and smartphone display).
   e. Describe the similarities and differences between the three screenshots.
   f. Describe two ways in which the display has been modified for smartphones.
   g. Does the website meet the needs of its target audience in all three display modes? Why or why not? Justify your answer.

Focus on Web Design

Choose two sites that are similar in nature or have a similar target audience, such as the following:

- https://amazon.com and https://bn.com
- https://chicagobears.com and https://greenbaypackers.com
- https://cnn.com and https://msnbc.com

a. Describe how the two sites that you chose to review exhibit the design principles of repetition, contrast, alignment, and proximity.

b. Describe how the two sites that you chose to review exhibit web design best practices. How would you improve these sites? Recommend three improvements for each site.

Web Project Case Study

The purpose of this Web Project Case Study is to design a website using recommended design practices. Your website might be about a favorite hobby or subject, your family, a church or club you belong to, a company that a friend owns, the company you work for, and so on. Your website will contain a home page and at least six (but no more than 10) content pages. The Web Project Case Study provides an outline for a semester-long project in which you design, create, and publish an original website.

Project Milestones

- Web Project Topic Approval (must be approved before moving on to other milestones)
- Web Project Planning Analysis Sheet
- Web Project Site Map
- Web Project Page Layout Design
- Web Project Update 1
- Web Project Update 2
- Publish and Present Project

1. **Web Project Topic Approval.** The topic of your website must be approved by your instructor. Write a one-page paper with a discussion of the following items:

   - What is the name and purpose of the site?

     List the website name and the reasons you are creating the site.

   - What do you want the site to accomplish?

     Explain the goal you have for the site. Describe what needs to happen for you to consider your site a success.

   - Who is your target audience?

     Describe your target audience by age, gender, socioeconomic characteristics, and so on.

   - What opportunity or issue is your site addressing?

     Note: Your site might be addressing the opportunity of providing information about a topic to others, creating an initial web presence for a company, and so on.

   - What type of content might be included in your site?

     Describe the type of text, graphics, and media you will need for the site.

   - List at least two related or similar sites found on the Web.

2. **Web Project Planning Analysis Sheet.** Write a one-page paper with a discussion of the following items. Include the following headings:

   Website Goal

   List the website name and describe the goal of your site in one or two sentences.

What results do I want to see?

List the working title of each page on your site. A suggested project scope is 7–11 pages.

What information do I need?

List the sources of the content (facts, text, graphics, sounds and video) for the web pages you listed. While you should write the text content yourself, you may use outside sources for royalty-free images and multimedia. Review copyright considerations (see Chapter 1).

3. **Web Project Site Map.** Use the drawing features of a word processing program, a graphic application, or paper and pencil to create a site map of your website that shows the hierarchy of pages and relationships between pages. Unless otherwise directed by your instructor, use the style for a site map shown in Figure 3.3.

4. **Web Project Page Layout Design.** Use the drawing features of a word processing program, a graphic application, or paper and pencil to create wireframe page layouts for the home page and content pages of your site. Unless otherwise directed by your instructor, use the style for page layout composition shown in Figures 3.34–3.37. Indicate where the logo, navigation, text, and images will be located. Do not worry about exact wording or exact images.

5. **Project Update Meeting 1.** You should have at least three pages of your website completed by this time. If you have not done so already, your instructor will help you to publish your pages to the Web (see Chapter 12 for information about selecting a web host). Unless prior arrangements to meet are made, the Project Update Meeting will be held during class lab time. Bring the following items to discuss with your instructor:

   - The URL of your website
   - Source files of your web pages and images
   - Site map (revise as needed)

6. **Project Update Meeting 2.** You should have at least six pages of your website completed by this time. They should be published to the Web. Unless prior arrangements to meet are made, the Project Update Meeting will be held during class lab time. Prepare the following items to discuss with your instructor:

   - The URL of your website
   - Source files of your web pages and images
   - Site map (revise as needed)

7. **Publish and Present Project.** Finish publishing your project to your website. Be prepared to show your website to the class, explaining project goal, target audience, use of color, and any challenges you faced (and how you overcame them) while you completed the project.

Endnotes

1. "Desktop Browser Market Share Worldwide." *StatCounter Global Stats*, gs.statcounter.com/browser-market-share/desktop/worldwide.
2. "Mobile Browser Market Share Worldwide." *StatCounter Global Stats*, gs.statcounter.com/browser-market-share/mobile/worldwide.
3. "Screen Resolution Stats Worldwide." *StatCounter Global Stats*, gs.statcounter.com/screen-resolution-stats.
4. Cowan, Nelson. "Statement of Research Interests and Orientation." *Dr. Nelson Cowan | Working- Memory Laboratory, University of Missouri*, University of Missouri, 2019, memory.psych.missouri.edu/cowan.html.
5. "About the ICT Refresh." *About the ICT Refresh*, United States Access Board, www.access- board.gov/guidelines-and-standards/communications-and-it/about-the-ict-refresh.
6. "Web Content Accessibility Guidelines (WCAG) Overview." Edited by Shawn Lawton Henry, *Web Accessibility Initiative (WAI)*, W3C, 22 June 2018, www.w3.org/WAI/standards-guidelines/wcag/.
7. "Contrast Checker." *WebAIM: Web Accessibility in Mind*, Utah State University, webaim.org/resources/contrastchecker/.
8. "Deque Color Contrast Analyzer." *Deque University*, dequeuniversity.com/color-contrast.
9. Holst, Christian. "Readability: the Optimal Line Length." *Readability: the Optimal Line Length - Articles - Baymard Institute*, 2010, baymard.com/blog/line-length-readability.
10. "Colour Blindness." *Colour Blind Awareness*, www.colourblindawareness.org/colour-blindness/.
11. Horvath, Meghan. "How to Design for Color Blindness." *The Latest Voice of Customer and CX Trends | Usabilla Blog*, Usabilla by Survey Monkey, 13 Sept. 2018, usabilla.com/blog/how-to- design-for-color-blindness/.
12. Nielsen, Jakob. "How Long Do Users Stay on Web Pages?" Nielsen Norman Group, 11 Sept. 2011, www.nngroup.com/articles/how-long-do-users-stay-on-web-pages/.
13. "Demographics of Internet and Home Broadband Usage in the United States." *Pew Research Center: Internet, Science & Tech*, Pew Research Center, 2020, www.pewresearch.org/internet/fact-sheet/internet-broadband/.
14. "Demographics of Mobile Device Ownership and Adoption in the United States." *Pew Research Center: Internet, Science & Tech*, Pew Research Center, 2020, www.pewresearch.org/internet/fact-sheet/mobile/.
15. "Desktop vs Mobile vs Tablet Market Share Worldwide." *StatCounter Global Stats*, gs.statcounter.com/platform-market-share/desktop-mobile-tablet.
16. "Mobile Web Application Best Practices." Edited by Adam Connors and Bryan Sullivan, *W3C*, W3C, 14 Dec. 2010, www.w3.org/TR/mwabp/.
17. Marcotte, Ethan, et al. "Responsive Web Design." *A List Apart*, 1 May 2019, alistapart.com/article/responsive-web-design/.
18. "Media Queries." Edited by Eivind Uggedal, *Media Queries*, mediaqueri.es/.

Cascading Style Sheets Basics

Now that you have experience with configuring the structure and information on a web page with HTML, let's explore **Cascading Style Sheets (CSS)**. Web designers use CSS to separate the presentation style of a web page from the information on the web page. CSS is used to configure text, color, and page layout.

CSS first became a W3C Recommendation in 1996. Additional properties for positioning web page elements were introduced to the language with **CSS level 2 (CSS2)** in 1998 but did not reach official Recommendation status until 2011. **CSS level 3 (CSS3)** properties support features such as embedding fonts, rounded corners, and transparency. The CSS specification is divided into modules, each with a specific purpose. These modules move along the approval process independently. The W3C continues to evolve CSS, with proposals for many types of properties and functionality currently in draft form. This chapter introduces you to the use of CSS on the Web as you explore configuring color on web pages.

You'll learn how to...

- Describe the purpose of Cascading Style Sheets
- List advantages of using Cascading Style Sheets
- Configure color on web pages with Cascading Style Sheets
- Configure inline styles
- Configure embedded style sheets
- Configure external style sheets
- Configure web page areas with element name, class, id, and descendant selectors
- Describe the order of precedence in CSS
- Test your Cascading Style Sheets for valid syntax

Cascading Style Sheets Overview

For years, style sheets have been used in desktop publishing to apply typographical styles and spacing instructions to printed media; CSS provides this functionality (and much more) for web designers. CSS allows web designers to apply typographical styles (typeface, font size, and so on), color, and page layout instructions to a web page.

CSS is a flexible, cross-platform, standards-based language developed by the W3C.[1] Be aware that even though CSS has been in use for many years, new features are continually developed. Be aware that different browsers may not support newer features of CSS in exactly the same way. We concentrate on aspects of CSS that are well supported by modern browsers.

Advantages of Cascading Style Sheets

There are several advantages to using CSS (see Figure 4.1):

▶ **Typography and page layout can be better controlled.**   These features include font size, line spacing, letter spacing, indents, margins, and element positioning.

▶ **Style is separate from structure.**   The format of the text and colors used on the page can be configured and stored separately from the body section of the web page document.

▶ **Styles can be stored.**   You can store styles in a separate document and associate them with the web page. When the styles are modified, the web page code remains intact. This means that if your client decides to change the background color from red to white, you only need to change one file that contains the styles, instead of each web page document.

▶ **Documents are potentially smaller.**   The formatting is separate from the document; therefore, the actual documents should be smaller.

▶ **Site maintenance is easier.**   Again, if the styles need to be changed, it's possible to complete the modifications by changing only the style sheet file.

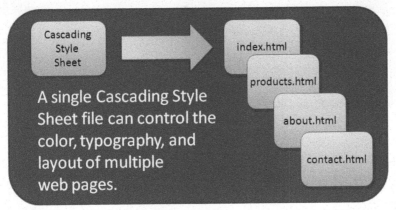

A single Cascading Style Sheet file can control the color, typography, and layout of multiple web pages.

Cascading Style Sheet

index.html
products.html
about.html
contact.html

FIGURE 4.1 *The power of a single CSS file.*

Methods of Configuring Cascading Style Sheets

Web designers use four methods to incorporate CSS technology in a website: inline, embedded, external, and imported.

- **Inline styles** are coded in the body of the web page as an attribute of an HTML tag. The style only applies to the specific element that contains it as an attribute.

- **Embedded styles** are defined in the head section of a web page. These style instructions apply to the entire web page document.

- **External styles** are coded in a separate text file, called an external style sheet. This text file is associated with a web page by coding a link element in the head section.

- **Imported styles** are similar to external styles in that they can connect styles coded in a separate text file with a web page document. An external style sheet can be imported into embedded styles or into another external style sheet by using the `@import` directive.

The "Cascade" in Cascading Style Sheets

Figure 4.2 shows the "cascade" (**order of precedence**) that applies the styles in order from outermost (external styles) to innermost (inline styles). This allows the site-wide styles to be configured with an external style sheet file but overridden when needed by more granular, page-specific styles (such as embedded or inline styles). The order the styles are coded in the web page matters. If styles are conflicting or apply to the same element, the last style rendered by the browser overrides earlier styles.

You'll learn to configure inline styles, embedded styles, and external styles in this chapter.

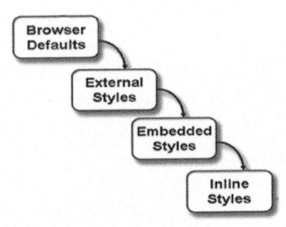

Browser Defaults → External Styles → Embedded Styles → Inline Styles

FIGURE 4.2 *The "cascade" of Cascading Style Sheets.*

CSS Selectors and Declarations

Style Rule Basics

Style sheets are composed of **rules** that describe the styling to be applied. Each **rule** has two parts: a **selector** and a **declaration**.

▶ **CSS Style Rule Selector**

There are several different types of selectors. The selector can be an HTML element name, a class name, or an id name. In this section, we'll focus on applying styles to element name selectors. You'll work with class selectors and id selectors later in this chapter.

▶ **CSS Style Rule Declaration**

The declaration indicates the CSS **property** you are setting (such as color) and the value you are assigning to the property.

For example, the CSS rule shown in Figure 4.3 would set the color of the text used on a web page to blue. The selector represents the body element, and the declaration sets the color property to the value of blue.

FIGURE 4.3 *Using CSS to set the text color to blue.*

The `background-color` Property

The purpose of the CSS **background-color** property is to configure the background color of an element. The following style rule will configure the background color of a web page to be yellow:

```
body { background-color: yellow }
```

Notice how the declaration is enclosed within braces and how the colon symbol (:) separates the declaration property and the declaration value.

The `color` Property

The CSS property to configure the text color of an element is **`color`**. The following CSS style rule will configure the text color of a web page to be blue:

```
body { color: blue }
```

Configure Background and Text Color

To configure more than one property for a selector, use a semicolon (;) to separate the declarations. The following CSS style rule configures the web page in Figure 4.4 with white text and an orchid background:

```
body { color: white; background-color: orchid; }
```

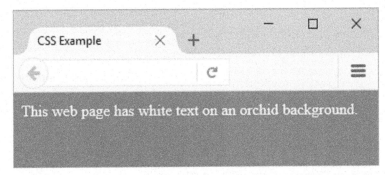

FIGURE 4.4 *A web page with orchid background color and white text color.*

You might be asking how you would know what properties and values are allowed to be used. See the CSS Cheat Sheet in the Appendix for a detailed list of CSS properties. This chapter introduces you to the CSS properties commonly used to configure color, shown in Table 4.1.

TABLE 4.1 *CSS Properties Used in This Chapter*

Property	Description	Value
background-color	Background color of an element	Any valid color value
color	Foreground (text) color of an element	Any valid color value

CSS Syntax for Color Values

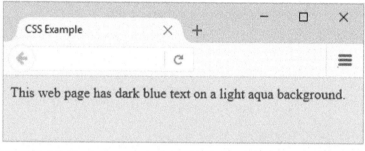

FIGURE 4.5 *The color was configured using hexadecimal color values.*

The previous section used color name keywords to configure color with CSS. There are a limited number of color name keywords.[2] For more flexibility and control, use a numerical color value, such as the hexadecimal color values introduced in Chapter 3 (in the Web Color Palette section). The Web Safe Color Palette Appendix provides examples of colors created with hexadecimal values.

A style rule to configure the web page displayed in Figure 4.5 with dark blue text (#000066) on a light aqua background (#CCFFFF) is

```
body { color: #000066; background-color: #CCFFFF; }
```

The spaces in these declarations are optional. The ending semicolon (;) is also optional but useful in case you need to add additional style rules at a later time. The following alternative versions of the code above are also valid:

EXAMPLE 1:
```
body {color:#000066;background-color:#CCFFFF}
```

EXAMPLE 2:
```
body { background-color:#000066; color:#CCFFFF; }
```

EXAMPLE 3:
```
body {
color: #000066;
background-color: #CCFFFF;
}
```

EXAMPLE 4:
```
body { color: #000066;
       background-color: #CCFFFF;
}
```

The W3C's CSS Color Module[3] provides a variety of ways to configure colors:

- color name
- hexadecimal color value
- hexadecimal shorthand color value
- decimal color value (RGB triplet)
- HSL (hue, saturation, and lightness) color value notation new to CSS3 (introduced in Chapter 6)

We'll typically use hexadecimal color values in this book. Table 4.2 shows a variety of CSS syntax examples that configure a paragraph with red text.

TABLE 4.2 *Syntax to Configure a Paragraph with Red Text*

CSS Syntax	Color Type
`p { color: red }`	Color name
`p { color: #FF0000 }`	Hexadecimal color value
`p { color: #F00 }`	Shorthand hexadecimal (one character for each hexadecimal pair—only used with web-safecolors)
`p { color: rgb(255,0,0) }`	Decimal color value (RGB triplet)
`p { color: hsl(0, 100%, 50%) }`	HSL color values

Although hexadecimal color notation is commonly used on most web pages, the W3C developed a color notation called HSL (hue, saturation, and lightness) to provide a more intuitive way to describe color on web pages.[4] The hue is the actual color which is represented by numeric values ranging from 0 to 360 (like the 360 degrees in a circle). For example, red is represented by both the values 0 and 360, green is represented by 120, and blue is represented by 240. Saturation is indicated by a percentage value (full color saturation = 100% and gray = 0%). A percentage value is also used to configure lightness (normal color = 50%, white = 100%, and black = 0%). Table 4.2 includes the HSL representation for the color red. A dark blue color could be represented by `hsl(240, 100%, 25%)`. Explore the following color tools:

- ColorHexa
  https://www.colorhexa.com/
- Color Designer
  https://colordesigner.io/

Are there other methods to configure color with CSS?

Yes, the CSS Color Module provides a way for web designers to configure not only color but also the transparency of the color with RGBA (Red, Green, Blue, Alpha) color and HSLA (Hue, Saturation, Lightness, Alpha) color. The CSS opacity property and CSS gradient backgrounds can be also be used to configure color on web pages. You'll explore these techniques in Chapter 6.

Configure Inline CSS

There are four methods for configuring CSS: inline, embedded, external, and imported. In this section, we focus on inline CSS.

The `style` Attribute

Inline styles are coded as an attribute on an HTML tag using the **`style` attribute**. The value of the `style` attribute is set to the style rule declaration that you need to configure. Recall that a declaration consists of a property and a value. Each property is separated from its value with a colon (:). The following code will set the text color of an `<h1>` tag to a shade of red:

```
<h1 style="color:#cc0000">This is displayed as a red heading</h1>
```

If there is more than one property, each is separated by a semicolon (;). The following code configures the heading with a red text color and a gray background color:

```
<h1 style="color:#cc0000;background-color:#cccccc;">
This is displayed as a red heading on a gray background</h1>
```

 Hands-On Practice 4.1

In this Hands-On Practice, you will configure a web page with inline styles. The inline styles will specify the following:

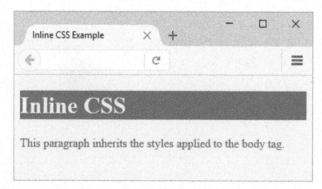

▶ Global body tag styles for an off-white background with teal text. These styles will be inherited by other elements within the body of the web page by default.

▶ Styles for an h1 element with a teal background with off-white text. This will override the global styles configured on the body element.

A sample is shown in Figure 4.6. Launch a text editor and open the template.html file from the chapter4 folder in the student files.

FIGURE 4.6 *Web page using inline styles.*

Modify the title element and add heading tag, paragraph tags, style attributes, and text to the body section as indicated by the following code:

```html
<!DOCTYPE html>
<html lang="en">
<head>
<title>Inline CSS Example</title>
<meta charset="utf-8">
</head>
<body style="background-color:#F5F5F5;color:#008080;">
  <h1 style="background-color:#008080;color:#F5F5F5;">Inline CSS</h1>
  <p>This paragraph inherits the styles applied to the body tag.</p>
</body>
</html>
```

Save the document as inline.html on your hard drive or flash drive. Launch a browser to test your page. It should look similar to the page shown in Figure 4.6. Note that the inline styles applied to the body tag are inherited by other elements on the page (such as the paragraph) unless more-specific styles are specified (such as those coded on the <h1> tag). You can compare your work with the solution found in the student files (chapter4/4.1/inline.html).

Let's continue and add another paragraph with the text color configured to be dark gray.

```html
<p style="color:#333333"> This paragraph overrides the text color
style applied to the body tag.</p>
```

Save the document as inline2.html. It should look similar to the page shown in Figure 4.7. You can compare your work with the solution at chapter4/4.1/inline2.html in the student files.

Note that the inline styles applied to the second paragraph override the global styles applied to the body of the web page. What if you had ten paragraphs that needed to be configured in this manner? You'd have to code an inline style on *each* of the 10 paragraph tags. This would add quite a bit of redundant code to the page. For this reason, inline styles are not the most efficient way to use CSS. In the next section, you'll learn how to configure embedded styles, which can apply to the entire web page document.

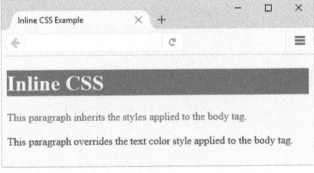

FIGURE 4.7 *The second paragraph's inline styles override the global styles configured on the body tag.*

While inline styles can sometimes be useful, you'll find that you won't use this technique much in practice—it is inefficient, adds extra code to the web page document, and is inconvenient to maintain. However, inline styles can be quite handy in some circumstances, such as when you post an article to a content management system or blog and need to tweak the site-wide styles a bit to help get your point across.

Configure Embedded CSS

The Style Element

The opening `<style>` tag begins the area with embedded style rules, and the closing `</style>` tag ends the area containing embedded style rules. Embedded styles apply to the entire document and are typically placed within a **style element** located in the head section of a web page. Note that while HTML 5.2 also allows the style element in the body section, we'll always code the style element in the head section.

FIGURE 4.8 *Web page using embedded styles.*

The web page in Figure 4.8 uses embedded styles to set the text color and background color of the web page document with the body element selector. See the example in the student files at chapter4/embed.html.

```
<!DOCTYPE html>
<html lang="en">
<head>
<title>Embedded CSS Example</title>
<meta charset="utf-8">
<style>
body { background-color: #E6E6FA;
       color: #191970;
}
</style>
</head>
<body>
  <h1>Embedded CSS</h1>
  <p>This page uses embedded styles.</p>
</body>
</html>
```

Notice the way the style rules were coded with each rule on its own line. This makes the styles more readable and easier to maintain than one long row of text. The styles are in effect for the entire web page document because they were applied to the `<body>` tag using the body element selector.

 Hands-On Practice 4.2 ———————————————————————————

Launch a text editor and open the starter.html file from the chapter4 folder in the student files. Save your page as embedded.html and test it in a browser. Your page should look similar to the one shown in Figure 4.9.

In this Hands-On Practice, you'll code embedded styles to configure selected background and text colors. You'll use the body element selector to configure the default background color (#F9F0FE) and default text color (#5B3256) for the entire page. You'll also use the h1 and h2 element selectors to configure different background and text colors for the heading areas.

Edit the embedded.html file in a text editor and add the following code in the head section above the closing </head> tag:

```
<style>
body { background-color: #F9F0FE;
       color: #5B3256; }
h1 { background-color: #833B83;
     color: #F9F0FE; }
h2 { background-color: #AD77C3;
     color: #F9F0FE; }
</style>
```

Save and test your file in a browser. Figure 4.10 displays the web page along with color swatches. A monochromatic color scheme was chosen. Notice how the repetition of a limited number of colors adds interest and unifies the design of the web page.

View the source code for your page and review the CSS and HTML code. See chapter4/4.2/ embedded.html in the student files for an example of this web page. Note that all the styles are in a single location on the web page. Since embedded styles are coded in a specific location, they are easier to maintain over time than inline styles. Also notice that you only needed to code the styles for the h2 element selector once (in the head section) and both of the <h2> tags applied the h2 style. This is more efficient than coding the same inline style on each <h2> tag.

However, it's uncommon for a website to have only one page. Repeating the CSS in the head section of each web page file is inefficient and difficult to maintain. In the next section, you'll use a more efficient approach—configuring an external style sheet.

FIGURE 4.9 *The web page without any styles.*

FIGURE 4.10 *The web page after embedded styles are configured.*

Configure
External CSS

VideoNote
External Style Sheets

The flexibility and power of CSS are best utilized when the CSS is external to the web page document. An external style sheet is a text file with a .css file extension that contains CSS style rules. The external style sheet file is associated with a web page using the link element. This provides a way for multiple web pages to be associated with the same external style sheet file. The external style sheet file does not contain any HTML tags—it only contains CSS style rules.

The advantage of external CSS is that styles are configured in a single file. This means that when styles need to be modified, only one file needs to be changed, instead of multiple web pages. On large sites, this can save a web developer much time and increase productivity. Let's get some practice with this useful technique.

The Link Element

The **link element** associates an external style sheet with a web page. It is placed in the head section of the page. The link element is a stand-alone, void tag. When coding in HTML5, two attributes are used with the link element: `rel` and `href`.

- ▶ The value of the **rel attribute** is `"stylesheet"`.
- ▶ The value of the **href attribute** is the name of the style sheet file.

Code the following in the head section of a web page to associate the document with the external style sheet named color.css:

```
<link rel="stylesheet" href="color.css">
```

Hands-On Practice 4.3

Let's practice using external styles. First, you'll create an external style sheet. Then you'll configure a web page to be associated with the external style sheet.

Create an External Style Sheet. Launch a text editor and type style rules to set the background color of a page to blue and the text color to white. Save the file as color.css. The code follows:

```
body { background-color: #0000FF;
       color: #FFFFFF; }
```

Figure 4.11 shows the external color.css style sheet displayed in Notepad. Notice that there is no HTML in this file. HTML tags are not coded within an external style sheet. Only CSS rules (selectors, properties, and values) are coded in an external style sheet.

FIGURE 4.11 *The external style sheet color.css.*

Configure the Web Page. To create the web page shown in Figure 4.12, launch a text editor and open the template.html file from the chapter4 folder in the student files. Modify the title element, add a link tag to the head section, and add a paragraph to the body section as indicated by the following code:

```
<!DOCTYPE html>
<html lang="en">
<head>
<title>External Styles</title>
<meta charset="utf-8">
<link rel="stylesheet" href="color.css">
</head>
<body>
<p>This web page uses an external style sheet.</p>
</body>
</html>
```

Save your file as external.html in the same folder as your color.css file. Launch a browser and test your page. It should look similar to the page shown in Figure 4.12. You can compare your work with the solution in the student files (chapter4/4.3/external.html).

The color.css style sheet can be associated with any number of web pages. If you ever need to change the style of formatting, you only need

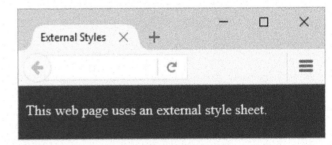

FIGURE 4.12 *This page is associated with an external style sheet.*

to change a single file (color.css) instead of multiple files (all of the web pages). As mentioned earlier, this technique can boost productivity on a large site. This is a simple example, but the advantage of having only a single file to update is significant for both small and large websites.

CSS Selectors: Class, Id, and Descendant

The Class Selector

Use a CSS **class selector** to apply a CSS declaration to one or more areas on a web page. When setting a style for a class, configure the class name as the selector. Place a dot or period (.) in front of the class name in the style sheet. A class name must begin with a letter and may contain numbers, hyphens, and underscores. Class names may not contain spaces. The following code configures a class called `feature` in a style sheet with a foreground (text) color set to red: `.feature { color: #FF0000; }`

The styles set in the new class can be applied to any element you wish. You do this by using the class attribute, such as `class="feature"`. The following code will apply the feature class styles to a `<li>` element: `<li class="feature">Usability Studies</li>`

The Id Selector

Use an **id selector** to identify and apply a CSS rule uniquely to a *single area* on a web page. Unlike a class selector which can be applied multiple times on a web page, an id may only be applied once per web page. When setting a style for an id, place a hash mark (#) in front of the id name in the style sheet. An id name may contain letters, numbers, hyphens, and underscores. Id names may not contain spaces. The following code will configure an id called `content` in a style sheet:

```
#content { color: #333333; }
```

The styles set in the `content` id can be applied to the element you wish by using the id attribute, `id="content"`. The following code will apply the content id styles to a div tag:

```
<div id="content">This sentence will be displayed using styles
configured in the content id.</div>
```

The Descendant Selector

Use a **descendant selector** to specify an element within the context of its container (parent) elements. Using descendant selectors can help you to reduce the number of different classes and ids but still allow you to configure CSS for specific areas on the web page. To configure a descendant selector, list the container selector (which can be an element selector, class, or id) followed by the specific selector you are styling. For example, to specify a green text color for paragraphs located *within* the `main` element, code the following style rule:

```
main p { color: #00ff00; }
```

Hands-On Practice 4.4

In this Hands-On Practice, you will modify the Trillium Media Design page while you practice configuring a class and an id. Launch a text editor and open the embedded.html file from the chapter4/4.2 folder in the student files. Save the file as classid.html.

Configure the CSS. Edit the embedded CSS in the head section of the web page as you configure a class named `feature` and an id named `new`.

1. Create a class named `feature` that configures red (#B33939) text. Add the following code to the embedded styles in the head section of the web page:

   `.feature { color: #B33939; }`

2. Create an id named `new` that configures a medium blue text color. Add the following code to the embedded styles in the head section of the web page:

   `#new { color: #227093; }`

Configure the HTML. Associate HTML elements with the class and id you just created.

FIGURE 4.13 *CSS class and id selectors are used on this page.*

1. Modify the last two `<li>` tags in the unordered list. Add a class attribute that associates the `<li>` with the `feature` class as follows:

   `<li class="feature">Usability Studies</li>`
   `<li class="feature">Search Engine Optimization</li>`

2. Modify the second opening paragraph tag. Add an id attribute that associates the paragraph with the id named `new`:

   `<p id="new">`

Save your classid.html file and test it in a browser. Your page should look similar to the image shown in Figure 4.13. Notice how the class and id styles are applied. The student files contain a sample solution at chapter4/4.4/classid.html.

 For maximum compatibility, choose your class and id names carefully. Always begin with a letter. Do not use any blank spaces. Feel free to use numerals, the dash character, and the underscore character in addition to letters.

Span Element

The Span Element

The inline **span element** defines a section on a web page that is displayed inline without empty space above and below. A span element begins with a `<span>` tag and ends with a `</span>` tag. Use the span element when you need to format an area that is contained within another, such as within a `<p>`, `<blockquote>`, or `<div>` element.

 Hands-On Practice 4.5 ─────────────────────────────

In this Hands-On Practice, you will experiment with span elements in the Trillium Media Design home page. Launch a text editor and open the starter.html file from the chapter4 folder in the student files. Save your file as span.html and test it in a browser. Your page should look similar to the one shown in Figure 4.9.

FIGURE 4.14 *This page uses the span element.*

Open span.html in a text editor and view the source code. In this Hands-On Practice, you'll code embedded styles to configure selected background and text colors. You'll also add `<span>` tags to the web page. When you are finished with the first part of this Hands-On Practice, your web page will be similar to Figure 4.14.

Part 1

Configure the Embedded Styles. Edit span2.html in a text editor and add embedded styles in the head section above the closing `</head>` tag. You will configure styles for a class named `companyname` and for the body, h1, h2, nav, and footer element selectors. The code is

```
<style>
body { background-color: #F7F7F7;
      color: #191970; }
h1 { background-color: #833B83;
     color: #F9F0FE; }
h2 { color: #AD77C3; }
nav { background-color: #EAEAF2; }
footer { color: #666666; }
.companyname { color: #833B83; }
</style>
```

Configure the Company Name. View Figure 4.14 and notice that the company name, Trillium Media Design, is displayed in a different color than the other text within the first paragraph. You've already created a class named `companyname` in the CSS. You'll use a span to apply this formatting. Find the text "Trillium Medium Design" in the first paragraph. Configure a span element to contain this text. Assign the span to the class named `companyname`. A sample code excerpt follows:

```
<p><span class="companyname">Trillium Media Design</span> will bring
```

Save your file and test in a browser. Your page should look similar to the one shown in Figure 4.14. The student files contain a sample solution at chapter4/4.5/span.html.

Part 2

As you review your web page and Figure 4.14, notice the empty space between the h1 element and the navigation area—the empty space is the default bottom margin of the h1 element. The margin is one of the components of the CSS box model with which you'll work in Chapter 6. One technique that will cause the browser to collapse this empty space is to configure the margin between the elements. Add the following style to the h1 element selector in the embedded CSS: `margin-bottom: 0;`

Save the file and launch in a browser. Your web page should now be similar to Figure 4.15. Notice how the display of the h1 and navigation area has changed. The student files contain a sample solution at chapter4/4.5/rework.html.

FIGURE 4.15 *The new header area.*

 How do I know when to use an id, a class, or a descendant selector?

The most efficient way to configure CSS is to use HTML elements as selectors. However, sometimes you need to be more specific—that's when other types of selectors are useful. Create a class when you need to configure one or more specific objects on a web page in the same way. A class can be applied more than once per web page. An id is similar to a class, but be mindful that it is not valid to apply an id more than once on a web page. To repeat: an id can be used once and only once on each web page. Use an id for a unique item, such as the navigation hyperlink that indicates the current page. As you become more comfortable with CSS, you'll begin to see the power and efficiency of descendant selectors, which allow you to target elements within a specific context (such as all paragraphs in the footer area) without the need to code additional classes or ids within the HTML code.

Practice with CSS

Hands-On Practice 4.6

In this Hands-On Practice, you'll continue to gain experience using external style sheets as you modify the Trillium Media Design website to use an external style sheet. You'll create the external style sheet file named trillium.css, modify the home page (index.html) to use external styles instead of embedded styles, and associate a second web page with the trillium.css style sheet.

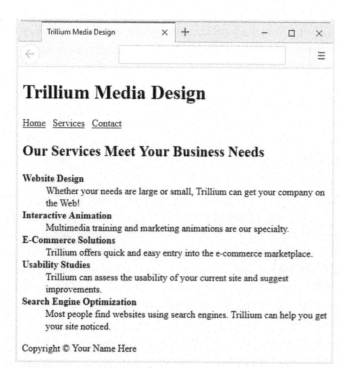

FIGURE 4.16 *The external style sheet named trillium.css.*

You'll use the span.html file from Hands-On Practice 4.5 shown in Figure 4.14 as a starting point.

Launch a text editor and open the span.html file from the chapter4/4.5 folder in the student files. Save the file as index.html in a folder named trillium.

Convert the Embedded CSS to External CSS Edit the index.html file and select the CSS rules (all the lines of code between, but not including, the `<style>` and `</style>` tags). Select Edit > Copy to copy the CSS code to the clipboard. You will place the CSS in a new file. Launch a text editor, select File > New to create a new file, paste the CSS style rules by selecting Edit > Paste, and save the file as trillium.css in the trillium folder. See Figure 4.16 for a screenshot of the new trillium.css file in the Notepad text editor. Notice that there are no HTML elements in trillium.css—not even the `<style>` element. The file contains CSS rules only.

Trillium Media Design

Home   Services   Contact

Our Services Meet Your Business Needs

Website Design
Whether your needs are large or small, Trillium can get your company on the Web!
Interactive Animation
Multimedia training and marketing animations are our specialty.
E-Commerce Solutions
Trillium offers quick and easy entry into the e-commerce marketplace.
Usability Studies
Trillium can assess the usability of your current site and suggest improvements.
Search Engine Optimization
Most people find websites using search engines. Trillium can help you get your site noticed.

Copyright © Your Name Here

FIGURE 4.17 *The services.html page is not yet associated with a style sheet.*

Associate the Web Page with the External CSS File Next, edit the index.html file in a text editor. Delete the CSS code you just copied. Delete the closing `</style>` tag. Replace the opening `<style>` tag with a `<link>` element to associate the style sheet named trillium.css. The `<link>` element code follows:

```
<link href="trillium.css"
rel="stylesheet">
```

Save the file and test it in a browser. Your web page should look just like the one shown in Figure 4.14. Although it looks the same, the difference is in the code—the page now uses external instead of embedded CSS.

Now, for the fun part—you'll associate a second page with the style sheet. The student files contain a services.html page for Trillium at chapter4/services.html. When you display this page in a browser, it should look similar to the one shown in Figure 4.17. Notice that although the structure of the page is similar to the home page, the styling of the text and colors is absent.

Launch a text editor to edit the services.html file. Code a `<link>` element to associate the services.html web page with the trillium.css external style sheet. Place the following code in the head section above the closing `</head>` tag:

```
<link href="trillium.css"
rel="stylesheet">
```

Save your file in the trillium folder and test in a browser. Your page should look similar to Figure 4.18—the CSS rules have been applied!

If you click the "Home" and "Services" hyperlinks, you can move back and forth between the index.html and services.html pages in the browser. The student files contain a sample solution in the chapter4/4.6 folder.

FIGURE 4.18 *The services.html page has been associated with trillium.css.*

Notice that when using an external style sheet, if the style rule declarations need to be changed in the future, you'll typically only have to modify *one* file—the external style sheet. Think about how this can improve productivity on a site with many pages. Instead of modifying potentially hundreds of pages to make a color or font change, only a single file—the CSS external style sheet—needs to be updated. Becoming comfortable with CSS will be important as you develop your skills and increase your technical expertise.

FAQ

My CSS doesn't work; what can I do?

Coding CSS is a detail-oriented process. There are several common errors that can cause the browser not to apply CSS correctly to a web page. With a careful review of your code and the following tips, you should get your CSS working:

▶ Verify that you are using the colon ":" and semicolon ";" symbols in the right spots—they are easy to confuse. The : symbol should separate the properties from their values. The ; symbol should be placed between each property : value configuration.

▶ Check that you are not using = signs instead of : between each property and its value.

▶ Verify that the { and } symbols are properly placed around the style rules for each selector.

▶ Check the syntax of your selectors, their properties, and property values for correct usage.

▶ If part of your CSS works and part doesn't, read through the CSS and check to determine the first rule that is not applied. Often the error is in the rule *above* the rule that is not applied.

▶ Use the W3C's CSS validator at http://jigsaw.w3.org/css-validator to help you find syntax errors.

The Cascade

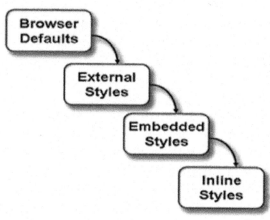

FIGURE 4.19 *The cascade.*

Figure 4.19 shows the "cascade" (**order of precedence**) that applies the styles from outermost (external styles) to innermost (inline styles).

This set of rules allows the site-wide styles to be configured but overridden when needed by more granular page-specific styles (such as embedded or inline styles).

External styles can apply to multiple pages. The order the styles are coded in the web page matters. When using both external and embedded styles, it is a typical practice to code the link element (for external styles) before the style element (for embedded styles). So, when a web page contains both an association with an external style sheet and embedded styles, the external styles will be applied first, and then the embedded styles will be applied. This approach allows a web developer to override global external styles on selected pages.

If a web page also contains inline styles, any external and embedded styles are applied first as just described, and then the inline styles are applied. This approach allows a web developer to override page-wide styles for particular HTML tags or classes.

Note that an HTML tag or attribute will override styles. For example, a `<strong>` tag will override corresponding font-related styles configured for an element. If no attribute or style is applied to an element, the browser's default is applied. However, the appearance of the browser's default may vary by browser, and you might be disappointed with the result. Use CSS to specify the properties of your text and web page elements. Avoid depending on the browser's default.

In addition to the general cascade of CSS types described previously, the style rules themselves follow an order of precedence. Style rules applied to more local elements (such as a paragraph) take precedence over those applied to more global elements (such as a `<div>` that contains the paragraph).

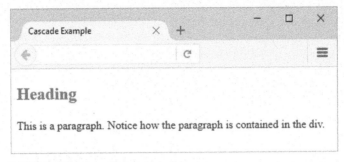

FIGURE 4.20 *Inheritance in action.*

Let's look at the code for the page shown in Figure 4.20 (also found in the student files at chapter4/cascade1.html). Consider the following CSS code:

```
.special { color: red; }
p { color: blue; }
```

The CSS has two style rules: a rule creating a class named special that configures red text and a rule configuring all paragraphs to display blue text.

The HTML on the page contains a <div> with multiple elements, such as headings and paragraphs, as shown in the following code:

```
<div class="special">
<h2>Heading</h2>
<p>This is a paragraph. Notice how the paragraph is contained in the div.</p>
</div>
```

As shown in Figure 4.20, here is how the browser would render the code:

1. The text within the heading is displayed using the color red because it is part of the <div> assigned to the special class. It inherits the properties from its parent (<div>) class. This is an example of **inheritance,** in which certain CSS properties are passed down to elements nested within a container element, such as a <div> or <body> element.

2. The text within the paragraph is displayed using the color blue because the browser applies the styles associated with the most local element (the paragraph). Even though the paragraph is within (and is considered a child of) the special class, the local paragraph style rules take precedence and are applied by the browser.

Don't worry if inheritance and order of precedence seem a bit overwhelming at this point. CSS definitely becomes easier with practice. You will get a chance to practice with the "cascade" as you complete the next Hands-On Practice.

Practice with the Cascade

 Hands-On Practice 4.7 ————————————————————————

You will experiment with the "cascade" in this Hands-On Practice as you work with a web page that uses external, embedded, and inline styles.

1. Create a new folder named mycascade.

2. Launch a text editor. Open a new file. Save the file as site.css in the mycascade folder. You will create an external style sheet that sets the background color of the web page to a shade of yellow (#FFFFCC) and the text color to black (#000000). The code follows:

```
body { background-color: #FFFFCC; color: #000000; }
```

Save and close the site.css file.

3. Open a new file in the text editor and save it as index.html in the mycascade folder. The web page will be associated with the external style sheet site.css, use embedded styles to set the global text color to blue, and use inline styles to configure the text color of the second paragraph. The file index.html will contain two paragraphs of text. The code for index.html follows:

```
<!DOCTYPE html>
<html lang="en">
<head>
<title>The Cascade in Action</title>
<meta charset="utf-8">
<link rel="stylesheet" href="site.css">
<style>
body { color: #0000FF; }
</style>
</head>
<body>
<p>This paragraph applies the external and embedded styles &mdash;
note how the blue text color that is configured in the embedded
styles takes precedence over the black text color configured in
the external stylesheet.</p>
<p style="color: #FF0000">Inline styles configure this paragraph
to have red text and take precedence over the embedded and external
styles.</p>
</body>
</html>
```

4. Save index.html and display it in a browser. Your page should look similar to the sample shown in Figure 4.21. The student files contain a sample solution at chapter4/4.7/index.html.

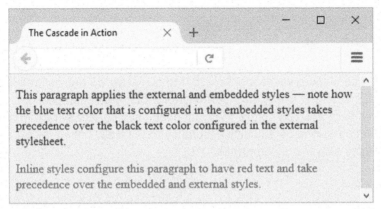

FIGURE 4.21 *The cascade in action.*

Take a moment to examine the index.html web page and compare it with its source code. The web page picked up the yellow background from the external style sheet. The embedded style configured the text to be the color blue, which overrides the black text color in the external style sheet. The first paragraph in the web page does not contain any inline styles, so it inherits the style rules in the external and embedded style sheets. The second paragraph contains an inline style of red text color; this setting overrides the corresponding external and embedded styles.

?FAQ Is it always better to use external CSS?

The answer is it depends. If you are creating a stand-alone web page (like some of the practice pages in this chapter), it is easier to work with a single file and code embedded CSS in the head section than to work with two files (the web page and the external CSS file). However, if you are creating a website, the best approach is to place all the CSS in an external CSS file. Later on if you need to change the styles, you'll only need to edit the CSS file!

CSS Syntax Validation

VideoNote
CSS
Validation

The W3C has a free Markup Validation Service (http://jigsaw.w3.org/css-validator) that will validate your CSS code and check it for syntax errors. CSS validation provides students with quick self-assessment—you can prove that your code uses correct syntax. In the working world, **CSS validation** serves as a quality assurance tool. Invalid code may cause browsers to render the pages slower than otherwise.

 Hands-On Practice 4.8

In this Hands-On Practice, you will use the W3C CSS Validation Service to validate an external CSS style sheet. This example uses the color.css file completed in Hands-On Practice 4.3 (student files chapter4/4.3/color.css). Locate color.css and open it in a text editor. We will add an error to the color.css file. Find the body element selector style rule and delete the first "r" in the `background-color` property. Remove the # from the `color` property value. Save the file.

Next, attempt to validate the color.css file. Visit the W3C CSS Validation Service page at http://jigsaw.w3.org/css-validator and select the "By file upload" tab. Click the "Browse" button and select the color.css file from your computer. Click the "Check" button. Your display should be similar to that shown in Figure 4.22. Notice that two errors were found. The selector is listed, followed by the reason an error was noted.

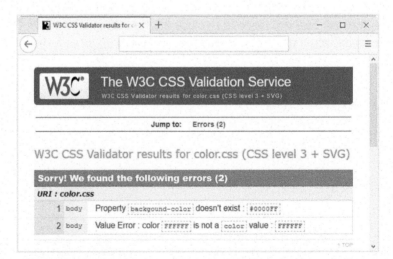

FIGURE 4.22 *The validation results indicate errors. Screenshots of W3C. Courtesy of W3C (World Wide Web Consortium)*

Notice that the first message in Figure 4.22 indicates that the "backgound-color" property does not exist. This is a clue to check the syntax of the property name. Edit color.css and correct the error. Test and revalidate your page. Your browser should now look similar to the one shown in Figure 4.23 and report only one error.

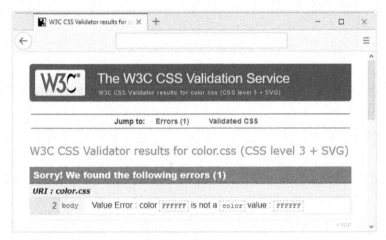

FIGURE 4.23 *The valid CSS is displayed below the errors (and warnings, if any). Screenshots of W3C. Courtesy of W3C (World Wide Web Consortium)*

The error reminds you that FFFFFF is not a color value—the validator expects you to already know that you need to add a "#" character to code a valid color value, #FFFFFF. Notice how any valid CSS rules are displayed below the error messages. Correct the color value, save the file, and test again.

Your results should look similar to those shown in Figure 4.24. There are no errors listed. This means that your file passed the CSS validation test. Congratulations, your color.css file contains valid CSS syntax! It's a good practice to validate your CSS style rules. The CSS validator can help you to identify code that needs to be corrected quickly and indicate which style rules a browser is likely to consider valid. Validating CSS is one of the many productivity techniques that web developers commonly use.

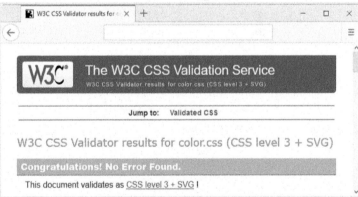

FIGURE 4.24 *The CSS is valid! Screenshots of W3C. Courtesy of W3C (World Wide Web Consortium)*

Review and Apply

Review Questions

1. Which type of CSS is coded in the body of the web page as an attribute of an HTML tag?
   - a. embedded
   - b. inline
   - c. external
   - d. imported

2. Which of the following can be a CSS selector?
   - a. an HTML element name
   - b. a class name
   - c. an id name
   - d. all of the above

3. Which of the following is the CSS property used to set the background color?
   - a. bgcolor
   - b. color
   - c. bcolor
   - d. background-color

4. Which of the following describes two components of CSS rules?
   - a. selectors and declarations
   - b. properties and declarations
   - c. selectors and attributes
   - d. none of the above

5. Which of the following associates a web page with an external style sheet?
   - a. `<style rel="external" href="style.css">`
   - b. `<style src="style.css">`
   - c. `<link rel="stylesheet" href="style.css">`
   - d. `<link rel="stylesheet" src="style.css">`

6. Which of the following configures a CSS class called news with red text (#FF0000) and light gray background (#EAEAEA)?
   - a. `news { color: #FF0000; background-color: #EAEAEA; }`
   - b. `.news { color: #FF0000; background-color: #EAEAEA; }`
   - c. `.news { text: #FF0000; background-color: #EAEAEA; }`
   - d. `#news {color: #FF0000; background-color: #EAEAEA; }`

7. An External Style Sheet uses the _____ file extension.
   - a. ess
   - b. css
   - c. htm
   - d. No file extension is necessary

8. Where do you place the code to associate a web page with an external style sheet?
   - a. in the external style sheet
   - b. in the DOCTYPE of the web page document
   - c. in the body section of the web page document
   - d. in the head section of the web page document

9. Which of the following configures a background color of #FFF8DC for a web page using CSS?
   - a. `body { background-color: #FFF8DC; }`
   - b. `document { background: #FFF8DC; }`
   - c. `body {bgcolor: #FFF8DC;}`
   - d. `body { color: #FFF8DC; }`

10. Which of the following do you configure to apply a style to more than one area on a web page?
    - a. id
    - b. class
    - c. group
    - d. link

Review Answers

1. b   2. d   3. d   4. a   5. c   6. b   7. b   8. d   9. a   10. b

Hands-On Exercise

Practice with External Style Sheets. In this exercise, you will create two external style sheet files and a web page. You will experiment with linking the web page to the external style sheets and note how the display of the page is changed.

a. Create an external style sheet (call it format1.css) to format as follows: document background color of white and document text color of #000099.

b. Create an external style sheet (call it format2.css) to format as follows: document background color of yellow and document text color of green.

c. Create a web page about your favorite movie that displays the movie name in an `<h1>` tag, a description of the movie in a paragraph, and an unordered (bulleted) list of the main actors and actresses in the movie. The page should also include a hyperlink to a website about the movie and an e-mail link to yourself. This page should be associated with the format1.css file. Save the page as moviecss1.html. Be sure to test your page in more than one browser.

d. Modify the moviecss1.html page to be associated with the format2.css external style sheet instead of the format1.css file. Save the page as moviecss2.html and test it in a browser. Notice how different the page looks!

Focus on Web Design

In this activity, you will design a color scheme, code an external CSS file for the color scheme, and code an example web page that applies the styles you configured. Use any of the following sites to help you get started with color and web design ideas:

Psychology of Color

- Infoplease: https://www.infoplease.com/spot/colors1.html
- Empowered by Color: https://www.empower-yourself-with-color-psychology.com/meaning-of-colors.html
- Designzzz: https://www.designzzz.com/infographic-psychology-color-web-designers/

Color Scheme Generators

- Color Blender: https://meyerweb.com/eric/tools/color-blend
- Color.org: http://www.colr.org
- Colors on the Web: http://www.colorsontheweb.com/Color-Tools/Color-Wizard
- Adobe Color: https://color.adobe.com/create/color-wheel
- Paletton.com: http://paletton.com

You have the following tasks:

a. Design a color scheme. List three hexadecimal color values in addition to white (#FFFFFF) or black (#000000) in your design.

b. Describe the process you went through as you selected the colors. Describe why you chose these colors. What type of website would they be appropriate for? List the URLs of any resources you used.

c. Create an external CSS file name color1.css that configures text color and background color selections for the document, h1 element selector, p element selector, and footer element selector using the colors you have chosen.

d. Create a web page named color1.html that shows examples of the CSS style rules.

Case Study

You will continue the case studies from Chapter 2 as you configure the websites to use external style sheets.

Pacific Trails Resort Case Study

In this chapter's case study, you will use the existing Pacific Trails (Chapter 2) website as a starting point while you create a new version of the website that uses an external style sheet to configure color (see Figure 4.25).

FIGURE 4.25 *New Pacific Trails Resort Home page with color swatches.*

You have five tasks in this case study:

1. Create a new folder for the Pacific Trails Resort website.

2. Create an external style sheet named pacific.css.

3. Update the Home page: index.html.

4. Update the Yurts page: yurts.html.

5. Update the pacific.css style sheet.

Task 1: Create a folder called ch4pacific to contain your Pacific Trails Resort website files. Copy the index.html and yurts.html files from the Chapter 2 Case Study pacific folder.

Task 2: The External Style Sheet. Launch a text editor. You will create an external style sheet named pacific.css. A sample wireframe is shown in Figure 4.26.

Code CSS to configure the following:

- Global styles for the document (use the body element selector) with background color white (#FFFFFF) and text color dark gray (#555555).
- Style rules for the header element selector that configure background color (#002171) and text color (#FFFFFF).
- Styles for the nav element selector that configure sky blue background color (#BBDEFB).
- Styles for the h2 element selector that configure medium blue text color (#1976D2).
- Styles for the dt element selector that configure dark blue text color (#002171).
- Styles for a class named `resort` that configure medium blue text color (#1976D2).

FIGURE 4.26 *The wireframe for the Pacific Trails Resort Home page.*

Save the file as pacific.css in the ch4pacific folder. Check your syntax with the CSS validator at http://jigsaw.w3.org/css-validator. Correct and retest if necessary.

Task 3: The Home Page. Launch a text editor and open the home page, index.html.

a. Associate the pacific.css external style sheet. Add a <link> element in the head section to associate the web page with the pacific.css external style sheet file.

b. Find the company name (Pacific Trails Resort) in the first paragraph below the h2. Configure a span that contains this text. Assign the span tag to the `resort` class.

c. Look for the company name (Pacific Trails Resort) directly above the street address. Configure a span that contains this text. Assign the span tag to the `resort` class.

d. Assign the div that contains the address and phone information to an id named `contact`. We'll configure CSS for this id in a future case study.

Save and test your index.html page in a browser. It should be similar to the page shown in Figure 4.27, and you'll notice that the styles you configured in the external CSS file are applied!

FIGURE 4.27 *First version of the new index.html page.*

Task 4: The Yurts Page. Launch a text editor and open the yurts.html file. An example of the new version of the web page is shown in Figure 4.28.

    a. Add a `<link>` element in the head section to associate the web page with the pacific.css external style sheet file.

Save and test your new yurts.html page. It should look similar to the one shown in Figure 4.28.

FIGURE 4.28 *First version of the new yurts.html page.*

Task 5: Update the CSS. You may notice an empty space between the header area and the navigation area. The empty space is the default bottom margin of the h1 element. Refer back to Hands-On Practice 4.5 (Part 2) and recall that a technique to cause the browser to collapse this empty space is to configure the margin. To set the bottom margin of the h1 element to 0, code the following style for an h1 element selector in the pacific.css file: `margin-bottom: 0;`

Save the pacific.css file. Launch a browser and test your index.html and yurts.html pages. The gap between the h1 element and the navigation area should be gone. Your home page should now display similar to Figure 4.25. Click the navigation link to display the yurts.html page; it should also render with the new styling from the pacific.css external style sheet.

This case study demonstrated the power of CSS. Just a few lines of code have transformed the display of the web pages in the browser.

Path of Light Yoga Studio Case Study

In this chapter's case study, you will use the existing Path of Light Yoga Studio (Chapter 2) website as a starting point while you create a new version of the website that uses an external style sheet to configure color (see Figure 4.29).

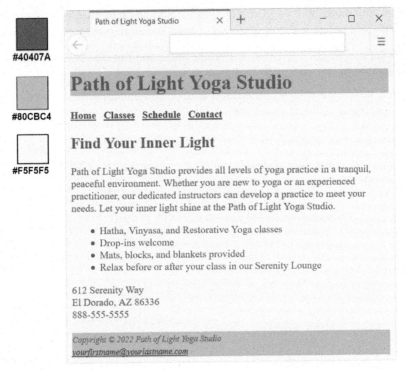

FIGURE 4.29 *New Path of Light Yoga Studio Home page with color swatches.*

You have four tasks in this case study:

1. Create a new folder for the Path of Light Yoga Studio website.
2. Create an external style sheet named yoga.css.
3. Update the Home page: index.html.
4. Update the Classes page: classes.html.

Task 1: Create a folder called ch4yoga to contain your Path of Light Yoga Studio website files. Copy the index.html and classes.html files from the Chapter 2 Case Study yoga folder.

Task 2: The External Style Sheet. Launch a text editor. You will create an external style sheet named yoga.css. A sample wireframe is shown in Figure 4.30.

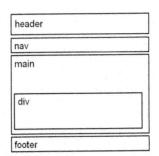

FIGURE 4.30 *The wireframe for the Path of Light Yoga Studio Home page.*

Code CSS to configure the following:

- Global styles for the document (use the body element selector) with an off-white background color (#F5F5F5) and violet text color (#40407A).
- Styles for the header element selector that configure a background color (#80CBC4).
- Styles for the footer element selector that configure a background color (#80CBC4).

Save the file as yoga.css in the ch4yoga folder. Check your syntax with the CSS validator at http://jigsaw.w3.org/css-validator. Correct and retest if necessary.

Task 3: The Home Page. Launch a text editor and open the home page, index.html.

    a. Associate the yoga.css external style sheet. Add a `<link>` element in the head section to associate the web page with the yoga.css external style sheet file.

Save and test your index.html page in a browser. It should be similar to the page shown in Figure 4.29, and you'll notice that the styles you configured in the external CSS file are applied!

Task 4: The Classes Page. Launch a text editor and open the classes.html file. An example of the new version of the web page is shown in Figure 4.31.

    a. Code a `<link>` element in the head section to associate the web page with the yoga.css external style sheet file.

Save and test your new classes.html page. It should look similar to Figure 4.31.

This case study demonstrated the power of CSS. Just a few lines of code have transformed the display of the web pages in the browser.

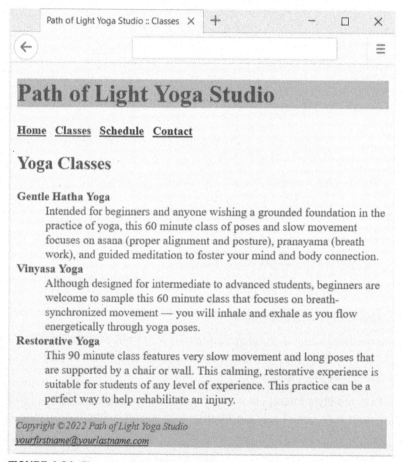

FIGURE 4.31 *The new classes.html page.*

Endnotes

1.   "Cascading Style Sheets Home Page." *W3C*, www.w3.org/Style/CSS/.

2.   "CSS/Properties/Color/Keywords." *CSS/Properties/Color/Keywords - W3C Wiki*, W3C, 21 Aug. 2011, www.w3.org/wiki/CSS/Properties/color/keywords.

3.   "CSS Color Module Level 3." Edited by Tantek Çelik et al., *W3C*, W3C, 19 June 2018, www.w3.org/TR/css-color-3/.

4.   "CSS Color Module Level 3: 4.2.4 HSL Color Values." Edited by Tantek Çelik et al., *W3C*, W3C, 19 June 2018, www.w3.org/TR/css-color-3/#hsl-color.

Graphics & Text Styling Basics

A key component of a compelling website is the use of interesting and appropriate graphics. This chapter introduces configuring text with Cascading Style Sheets (CSS) as you work with visual elements on web pages. When you include images on your website, it is important to remember that not all web users are able to view them. Some users may have vision problems and need assistive technology such as a screen reader application that reads the web page to them. In addition, search engines send out spiders and robots to walk the web and catalog pages for their indexes and databases; such programs do not access your images. Some of your visitors may be using a mobile device that may not display your images. As a web designer, strive to create pages that are enhanced by graphical elements but that are usable without them.

You'll learn how to...

- Describe types of graphics used on the Web
- Apply the img, figure, and figcaption elements to add graphics to web pages
- Configure images as backgrounds on web pages
- Configure images as hyperlinks
- Configure image maps
- Configure bullets in unordered lists with images
- Configure multiple background images with CSS
- Configure text typeface, size, weight, and style with CSS
- Align and indent text with CSS

Graphics on the Web

Graphics can make web pages compelling and engaging. This section introduces features of graphic files commonly used on the Web: GIF, JPEG, PNG, and WebP.

FIGURE 5.1 *This logo is a GIF.*

Graphic Interchange Format (GIF)

The **GIF** image format is best used for line drawings containing mostly solid tones and simple images such as clip art. The maximum number of colors in a GIF file is 256. GIF images have a .gif file extension. Figure 5.1 shows a logo image created in GIF format.

Optimization. To avoid slow-loading web pages, graphic files should be optimized for the Web. **Image optimization** is the process of creating an image with the lowest file size that still renders a good-quality image—balancing image quality and file size. GIF images are typically optimized by using a graphics application to reduce the number of colors in the image.

FIGURE 5.2 *Transparent and nontransparent GIF images.*

Transparency. The format GIF89A used by GIF images supports image **transparency**. In a graphics application, such as the open-source GIMP, one color (typically the background color) of the image can be set to be transparent. The background color (or background image) of the web page shows through the transparent area in the image. Figure 5.2 displays two GIF images on a blue texture background.

Animation. An **animated GIF** consists of multiple images or frames, each of which is slightly different. When the frames display on the screen in order, the image appears animated.

Compression. **Lossless compression** is used when a GIF is saved. This means that nothing in the original image is lost and that the compressed image, when rendered by a browser, will contain the same pixels as the original.

Interlacing. Browsers render, or display, web page documents in order, line by line, starting at the top of the document. They display standard images as the files are read in order from top to bottom. The top of a standard image begins to display after 50% of the image has been read by a browser. When a GIF graphic file is created, it can be configured as interlaced.

An **interlaced image** progressively displays and seems to fade in as it downloads. The image first appears fuzzy but gradually becomes clearer and sharper, which can help to reduce the perceived load time of your web page.

Joint Photographic Experts Group (JPEG)

The **JPEG** image format is best used for photographs. In contrast to a GIF image, a JPEG image can contain 16.7 million colors. However, JPEG images cannot be made transparent, and they cannot be animated. JPEG images usually have a .jpg or .jpeg file extension.

JPEG images are saved using **lossy compression**. This means that some pixels in the original image are lost or removed from the compressed file. When a browser renders the compressed image, the display is similar to but not exactly the same as the original image.

FIGURE 5.3 *JPEG saved at 80% quality (55KB file size).*

There are trade-offs between the quality of the image and the amount of compression. An image with less compression will have higher quality and result in a larger file size. An image with more compression will have lower quality and result in a smaller file size.

When you take a photo with a digital camera, the file size is too large for optimal display on a web page. Figure 5.3 shows an optimized version of a digital photo with an original file size of 250KB. The image was optimized using a graphics application set to 80% quality, is now only 55KB, and displays well on a web page.

Figure 5.4 was saved with 20% quality and is only 19KB, but its quality is unacceptable. The quality of the image degrades as the file size decreases. The square blockiness you see in Figure 5.4 is called **pixelation** and should be avoided.

FIGURE 5.4 *JPEG saved at 20% quality (19KB file size).*

Another technique used with web graphics is to display a small version of the image, called a **thumbnail image**. Often, the thumbnail is configured as an image hyperlink to display the larger image. Figure 5.5 shows a thumbnail image.

Progressive JPEG. When a JPEG file is created, it can be configured as progressive. A **progressive JPEG** is similar to an interlaced GIF in that the image progressively displays and seems to fade in as it downloads.

FIGURE 5.5 *Thumbnail image (5KB).*

Portable Network Graphics (PNG)

The Portable Network Graphics (**PNG**) image format was initially specified by the W3C as a replacement for the GIF image format.[1] PNG images have a .png file extension and are well-supported by modern browsers.

FIGURE 5.6 *The PNG file format works well for both line art and photographs.*

PNG (pronounced "ping") graphics work well for both line art and photographs (shown in Figure 5.6) because the file format offers support for selected features of GIFs and JPEGs. Similar to GIF images, PNG graphics use lossless compression, so the image can be perfectly reconstructed after being compressed to reduce file size. PNG images also support interlacing and transparency (including variable transparency levels). Like JPG images, PNG graphics can support millions of colors but cannot be animated.

FIGURE 5.7 *The first frame of an APNG image.*

Animated Portable Network Graphics (APNG). While PNG graphics do not support animation, Mozilla created an extension of the PNG file format, called **APNG**, that supports animation.[2] APNG images use a .png or .apng file extension. Although well- supported by modern browsers, APNG images are backward compatible in that a non-supporting browser will show the first frame of the animation. Advantages of APNG animated image over animated GIFs include support for more colors, additional levels of transparency, and potentially smaller file sizes. Figure 5.7 shows the first frame of an APNG image that is 10K smaller than the original animated GIF images. APNG images can be created with the GIMP software application and by tools that convert animated GIFs to APNG images.[3]

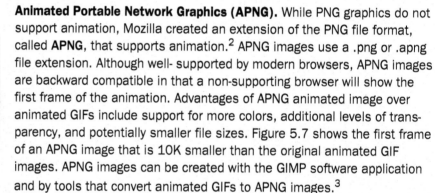

FIGURE 5.8 *The file size of this WebP image is over 25% smaller than the original JPEG.*

WebP Image Format

The relatively new **WebP** image format (shown in Figure 5.8) created by Google offers improved file compression over other types of images used on the Web. The WebP (pronounced "weppy") image format supports both lossy (like JPG) and lossless (like PNG) compression. In a Google comparison test, WebP images were 26% in smaller file size than comparable PNG images and 25–34% smaller in file size than comparable JPG images.[4]

WebP graphics can support millions of colors, transparency, and animation. WebP images do not support interlacing.

Initially, only the Chrome, Opera, and Android browsers displayed WebP images. Firefox and Edge were the next browsers to add support for WebP. At the time this was written, the Safari browser only offered experimental support.[5] Since some browsers, such as Internet Explorer, do not support WebP, special consideration is needed for backward compatibility (see Chapter 8).

Table 5.1 summarizes the characteristics of GIF, JPEG, PNG, and WebP file formats. In the next section, you'll begin coding the HTML img element to display GIF, JPEG, and PNG images on web pages. You will work with WebP images in Chapter 8.

TABLE 5.1 *Overview of Common Web Graphic File Formats*

Image Type	File Extension	Compression	Transparency	Animation	Colors	Progressive Display
GIF	.gif	Lossless	Yes	Yes	256	Interlacing
JPEG	.jpg or .jpeg	Lossy	No	No	Millions	Progressive
PNG	.png	Lossless	Yes	No	Millions	Interlacing
WebP	.webp	Lossy and Lossless	Yes	Yes	Millions	None

Popular Graphics Applications

A variety of graphics applications can be used to edit, optimize, and export images in GIF, JPEG, and PNG file formats such as:

- Adobe Photoshop
  https://www.adobe.com/products/photoshop.html
- GIMP
  https://www.gimp.org
- Sketch
  https://www.sketchapp.com

There are several options for converting a GIF, JPG, or PNG image to WebP format, including websites with online converter tools,[6,7] Sketch, and GIMP.

? FAQ   How do I choose which type of image to use?

GIF, JPEG, and PNG images are well-supported on the Web. GIFs are best suited for line art or clip art. JPEGs are best suited for photographs. PNGs can be used for either purpose, so you may want to consider the file size and quality of the images when you make your decision. As time goes on, support for WebP by modern browsers will increase.

Img Element

The **img element** (often referred to as the image element) configures graphics on a web page, such as photographs, banners, company logos, and navigation buttons. The img element is a void element and is not coded as a pair of opening and closing tags. The following code example configures an image named logo.gif, which is located in the same folder as the web page:

```
<img src="logo.gif" height="200" width="500" alt="My Company Name">
```

The **src attribute** specifies the file name of the image. The **alt attribute** provides a text replacement, typically a text description, of the image. The browser reserves the correct amount of space for your image if you use the height and width attributes with values either equal to or approximately the size of the image. Table 5.2 lists the img element attributes and their values. Commonly used attributes are shown in bold.

TABLE 5.2 *Attributes of the Img Element*

Attribute	Value
align	right, left (default), top, middle, bottom; obsolete in HTML5, use CSS layout techniques such as float (Chapter 7), flexbox (Chapter 8), or grid (Chapter 8)
alt	Text phrase that describes the image; required
height	Height of image in pixels or percentage
id	Text name, alphanumeric, beginning with a letter, no spaces—the value must be unique and not used for other id values on the same web page document
loading	eager (default, load image immediately), lazy (defer loading the image until the user scrolls near it); (Chapter 8)
longdesc	URL of a resource that contains an accessible description of a complex image
sizes	HTML 5.1 attribute supports the browser display of responsive images (Chapter 8)
src	The URL or file name of the image; required
srcset	HTML 5.1 attribute supports the browser display of responsive images (Chapter 8)
title	A text phrase containing advisory information about the image—typically more descriptive than the alt text
usemap	Value corresponding to the name and id of an associated map element
width	Width of image in pixels or percentage

? FAQ How should I name my image files?

Use all lowercase letters. Avoid punctuation symbols and spaces. Use hyphens to separate words. Do not change file extensions. Keep your file names short, but descriptive. For example: myphotowithmydogonmybirthday.jpg is too long, d1.jpg is probably too short, birthday-dog.jpg may be just about right.

In this Hands-On Practice, you will place a logo and a photograph on a web page. Create a new folder called kayakch5. The images used in this Hands-On Practice are located in the student files chapter5/starters folder. Copy the kayakdc.gif and hero.jpg files into your kayakch5 folder. A starter version of the KayakDoorCounty.net Home page is ready for you in the student files. Copy the chapter5/starter.html file into your kayakch5 folder. When you complete this Hands-On Practice, your page will look similar to the one shown in Figure 5.9—with two images. Launch a text editor and open the file.

1. Delete the text between the h1 opening and closing tags. Code an image tag for kayakdc.gif in this area. Remember to include the `src`, `alt`, `height`, and `width` attributes. Sample code follows:

```
<img src="kayakdc.gif"
alt="KayakDoorCounty.net"
width="500" height="60">
```

2. Code an image tag to display the hero.jpg image below the h2 element. The image is 500 pixels wide and 350 pixels high. Configure appropriate alt text for the image.

3. Save your page as index.html in the kayakch5 folder. Launch a browser and test your page. It should look similar to the one shown in Figure 5.9.

Note: If the images did not display on your web page, verify that you have saved the files inside the kayakch5 folder and that you have spelled the file names correctly in the `<img>` tags. The student files contain a sample solution in the chapter5/5.1 folder. Isn't it intriguing how images can add visual interest to a web page?

FIGURE 5.9 *A web page with images.*

Accessibility and the `alt` Attribute

Use the `alt` attribute to provide accessibility. Section 508 of the Rehabilitation Act requires the use of accessibility features for information technology (including websites) associated with the federal government. The `alt` attribute configures an alternative text description of the image. This alternative text may be used by the browser in two ways: The browser will show the text in the image area before the graphic is downloaded and displayed. Some browsers will also show the text as a tool tip whenever the web page visitor places a mouse over the image area.

Standard browsers such as Microsoft Edge and Mozilla Firefox are not the only type of application or user agent that can access your website. Major search engines run programs called spiders or robots; these programs index and categorize websites. They cannot process images, but some process the value of the alt attributes in image tags. Applications such as screen readers will read the text in the `alt` attribute out loud. A mobile browser may display the `alt` text instead of the image.

Image Hyperlinks

Writing the code to make an image function as a hyperlink is very easy. To create an **image link**, all you need to do is surround your `<img>` tag with anchor tags. For example, to place a link around an image called home.gif, use the following code:

```
<a href="index.html"><img src="home.gif" height="19" width="85"
alt="Home"></a>
```

A **thumbnail image link** is a small image configured as an image link with an href attribute value that points to another image file instead of to a web page. For example,

```
<a href="sunset.jpg"><img src="thumb.jpg" height="100" width="100"
alt="view a larger sunset"></a>
```

To see this in action, launch a browser and view chapter5/thumb.html in the student files.

 Hands-On Practice 5.2 ———————————————————————

You will add image links to the KayakDoorCounty.net Home page in this Hands-On Practice. You should already have the index.html, kayakdc.gif, and hero.jpg files in your kayakch5 folder. The new graphics used in this Hands-On Practice are located in the student files in the chapter5/starters folder. Copy the home.gif, tours.gif, reservations.gif, and contact.gif files into your kayakch5 folder. View Figure 5.10 to see how your page should look after you are done with this Hands-On Practice.

Let's get started. Launch a text editor and open index.html. Notice that the anchor tags are already coded—you'll just need to convert the text links to image links!

1. Whenever the main navigation consists of media, such as an image, some individuals may not be able to see the images (or may have images turned off in their browser). To provide navigation that is accessible to all, configure a set of plain text navigation links in the page footer area. Copy the `<nav>` element containing the navigation area to the lower portion of the page and paste it within the footer element, above the copyright line.

2. Locate the style tags in the head section and code the following style rule to configure a green background color for an id named bar:

   ```
   #bar { background-color: #152420; }
   ```

3. Now, focus on the top navigation area. Code `id="bar"` on the opening nav tag. Next, replace the text contained between each pair of anchor tags with an img element. Use home.gif for the link to index.html, tours.gif for the link to tours.html, reservations.gif for the link to reservations.html, and contact.gif for the link to contact.html. Be careful not to leave any extra spaces between the `<img>` tag and the opening and closing anchor tags. A sample follows:

```
<a href="index.html"><img src="home.gif"
alt="Home" width="90" height="35"></a>
```

As you code the `<img>` tags be mindful of the width of each image: home.gif (90 pixels), tours.gif (90 pixels), reservations.gif (190 pixels), and contact.gif (130 pixels).

4. Save your page as index.html. Launch a browser and test your page. It should look similar to the one shown in Figure 5.10.

The student files contain a sample solution in the chapter5/5.2 folder.

FIGURE 5.10 *The new Home page navigation with image links.*

Accessibility and Image Hyperlinks

When using an image for main navigation, there are two methods to provide for accessibility:

1. Add a row of plain text navigation hyperlinks in the page footer. These won't be noticed by most people but could be helpful to a person using a screen reader to visit your web page.
2. Configure the `alt` attribute for each image to contain the exact text that displays in the image. For example, code `alt="Home"` in the `<img>` tag for the Home button.

? FAQ **What if my images don't display?**

The following are common reasons for an image to not display on a web page:

▷ Is your image *really* in the website folder? Use Windows File Explorer or the Mac Finder to double check.

▷ Did you code the HTML and CSS correctly? Perform W3C CSS and HTML validation testing to find syntax errors that could prevent the image from displaying.

▷ Does your image have the exact file name that you have used in the CSS or HTML code? Attention to detail and consistency will be very helpful here.

Configure Background Images

Back in Chapter 4, you learned how to configure background color with the CSS `background-color` property. In addition to a background color, you can also choose to use an image for the background of an element.[8]

The `background-image` Property

Use the CSS **`background-image` property** to configure a background image. For example, the following CSS code configures the HTML body selector with a background using the graphic texture1.png, located in the same folder as the web page file:

```
body { background-image: url(texture1.png); }
```

Using Both Background Color and a Background Image

You can configure both a background color and a background image. The background color (specified by the `background-color` property) will display first. Next, the image specified as the background will be displayed as it is loaded by the browser.

By coding both a background color and a background image, you provide your visitor with a more pleasing visual experience. If the background image does not load for some reason, the background color will still have the expected contrast with your text color. If the background image is smaller than the web browser window and the web page is configured with CSS to not automatically tile (repeat), the page background color will display in areas not covered by the background image. The CSS for a page with both a background color and a background image is as follows:

```
body { background-color: #99CCCC;
       background-image: url(background.jpg); }
```

Browser Display of a Background Image

You may think that a graphic created to be the background of a web page would always be about the size of the browser window viewport. However, the dimensions of the background image are often much smaller than the typical viewport. The shape of a background image is typically either a long, thin rectangle, or a small rectangular block. Unless otherwise specified in a style rule, browsers repeat, or tile, these images to cover the page background, as shown in Figures 5.11 and 5.12. The images have small file sizes so that they download quickly.

Background Image

Web Page with Background Image

FIGURE 5.11 *A long, thin background image tiles down the page.*

Background Image

Web Page with Background Image

FIGURE 5.12 *A small square background is repeated to fill the web page window.*

The `background-attachment` Property

Use the **`background-attachment` property** to configure whether the background image remains fixed in place or scrolls along with the page in the browser viewport. Valid values for the `background-attachment` property include `fixed` and `scroll` (the default).

? FAQ **What if my images are in their own folder?**

It's a good idea to organize your website by placing all your images in a folder separate from your web pages. Notice that the CircleSoft website shown in Figure 5.13 has a folder called images, which contains GIF and JPEG files. To refer to these files in code, you also need to refer to the images folder. The following are some examples:

▸ The CSS code to configure the background.gif file from the images folder as the page background is as follows:

```
body { background-image: url(images/background.gif); }
```

▸ To configure a web page to display the logo.jpg file from the images folder, use the following code:

```
<img src="images/logo.jpg" alt="CircleSoft" width="588"
height="120">
```

FIGURE 5.13 *A folder named "images."*

Position Background Images

The `background-repeat` Property

The default behavior of a browser is to repeat, or tile, background images to cover the entire element's background. Figures 5.11 and 5.12 display examples of this type of tiling for a web page background. This tiling behavior also applies to other elements, such as backgrounds for headings, paragraphs, and so on. You can change automatic tiling of a background image with the CSS **background-repeat property**. The values for the `background-repeat` property include `repeat` (default), `repeat-y` (vertical repeat), `repeat-x` (horizontal repeat), and `no-repeat` (image does not repeat). Configure `background-repeat: no-repeat;` to display the background image only once. Figure 5.14 provides examples of the actual background image and the result of applying commonly used `background-repeat` property values. Additional values for the `background-repeat` property include `space` and `round`:

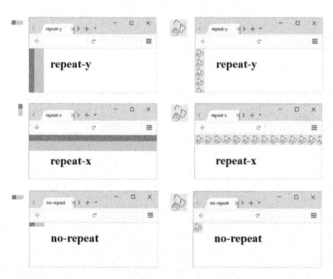

FIGURE 5.14 *Examples of the CSS* `background-repeat` *property.*

- `background-repeat: space;` Repeats the image in the background without clipping (or cutting off) parts of the image by adjusting empty space around the repeated images.
- `background-repeat: round;` Repeats the image in the background and scales (adjusts) the dimensions of the image to avoid clipping.

Positioning the Background Image

You can specify other locations for the background image besides the default top left location using the `background-position` property. Valid values for the `background-position` property include percentages; pixel values; or `left`, `top`, `center`, `bottom`, and `right`. The first value indicates horizontal position. The second value indicates vertical position. If only one value is provided, the second value defaults to center. In Figure 5.15, the flower image has been placed on the right side of the element using the following style rule:

```
h2 { background-image: url(flower.gif);
     background-position: right;
     background-repeat: no-repeat; }
```

FIGURE 5.15 *The flower background image was configured to display on the right side with CSS.*

Hands-On Practice 5.3

Let's practice using a background image. You will update the index.html file from Hands-On Practice 5.2 (shown in Figure 5.10). In this Hands-On Practice, you will configure the main element selector with a background image that does not repeat. Obtain the heroback.jpg image from the student files chapter5/starters folder. Copy the image into your kayakch5 folder. When you have completed this exercise, your page should look similar to the one shown in Figure 5.16. Launch a text editor and open index.html.

1. Locate the style tags in the head section. Code a new style rule for the main element selector to configure the `background-image` and `background-repeat` properties. Set the background image to be heroback.jpg. Set the background not to repeat. The main element selector style rules follow:

```
main { background-image: url(heroback.jpg);
       background-repeat: no-repeat; }
```

2. Remove the `<img>` tag that displays the hero.jpg image from the body of the web page.

3. Save your page as index.html. Launch a Firefox or Chrome browser and test your page. You may notice that the text within the main element is displayed over the background image. In this case, the page would look more appealing if the paragraph did not extend across the background image. Open index.html in a text editor and code a line break tag before the word "explore."

4. Save and test your page again. It should look similar to the page shown in Figure 5.16 if you are using a browser other than Internet Explorer. The student files contain a sample solution in the chapter5/5.3 folder. Internet Explorer does not support default styles for the HTML5 main element. If you are concerned about the display of your page in Internet Explorer, you can nudge this browser to comply by adding the `display: block;` declaration (see Chapter 7) to the styles for the main element selector. An example solution is in the student files (chapter5/5.3/iefix.html).

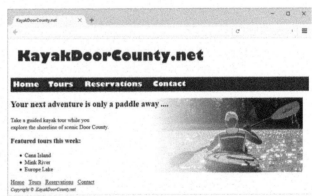

FIGURE 5.16 *The background image in the* `<main>` *area is configured with* `background-repeat: no-repeat.`

CSS Multiple Background Images

Now that you are familiar with background images, let's explore applying multiple background images to a web page. Figure 5.17 shows a web page with two background images configured on the body selector: a large photograph of a coffee cup on a table that displays across the entire web page and a small coffee cup drawing that displays once in lower left corner.

Use the CSS **background property** to configure multiple background images. Each image declaration is separated by a comma. You can optionally add property values to indicate the image's position and whether the image repeats. The `background` property uses a shorthand notation—just list the values that are needed for relevant properties such as `background-position` and `background-repeat`.

FIGURE 5.17 *The browser displays multiple background images.*

To provide for progressive enhancement when using multiple background images, first configure a separate `background-image` property with a single image (rendered by browsers that do not support multiple background images) prior to the background property configured for multiple images (to be rendered by supporting browsers and ignored by nonsupporting browsers).

Hands-On Practice 5.4

Let's practice configuring multiple background images. In this Hands-On Practice, you will configure the body element selector to display multiple background images on the web page. Create a new folder named coffee5. Copy all the files from the student files chapter5/coffeestarters folder into your coffee5 folder.

Launch a text editor and open coffee.html. You will add style rules for the body element selector. Configure the `background-image` property to display coffeepour.jpg. This style rule will be applied by browsers that do not support multiple background images. Configure a background property to display both the coffee.gif and the coffeepour.jpg image files. The coffee.gif image should be displayed in the lower left corner without repeating. The new code is shown in blue:

```
body { font-size: 150%; font-family: Arial; color: #992435;
       background-image: url(coffeepour.jpg);
       background-repeat: no-repeat;
       background: url(coffee.gif) no-repeat left bottom,
                   url(coffeepour.jpg) no-repeat;}
```

Save your file as index.html. Launch a browser and test your page in a modern browser. Your display should be similar to Figure 5.17. If the page is displayed in a browser that does not support multiple background images, only the large photograph will be displayed. The student files contain a sample solution in the chapter5/5.4 folder.

A **cinemagraph** is a type of animated GIF created by taking a video or series of photos with small changes (such as coffee pouring into a cup or hair waving in the wind), processing them in a graphics application such as Adobe Photoshop, and exporting the file as an animated GIF

FIGURE 5.18 *Multiple background images.*

or PNG. Figure 5.18 shows a web page with three background images: a large coffee cup cinemagraph GIF, a solid color rectangle, and a small sketch of a coffee cup.

 ## Hands-On Practice 5.5

In this Hands-On Practice, you will rework the example in Hands-on Practice 5.4 to display a cinemagraph as one of three background images on the web page. Use your files from Hands-On Practice 5.4 (see the chapter5/5.4 folder in the student files).

Launch a text editor and open index.html. You will modify the style rules for the body element selector. Configure the style rules to display the coffeepour.gif image instead of the coffeepour.jpg image. Edit the background property values to display a third image (coffeeback.gif) under the coffeelogo.gif and above the coffeepour.gif. The style rule for the body element selector follows:

```
body { font-size: 150%; font-family: Arial; color: #992435;
      background-image: url(coffeepour.gif);
      background-repeat: no-repeat;
      background: url(coffee.gif) no-repeat left bottom,
              url(coffeeback.gif) no-repeat,
              url(coffeepour.gif) no-repeat; }
```

Save your file as coffepour.html. Launch a browser and test your page in a modern browser. Your display should be similar to Figure 5.15. You will see an animation of coffee pouring into the cup. If the page is displayed in a browser that does not support multiple background images, only the large pouring coffee image will be displayed. The student files contain a sample solution in the chapter5/5.5 folder.

Fonts with CSS

The `font-family` Property

The **font-family property** configures font typefaces.[9] A web browser displays text with the fonts that have been installed on the user's computer. When a font that is not installed on your web visitor's computer is specified, the default font is substituted. Times New Roman is the default font displayed by most web browsers. Figure 5.19 shows font family categories and some common font typefaces.

Font Family Category	Font Family Description	Font Typeface Examples
serif	Serif fonts have small embellishments on the end of letter strokes; often used for headings.	Times New Roman, Georgia, Palatino
sans-serif	Sans-serif fonts do not have serifs; often used for web page text.	Arial, **Tahoma**, Helvetica, Verdana
monospace	Fixed-width font; often used for code samples.	Courier New, Lucida Console
cursive	Hand-written style; use with caution; may be difficult to read on a web page.	*Lucida Handwriting*, *Brush Script*, Comic Sans MS
fantasy	Exaggerated style; use with caution; sometimes used for headings; may be difficult to read on a web page.	Jokerman, **Impact**, Papyrus

FIGURE 5.19 *Common fonts.*

The Verdana, Tahoma, and Georgia font typefaces were specifically designed to display well on computer monitors. A common practice is to use a serif font (such as Georgia or Times New Roman) for headings and a sans-serif font (such as Verdana or Arial) for detailed text content. Not every computer has the same fonts installed. Create a built-in backup plan by listing multiple fonts and categories for the value of the `font-family` property. The browser will attempt to use the fonts in the order listed. The following CSS configures the p element selector to display text in Verdana (if installed) or Arial (if installed) or the default installed sans-serif font.

```
p { font-family: Verdana, Arial, sans-serif; }
```

 ## Hands-On Practice 5.6

In this Hands-On Practice, you will configure the `font-family` property. You will use your files from Hands-On Practice 5.3 (see the student files chapter5/5.3 folder) as a starting point. Launch a browser to display the index.html web page—notice that the text displays in the default browser font (typically Times New Roman). When you are finished with this Hands-On Practice, your page will look similar to the one shown in Figure 5.20.

Launch a text editor and open the index.html file. Configure the embedded CSS as follows:

1. Configure the body element selector to set global styles to use a sans-serif font typeface, such as Verdana or Arial. An example is

```
body { font-family: Verdana, Arial, sans-serif; }
```

2. Configure h2 and h3 element selectors to use a serif font typeface, such as Georgia or Times New Roman. You can configure more than one selector in a style rule by placing a comma before each new selector. Notice that "Times New Roman" is enclosed within quotation marks because the font name is more than a single word. Code the following style rule:

```
h2, h3 { font-family: Georgia,
         "Times New Roman", serif; }
```

Save your page as index.html in the kayakch5 folder. Launch a browser and test your page. It should look similar to the one shown in Figure 5.20. A sample solution is in the chapter5/5.6 folder.

FIGURE 5.20 *The new Home page.*

For many years, web designers were limited to a set of common fonts for text on web pages. CSS3 introduced `@font-face`, which can be used to "embed" other fonts within web pages although you actually provide the location of the font and the browser downloads it. For example, if you own the rights to freely distribute the font named MyAwesomeFont and it is stored in a file myawesomefont.woff in the same folder as your web page, the following CSS will make it available to your web page visitors:

```
@font-face { font-family: MyAwesomeFont;
             src: url(myawesomefont.woff) format("woff"); }
```

After you code the `@font-face` rule, you can apply that font to a selector in the usual way, such as in the following example that configures h1 elements:

```
h1 { font-family: MyAwesomeFont, Georgia, serif; }
```

Current browsers support `@font-face`, but there can be copyright issues. When you purchase a font to use on your own computer, you do not necessarily purchase the right to freely distribute it. Visit https://www.fontsquirrel.com to browse a selection of commercial-use free fonts available for download and use.

Google Web Fonts provides a collection of free hosted embeddable web fonts. Explore the fonts available at https://fonts.google.com. Once you choose a font, all you need to do is:

1. Copy and paste the link tag provided by Google in your web page document. (The link tag associates your web page with a CSS file that contains the appropriate `@font-face` rule.)

2. Configure your CSS `font-family` property with the Google web font name.

See the Getting Started guide at https://developers.google.com/webfonts/docs/getting_started for more information. Use web fonts judiciously to conserve bandwidth and avoid applying multiple web fonts to a web page. It's a good idea to use just one web font on a web page along with your typical fonts. This can provide you a way to use an uncommon font typeface in page headings and/or navigation without the need to create graphics for these page areas.

CSS Text Properties

Typography can be described as the style or arrangement of text. You will use CSS to configure the text on your web pages. In this section, you'll explore the `font-size`, `font-weight`, `font-style`, `line-height`, `text-align`, `text-decoration`, `text-indent`, `text-transform`, and `letter-spacing` properties.[10,11,12]

The `font-size` Property

The **font-size** property sets the size of the font. Table 5.3 lists several categories of font size values, characteristics, and recommended usage.

TABLE 5.3 *Configuring Font Size*

Value Category	Values	Notes
Text Value	`xx-small`, `x-small`, `small`, `medium` (default), `large`, `x-large`, `xx-large`	Scales well when text is resized; limited options for text size
Pixel Unit (px)	Numeric value with unit, such as `10px`	Pixel-perfect display depends on screen resolution; may not scale in every browser when text is resized
Point Unit (pt)	Numeric value with unit, such as `10pt`	Use to configure print version of web page (see Chapter 8); may not scale in every browser when text is resized
Em Unit (em)	Numeric value with unit, such as `.75em`	Recommended by W3C; scales well when text is resized in browser; many options for text size
Percentage Value	Numeric value with percentage, such as `75%`	Recommended by W3C; scales well when text is resized in browser; many options for text size

The **em unit** is a relative font unit that has its roots in the print industry back in the day when printers set type manually with blocks of characters. An em unit is the width of a square block of type (typically the uppercase M) for a particular font and type size. On web pages, an em unit corresponds to the width of the font and size used in the parent element (typically the body element). With this in mind, the size of an em unit is relative to the font typeface and default size. Percentage values work in a manner similar to em units. For example, `font-size: 100%;` and `font-size: 1em;` should render the same in a browser. To compare font sizes on your computer, launch a browser and view chapter5/fonts.html in the student files.

The `font-weight` Property

The **font-weight** property configures the boldness of the text. The CSS `font-weight: bold;` declaration has an effect similar to the `<b>` or `<strong>` HTML element. Sample CSS to configure bold text in the nav:

```
nav { font-weight: bold; }
```

The `font-style` Property

The **font-style** property is used to configure text displayed in italics. Valid values are `normal` (the default), `italic`, and `oblique`. The CSS `font-style: italic;` declaration has the same visual effect in the browser as an `<i>` or `<em>` HTML element. Sample CSS to configure italic text in the footer:

```
footer { font-style: italic; }
```

The `line-height` Property

The **line-height** property modifies the default height of a line of text and is often configured using a percentage value. Sample CSS to configure a paragraph with double-spaced lines:

```
p { line-height: 200%; }
```

The `text-align` Property

HTML elements are left-aligned by default—they begin at the left margin. The CSS **text-align** property configures the alignment of text and inline elements within block elements such as headings, paragraphs, and divs. The `left` (default), `center`, `right`, and `justify` values are valid. Sample CSS to configure centered text within an h1 element:

```
h1 { text-align: center; }
```

The `text-decoration` Property

The purpose of the CSS **text-decoration** property is to modify the display of text. Commonly used values include `none`, `underline`, `overline`, and `line-through`. Although hyperlinks are underlined by default, you can remove the underline with the text-decoration property. Sample CSS to remove the underline on a hyperlink:

```
a { text-decoration: none; }
```

The `text-indent` Property

The CSS **text-indent** property configures the indentation of the first line of text within an element. The value can be numeric (with a px, pt, or em unit) or a percentage. Sample CSS to configure a 5em indent for the first line of a paragraph:

```
p { text-indent: 5em; }
```

The `text-transform` Property

The **text-transform** property configures the capitalization of text. Valid values are `none` (default), `capitalize`, `uppercase`, or `lowercase`. Sample CSS to configure uppercase text within an h3 element:

```
h3 { text-transform: uppercase; }
```

The `letter-spacing` Property

The **letter-spacing** property configures the space between text characters. Valid values are `normal` (default) and a numeric pixel or em unit. Sample CSS to configure extra spacing within an h3 element:

```
h3 { letter-spacing: 3px; }
```

You'll get some practice using many of these new properties in the next section.

Practice with Graphics and Text

 Hands-On Practice 5.7 ───────────────────────────

You will apply your new skills with configuring images and text in this Hands-On Practice while you create the web page shown in Figure 5.21.

Create a folder named ch5practice.

Copy the starter.html file from the chapter5 folder in the student files into your ch5practice folder.

Copy the following files from the chapter5/starters folder into your ch5practice folder: hero.jpg, background.jpg, and headerbackblue.jpg.

FIGURE 5.21 *The new Home page.*

Launch a text editor, open the starter.html file, and save the file as index.html. Edit the code as follows:

1. Locate the style tags in the head section and code embedded CSS to style the following:

   a. Configure the body element selector to display background.jpg as the page background and set Verdana, Arial, or the default sans-serif font as the global font typeface.

   ```
   body { background-image:
       url(background.jpg);
       font-family: Verdana,
       Arial, sans-serif; }
   ```

   b. Configure the header element selector with a #000033 background color and to display the headerbackblue.jpg image in the background. Configure this image to display on the right and to not repeat. Also configure #FFFF99 text color, 400% line height, and a 1em text indent.

```
header { background-color: #000033;
         background-image: url(headerbackblue.jpg);
         background-position: right;
         background-repeat: no-repeat;
         color: #FFFF99;
         line-height: 400%;
         text-indent: 1em; }
```

c. Configure the h1, h2, and h3 element selectors with Georgia, Times New Roman, or the default serif font.

```
h1, h2, h3 { font-family: Georgia, "Times New Roman", serif; }
```

d. Configure the nav element selector with bold font that is 1.5em in size.

```
nav { font-weight: bold;
      font-size: 1.5em; }
```

e. Configure navigation anchor elements to not display an underline. Use a descendant selector.

```
nav a { text-decoration: none; }
```

f. Configure paragraph elements to be indented 2em units.

```
p { text-indent: 2em; }
```

g. Configure the footer element selector to be centered with italic font that is .80em in size.

```
footer { text-align: center;
         font-style: italic;
         font-size: .80em; }
```

2. Remove the small and i tags from the page footer area.

3. Code an img element after the h2 element to display the hero.jpg image. Set appropriate values for the alt, width, and height attributes.

```
<img src="hero.jpg" alt="tour guide paddling a kayak" width="500" height="350">
```

Save your file. Test your page in a browser. It should look similar to Figure 5.21. You can compare your work to the sample in the student files (chapter5/5.7).

Quick TIP
We used line-height and text-indent properties to configure empty space in this Hands-On Practice. However, there are other CSS properties which would be more appropriate to use for this purpose. You'll explore the box model in Chapter 6 and learn about how to configure empty space with the margin and padding properties.

FAQ
Is there a way to place a comment within CSS?

Comments are ignored by browsers and can be helpful to document or annotate (in human terms) the purpose of the code. An easy way to add a comment to CSS is to type "/*" before your comment and "*/" after your comment. For example,

```
/* Configure Footer */
footer { font-size: .80em; font-style: italic; text-align: center; }
```

Configure List Markers with CSS

The default display for an unordered list is to show a disc marker (often referred to as a bullet) in front of each list item. The default display for an ordered list is to show a decimal number in front of each list item. Use the **list-style-type property** to configure the marker for an unordered or ordered list. See Table 5.4 for common property values.

TABLE 5.4 *CSS Properties for Ordered and Unordered List Markers*[13]

Property	Description	Value	List Marker Display
list-style-type	Configures the style of the list marker	none	No list markers display
		disc	Circle ("bullet")
		circle	Open circle
		square	Square
		decimal	Decimal numbers
		upper-alpha	Uppercase letters
		lower-alpha	Lowercase letters
		lower-roman	Lowercase Roman numerals
list-style-image	Image replacement for the list marker	The url keyword with parentheses surrounding the file name or path for the image	The image displays in front of each list item
list-style-position	Configures placement of markers	inside	Markers are indented, text wraps under the markers
		outside (default)	Markers have default placement

The property `list-style-type: none` prevents the browser from displaying the list markers (you'll see a use for this when configuring navigation hyperlinks in Chapter 7). Figure 5.22 shows an unordered list configured with square markers using the following CSS:

```
ul { list-style-type: square; }
```

Figure 5.23 shows an ordered list configured with uppercase letter markers using the following CSS:

```
ol { list-style-type: upper-alpha; }
```

Configure an Image as a List Marker

Use the **list-style-image** property to configure an image as the marker in an unordered or ordered list. In Figure 5.24, an image named marker.gif was configured to replace the list markers using the following CSS:

```
ul {list-style-image: url(marker.gif); }
```

FIGURE 5.22
The unordered list markers are square.

- Specialty Coffee and Tea
- Gluten-free Pastries
- Organic Salads

FIGURE 5.23
The ordered list markers use uppercase letters.

A. Specialty Coffee and Tea
B. Gluten-free Pastries
C. Organic Salads

FIGURE 5.24
The list markers are replaced with an image.

Specialty Coffee and Tea
Gluten-free Pastries
Organic Salads

 Hands-On Practice 5.8

In this Hands-On Practice, you'll replace the list markers on a web page with an image file named marker.gif. You will use your files from Hands-On Practice 5.4 (see the student files chapter5/5.4 folder) as a starting point.

1. Launch a text editor and open index.html. Add the following style rule to the embedded CSS in the head section to configure the ul element selector with the `list-style-image` property:

   ```
   ul { list-style-image: url(marker.gif); }
   ```

2. Save your page as index.html. Launch in a browser and test your page. You should see a small coffee cup before each item in the unordered list as shown in Figure 5.24. The student files contain a sample solution in the chapter5/5.8 folder.

The Favorites Icon

Ever wonder about the small icon you sometimes see in the address bar or tab of a browser? That's a **favorites icon**, often referred to as a **favicon**, which is a square image (either 16 × 16 pixels or 32 × 32 pixels) associated with a web page. The favicon shown in Figure 5.25 may be displayed in the browser address bar, tab, or the favorites/bookmarks list.

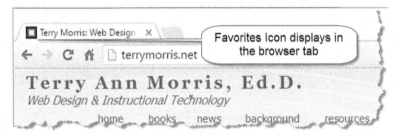

FIGURE 5.25 *The favorites icon displays in the browser tab.*

Configuring a Favorites Icon

Recall that in Chapter 4 you coded the `<link>` tag in the head section of a web page to associate an external style sheet file with a web page file. You can also use the `<link>` tag to associate a favorites icon with a web page. Three attributes are used to associate a web page with a favorites icon: `rel`, `href`, and `type`. The value of the `rel` attribute is `icon`. The value of the `href` attribute is the name of the image file. Recall from Chapter 1 that MIME types are used to indicate the format of data within a media file. The value of the `type` attribute describes the MIME type of the image—which defaults to `image/x-icon` for .ico files. The code to associate a favorites icon named favicon.ico to a web page follows:

```
<link rel="icon" href="favicon.ico" type="image/x-icon">
```

You may need to publish your files to the Web (see Chapter 12) in order for the favorites icon to display in Microsoft Edge and Internet Explorer. Other browsers, such as Firefox, display favicons more reliably and also support GIF, JPG, and PNG image formats. Be aware that if you use a .gif, .png, or .jpg file as a favorites icon, the MIME type should be image/ico. For example:

```
<link rel="icon" href="favicon.gif" type="image/ico">
```

Hands-On Practice 5.9 ————————————————————

Let's practice using a favorites icon. In this exercise, you will use your files from Hands-On Practice 5.3 (see the student files chapter5/5.3 folder) as a starting point and configure the favicon.ico file as a favorites icon. Obtain the favicon.ico file from the student files in the chapter5/starters folder and save it with your files.

1. Launch a text editor and open index.html. Add the following link tag to the head section of the web page:

   ```
   <link rel="icon" href="favicon.ico" type="image/x-icon">
   ```

2. Save your page as index.html. Launch a browser and test your page. You may notice the tiny kayaker in the browser tab as shown in Figure 5.26. The student files contain a sample solution in the chapter5/5.9 folder.

FIGURE 5.26 *The favorites icon displays in the Firefox browser tab.*

 How can I create my own favorites icon?

You can create your own favicon with a graphics application, such as GIMP, or with one of the following online tools:

▶ https://favicon.cc

▶ https://www.favicongenerator.com

▶ https://www.freefavicon.com

Image Maps

An **image map** is an image configured with multiple clickable or selectable areas that link to another web page or website. The selectable areas are called **hotspots**. Image maps can configure clickable areas in three shapes: rectangles, circles, and polygons. An image map requires the use of the img element, map element, and one or more area elements.

Map Element

The **map element** is a container tag that indicates the beginning and ending of the image map description. The **name attribute** is coded to associate the `<map>` tag with its corresponding image. The `id` attribute must have the same value as the `name` attribute. To associate a map element with an image, configure the `<img>` tag with the **usemap attribute** to indicate which `<map>` to use.

Area Element

The **area element** defines the coordinates or edges of the clickable area. It is a void tag that uses the `href`, `alt`, `title`, `shape`, and `coords` attributes. The `href` attribute identifies the web page to display when the area is clicked. The `alt` attribute provides a text description for screen readers. Use the **title attribute** to specify text that some browsers may display as a tooltip when the mouse is placed over the area. The **coords attribute** indicates the coordinate position of the clickable area. Table 5.5 describes the type of coordinates needed for each `shape` attribute value.

TABLE 5.5 *Shape Coordinates*

Shape	Coordinates	Meaning
rect	"x1,y1,x2,y2"	The coordinates at point (x1,y1) represent the upper-left corner of the rectangle. The coordinates at point (x2,y2) represent the lower-right corner of the rectangle
circle	"x,y,r"	The coordinates at point (x,y) indicate the center of the circle. The value of r is the radius of the circle, in pixels
poly	"x1,y1,x2,y2,x3,y3", etc.	The values of each (x,y) pair represent the coordinates of a corner point of the polygon

Exploring a Rectangular Image Map

We'll focus on a rectangular image map. For a rectangular image map, the value of the `shape` attribute is `rect`, and the coordinates indicate the pixel positions as follows:

- distance of the upper-left corner from the left side of the image
- distance of the upper-left corner from the top of the image
- distance of the lower-right corner from the left edge of the image
- distance to the lower-right corner from the top of the image.

Figure 5.27 shows an image of a fishing boat. This example is in the student files at chapter5/map.html.

- The dotted rectangle around the fishing boat indicates the location of the hotspot.
- The coordinates shown (24, 188) indicate that the top-left corner is 24 pixels from the left edge of the image and 188 pixels from the top of the image.
- The pair of coordinates in the lower-right corner (339, 283) indicates that this corner is 339 pixels from the left edge of the image and 283 pixels from the top of the image.

FIGURE 5.27 *A sample image map.*

The HTML code to create this image map follows:

```
<map name="boat" id="boat">
<area href="http://www.fishingdoorcounty.com"
  shape="rect" coords="24,188,339,283"
  alt="Door County Fishing Charter"
  title="Door County Fishing Charter">
</map>
<img src="fishingboat.jpg" usemap="#boat"
  alt="Door County" width="416" height="350">
```

Note the use of the `alt` attribute on the area element in the previous code sample. Configure a descriptive `alt` attribute for each area element associated with an image map to provide for accessibility.

Most web developers do not hand-code image maps. Web authoring tools, such as Adobe Dreamweaver, have features that help you to generate image maps. There are also free online image map generators available at:

- https://www.image-map.net
- https://imagemap.org
- https://mobilefish.com/services/image_map/image_map.php

Figure and Figcaption Elements

HTML5 introduced several elements that are useful to semantically describe the content. While you could use a generic div element to configure an area on a web page with an image and a caption, the figure and figcaption elements are more descriptive of the content. The div element is quite useful but very general in nature. When the figure and figcaption elements are used, the structure of the content is well defined.

The Figure Element

The block display **figure element** comprises a unit of content that is self-contained, such as an image, along with one optional figcaption element.

The Figcaption Element

The block display **figcaption element** provides a caption for the figure content.

 Hands-On Practice 5.10

In this Hands-On Practice, you will configure an area on a web page that contains an image with a caption by using the HTML5 figure and figcaption elements. Obtain the myisland.jpg file from the student files chapter5/starters folder. Save the myisland.jpg file in a folder named figure.

1. Launch a text editor. Open the file located at chapter5/template.html in the student files. Modify the title element. Add an image tag to the body section to display the myisland.jpg image as follows:

```
<img src="myisland.jpg" alt="Tropical
Island" height="480" width="640">
```

Save the file as index.html in the figure folder. Launch a browser to test your page. It should look similar to the page shown in Figure 5.28.

FIGURE 5.28 *The image is displayed on the web page.*

2. Configure a figure caption for the image. Launch a text editor and open the web page file. Add embedded CSS to the head section that configures the figcaption element selector to display bold, italic text with the Trajan Pro font typeface (or if Trajan Pro is not available, with Arial font). Configure the size of the font to be 1.5em; The code follows:

```
<style>
figcaption { font-weight: bold;
             font-style: italic
             font-family: Trajan Pro, Arial;
             font-size: 1.5em;

}
</style>
```

3. Edit the body section. Below the image, add a figcaption element that contains the following text: "Tropical Island Getaway." Configure a figure element that contains both the image and the figcaption. The code follows:

```
<figure>
<img src="myisland.jpg"
width="640" height="480"
alt="Tropical
Island">
<figcaption> Tropical
Island Getaway
</figcaption>
</figure>
```

FIGURE 5.29 *The HTML figure and figcaption elements were used in this web page.*

4. Save the file as index.html in the figure folder. Launch a browser to test your page. It should look similar to the page shown in Figure 5.29. The student files contain a sample solution in the chapter5/5.10 folder.

?FAQ **What if I don't know the height and width of an image?**

If you have a graphics application such as Adobe Photoshop or Microsoft Paint handy, launch the application and open the image. These applications include options that will display the properties of the image. It's also easy to use Windows Explorer to determine the dimensions of the image. First, display the folder containing the image and verify that the "Details pane" view is selected. Next, select the image file to display the dimensions, file size, and other image information.

Review and Apply

Review Questions

1. Which attribute specifies text that is available to browsers and other user agents that do not support graphics?

    a. alt

    b. text

    c. src

    d. accessibility

2. Which of the following creates an image link to the index.html page when the home.gif graphic is clicked?

    a. `<a href="index.html" src="home.gif" alt="Home"></a>`

    b. `<a href="index.html"><img src="home.gif" alt="Home"></a>`

    c. `<img src="home.gif" href="index.html" alt="Home">`

    d. `<a src="home.gif"><img src="home.gif" alt="Home"></a>`

3. Why should you include height and width attributes on an `<img>` tag?

    a. They are required attributes and must always be included.

    b. They help the browser reserve the appropriate space for the image.

    c. They help the browser display the image in its own window.

    d. None of the above.

4. Which declaration configures an unordered list item with a square list marker?

    a. `list-bullet: none;`

    b. `list-style-type: square;`

    c. `list-style-image: square;`

    d. `list-marker: square;`

5. Which CSS property will configure the font typeface?

    a. font-face   **b.** font-style

    c. font-family   **d.** typeface

6. Which configures a class called news with red text, large font, and Arial or a sans-serif font using CSS?

    a. `news { text: red;`
    `        font-size: large;`
    `        font-family: Arial,`
    `        sans-serif; }`

    b. `.news { text: red;`
    `        font-size: large;`
    `        font-family: Arial,`
    `        sans-serif; }`

    c. `#news { color: red;`
    `        font-size: large;`
    `        font-family: Arial,`
    `        sans-serif; }`

    d. `.news { color: red;`
    `        font-size: large;`
    `        font-family: Arial,`
    `        sans-serif; }`

7. Which of the following configures a graphic to repeat vertically down the side of a web page?

    a. `background-repeat: repeat-x;`

    b. `background-repeat: repeat;`

    c. `valign="left"`

    d. `background-repeat: repeat-y;`

8. Which CSS property configures the background image of an element?

    a. background-color   **b.** bgimage

    c. favicon   **d.** background-image

9. What is the process of creating an image with the lowest file size that still renders a good-quality image—balancing image quality and file size?

   a. pixelation

   b. optimization

   c. typography

   d. interlacing

10. Which of the following graphic file types do not support transparency?

   a. GIF

   b. JPG

   c. WebP

   d. PNG

Review Answers

1. a   2. b   3. b   4. b   5. c   6. d   7. d   8. d   9. b   10. b

Hands-On Exercises

1. Write the CSS code for an external style sheet file named mystyle.css that configures the text to be brown, 1.2em in size, and in Arial, Verdana, or a sans-serif font.

2. Write the HTML and CSS code for an embedded style sheet that configures a class called priority, which has bold and italic text.

3. Write the code to place an image called primelogo.gif on a web page. The image is 100 pixels high by 650 pixels wide.

4. Write the code to create an image hyperlink. The image is called schaumburgthumb.jpg. It is 100 pixels high by 150 pixels wide. The image should link to a larger image called schaumburg.jpg. There should be no border on the image.

5. Write the code to create a nav element containing three images used as navigation links. Table 5.6 provides information about the images and their associated links.

TABLE 5.6   *Image and Link Information*

Image Name	Link Page Name	Image Height	Image Width
homebtn.gif	index.html	50	200
productsbtn.gif	products.html	50	200
orderbtn.gif	order.html	50	200

6. Experiment with background images.

   a. Locate the twocolor.gif file in the student files chapter5/starters folder. Design a web page that uses this file as a background image that repeats down the left side of the browser window. Save your file as bg1.html.

   b. Locate the twocolor1.gif file in the student files chapter5/starters folder. Design a web page that uses this file as a background image that repeats across the top of the browser window. Save your file as bg2.html.

7. Design a new web page about your favorite movie. Name the web page movie5.html. Configure a background color for the page and either background images or background colors for at least two sections of the page. Search the Web for a photo of a scene from the movie, an actress in the movie, or an actor in the movie. Include the following information on your web page:

   • Title of the movie

   • Director or producer

- Leading actor
- Leading actress
- Rating (R, PG-13, PG, G, NR)
- A brief description of the movie
- An absolute link to a review about the movie

It is unethical to steal an image from another website. Some websites have a link to their copyright policy. Most websites will give permission for you to use an image in a school assignment. If there is no available policy, e-mail the site's contact person and request permission to use the photo. If you are unable to obtain permission, you may substitute with clip art or an image from a free site instead.

Focus on Web Design

Providing access to the Web for all people is an important issue. Visit the W3C's Web Accessibility Initiative and explore their WCAG 2.1 Quick Reference at https://w3.org/WAI/WCAG21/quickref. View additional pages at the W3C's site as necessary. Explore the checkpoints that are related to the use of color and images on web pages. Create a web page that uses color, uses images, and includes the information that you discovered.

Case Study

You will continue the case studies from Chapter 4 as you configure the websites to display images.

Pacific Trails Resort Case Study

In this chapter's case study, you will use the existing Pacific Trails (Chapter 4) website as a starting point to create a new version of the website that incorporates images. You will modify the design of the pages to display a large image on each page, as indicated in the wireframe in Figure 5.30. You will also create a new page, the Activities page.

You have five tasks in this case study:

1. Create a new folder for the Pacific Trails Resort website.

2. Update the pacific.css external style sheet file.

3. Update the Home page: index.html.

4. Update the Yurts page: yurts.html.

5. Create a new Activities page: activities.html.

wrapper

| header |
| nav |
| div with large image |
| main |
| div with contact info |
| footer |

Task 1: Create a folder called ch5pacific to contain your Pacific Trails Resort website files. Copy the files from the Chapter 4 Case Study ch4pacific folder. Copy the following files from the chapter5/casestudystarters/pacific folder in the student files and place them in your ch5pacific folder: coast.jpg, marker.gif, sunset.jpg, trail.jpg, and yurt.jpg.

Task 2: The External Style Sheet. Launch a text editor and open the pacific.css external style sheet file.

1. **The body element selector.** Add a declaration that configures Arial, Helvetica, or sans-serif font typeface.

FIGURE 5.30 *New Pacific Trails wireframe.*

2. **The header element selector.** Add declarations to display the background image named sunset.jpg on the right without any repeats. Also configure declarations to set 400% line-height and 1em text-indent.

3. **The nav element selector.** Add a declaration to configure bold text.

4. **The nav a element selector.** Code styles to eliminate the display of the underline for hyperlinks (hint: use the `nav a` descendant selector with `text-decoration: none;`)

5. **The h1 element selector.** Add a declaration to display text in Georgia, Times New Roman, or serif font typeface.

6. **The h2 element selector.** Add a declaration to display text in Georgia, Times New Roman, or serif font typeface.

7. **The h3 element selector.** Code styles to display text in Georgia, Times New Roman, or serif font typeface. Also configure #000033 text color.

8. **The ul element selector.** Code styles to display the marker.gif as the list marker (bullet).

9. **The footer element selector.** Code styles to configure 75% font size, italic font style, centered text, and Georgia, Times New Roman, or serif font typeface.

10. **The resort class selector.** Add a declaration to display bold text.

11. **The contact id selector.** Code styles to display text with 90% font size.

Save your pacific.css file. Check your syntax with the CSS validator (http://jigsaw.w3.org/css-validator). Correct and retest if necessary.

Task 3: The Home Page. Launch a text editor and open the home page, index.html. Remove the b, small, and i tags from the page. Code a div element with an `<img>` tag between the nav element and the main element. Configure the `<img>` tag to display the coast.jpg image. Configure the `alt`, `height`, and `width` attribute for the image. Note: In order for your page to look similar to Figure 5.31, use 100% for the value of the width attribute. A percentage width causes the image to fill a percentage of the width of the parent element. The W3C HTML validator may indicate that the percentage value is invalid. We will overlook the error for this case study. In Chapter 6, you'll learn to use CSS to configure width and height. Save and test your page in a browser. It should look similar to Figure 5.31.

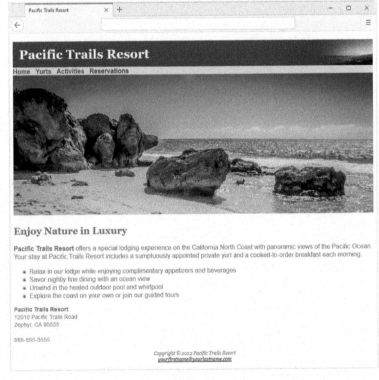

FIGURE 5.31 *Pacific Trails Resort Home page.*

FIGURE 5.32 *Pacific Trails Resort Yurts page.*

FIGURE 5.33 *New Pacific Trails Resort Activities page.*

Task 4: The Yurts Page. Launch a text editor and open the yurts.html file. Remove the b, small, and i tags from the page. Next, you will modify this file to display the yurt.jpg image in a similar manner as you configured the coast.jpg image on the home page. Save and test your new yurts.html page. It should look similar to the one shown in Figure 5.32.

Task 5: The Activities Page. Launch a text editor, open the yurts.html document, and save the file as activities.html—this is the start of your new activities page.

1. Modify the page title area as appropriate.

2. Modify the `<img>` tag to display the trail.jpg image.

3. Change the h2 text to the following: Activities at Pacific Trails.

4. Delete the description list.

5. Configure the following text using h3 tags for the headings and paragraph tags for the sentences.

   "Hiking

   Pacific Trails Resort has 5 miles of hiking trails and is adjacent to a state park. Go it alone or join one of our guided hikes.

   Kayaking

   Ocean kayaks are available for guest use.

   Bird Watching

   While anytime is a good time for bird watching at Pacific Trails, we offer guided bird-watching trips at sunrise several times a week."

   Save your activities.html file. Launch a browser and test your new activities.html page. It should look similar to Figure 5.33.

   As you view your web pages in a browser, notice how the use of graphics and typography creates a more appealing web presence for Pacific Trails Resort.

Path of Light Yoga Studio Case Study

In this chapter's case study, you will use the existing Path of Light Yoga Studio (Chapter 4) website as a starting point while you create a new version of the website that incorporates images.

You have five tasks in this case study:

1. Create a new folder for the Path of Light Yoga Studio website.

2. Update the yoga.css external style sheet.

3. Update the Home page: index.html.

4. Update the Classes page: classes.html.

5. Create a new Schedule page: schedule.html.

Task 1: Create a folder called ch5yoga to contain your Path of Light Yoga Studio website files. Copy the files from the Chapter 4 Case Study ch4yoga folder and place them in your ch5yoga folder. Locate the chapter5/casestudystarters/yoga folder in the student files. Copy the following files to your ch5yoga folder: lilyheader.jpg, yogadoor.jpg, yogalounge.jpg, and yogamat.jpg.

Task 2: The External Style Sheet. Launch a text editor and open the yoga.css external style sheet file.

1. **The body element selector.** Add a declaration that configures Verdana, Arial, or sans-serif font typeface.

2. **The header element selector.** Add declarations to configure lilyheader.jpg as a background image that displays on the right without repeating.

3. **The nav element selector.** Code styles to configure centered, bold text.

4. **The nav a element selector.** Code styles to eliminate the display of the underline for hyperlinks (hint: use the `nav a` descendant selector with `text-decoration: none;`).

5. **The h1 element selector.** Code styles to display 400% line height and 1em text-indent.

6. **The footer selector.** Add declarations to configure small, italic, and centered text.

7. **The li element selector and dd element selector.** Code styles to configure 90% font size.

Save your file. Use the CSS Validator (http://jigsaw.w3.org/css-validator) to check your syntax. Correct and retest if necessary.

FIGURE 5.34 *Path of Light Yoga Studio Home page.*

wrapper

header

nav

main

div with large image

footer

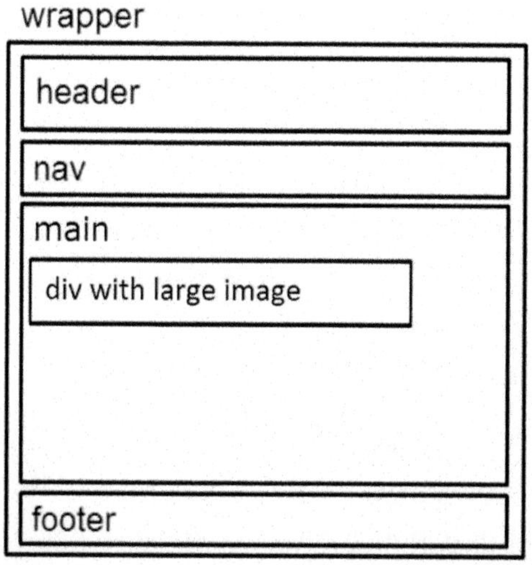

FIGURE 5.35 *Wireframe for Classes and Schedule pages.*

Task 3: The Home Page. Launch a text editor and open the home page, index.html. Remove the b, small, and i tags from the page. Add an `<img>` tag above the h2 element. Configure the `<img>` tag to display the yogadoor.jpg image. Configure the `alt`, `height`, and `width` attributes for the image. Also configure the image to appear to the right of the text by coding the `align="right"` attribute on the `<img>` tag. Note: The W3C HTML validator will indicate that the align attribute is invalid. We'll overlook the error for this case study. In Chapter 7, you'll learn to use the CSS float property (instead of the align attribute) to configure this type of layout. Save and test your page in a browser. It should look similar to Figure 5.34.

Task 4: The Classes Page. It's common for the content pages of a website to have a slightly different layout than the home page. The wireframe shown in Figure 5.35 depicts the layout of the Classes and Schedule pages.

Launch a text editor and open classes.html. Remove the b, small, and i tags from the page. Configure a div element to display the yogamat.jpg image. As shown in the wireframe in Figure 5.35, this div is located at the top of the main element. The div element contains a line break tag followed by an `<img>` tag. Display the yogamat.jpg image within the `<img>` tag. Configure the alt, height, and width attributes for the image. Note: for a more pleasing page display, configure the image's width at 100%. A percentage width causes the image to fill a percentage of the width of the parent element. The W3C HTML validator may indicate that the percentage value is invalid. We will overlook the error for this case study. In Chapter 6, you'll learn to use CSS to configure width and height. Save and test your page in a browser. It should look similar to Figure 5.36.

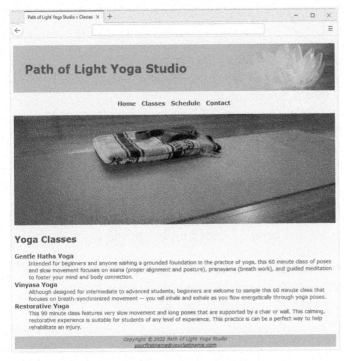

FIGURE 5.36 *Path of Light Yoga Studio Classes page.*

FIGURE 5.37 *Path of Light Yoga Studio Schedule page.*

Task 5: The Schedule Page. Use the Classes page as the starting point. Launch a text editor, open classes.html, and save the file as schedule.html.

Modify the schedule.html file to look similar to the Schedule page, as shown in Figure 5.37:

1. Change the page title to an appropriate phrase.

2. Modify the `<img>` tag to display the yogalounge.jpg image.

3. Replace the text within the h2 element with Yoga Schedule.

4. Delete the description list from the page.

5. Configure a paragraph below the h2 element:

   Mats, blocks, and blankets provided. Please arrive 10 minutes before your class begins. Relax in our Serenity Lounge before or after your class.

6. Configure an h3 element with Monday—Friday.

7. Configure an unordered list with

   9:00am Gentle Hatha Yoga

   10:30am Vinyasa Yoga

   5:30pm Restorative Yoga

   7:00pm Gentle Hatha Yoga

8. Configure an h3 element with Saturday & Sunday.

9. Configure an unordered list

   10:30am Gentle Hatha Yoga

   Noon Vinyasa Yoga

   1:30pm Gentle Hatha Yoga

   3:00pm Vinyasa Yoga

   5:30pm Restorative Yoga

Save the schedule.html file. When you test your page in a browser, it should look similar to Figure 5.37. As you view your web pages in a browser, notice how the use of graphics and typography creates a more appealing web presence for the yoga studio.

Endnotes

1. Adler, Mark, et al. "Portable Network Graphics (PNG) Specification (Second Edition)." Edited by David Duce, *W3C*, 10 Nov. 2003, www.w3.org/TR/PNG/.
2. Parmenter, Stuart, et al. "APNG Specification." *APNG Specification – MozillaWiki*, 26 Sept. 2015, wiki.mozilla.org/APNG_Specification.
3. "GIF to Animated PNG Converter." *Online Animated GIF Tools*, ezgif.com/gif-to-apng.
4. "A New Image Format for the Web | WebP | Google Developers." *Google*, Google, developers.google.com/speed/webp/.
5. Deveria, Alexis "Can I Use... WebP Image Format?" *Can I Use... Support Tables for HTML5, CSS3, Etc*, caniuse.com/#feat=webp.
6. "Online JPG to WebP Converter." *Online Animated GIF Tools*, ezgif.com/jpg-to-webp.
7. "Convert Images to WebP." *Online-Convert*, image.online-convert.com/convert-to-webp.
8. "CSS Backgrounds and Borders Module Level 3." Edited by Bert Bos et al., *W3C*, 17 Oct. 2017, www.w3.org/TR/css-backgrounds-3/.
9. "CSS Fonts Module Level 3: 3.1. Font family: the font-family property." Edited by John Daggett et al., *W3C*, 20 Sept. 2018, www.w3.org/TR/css-fonts-3/#propdef-font-family.
10. "CSS Fonts Module Level 3." Edited by John Daggett et al., *W3C*, 20 Sept. 2018, www.w3.org/TR/css-fonts-3/.
11. "CSS Text Module Level 3." Edited by Elika J. Etemad et al., *W3C*, 13 Nov. 2019, www.w3.org/TR/css-text-3/.
12. "CSS Text Decoration Module Level 3." Edited by Elika J. Etemad and Koji Ishii, *W3C*, 13 Aug. 2019, www.w3.org/TR/css-text-decor-3/.
13. "CSS Lists Module Level 3." Edited by Elika J. Etemad and Tab Atkins, *W3C*, 17 Aug. 2019, www.w3.org/TR/css-lists-3/.

CHAPTER 6

More CSS Basics

You'll add to your Cascading Style Sheets (CSS) skill set in this chapter. You will begin to work with the CSS box model and configure margin, border, and padding. You'll also explore CSS properties to round corners, apply shadow, adjust display of background images, and configure color and opacity.

You'll learn how to...

- Describe and apply the CSS box model
- Configure width and height with CSS
- Configure margin, border, and padding with CSS
- Center web page content with CSS
- Apply shadows with CSS
- Configure rounded corners with CSS
- Apply CSS properties to background images
- Configure opacity, RGBA color, HSLA color, and gradients with CSS

Width and Height with CSS

There are many ways to configure width and height with CSS.[1] This section introduces you to the width, min-width, max-width, and height properties. Table 6.1 lists commonly used width and height units and their purpose.

TABLE 6.1   *Unit Types and Purpose*

Unit	Purpose
px	px Configures a fixed number of pixels as the value
em	Configures a value relative to the font size
%	Configures a percentage value of the parent element
vh	Configures a value relative to 1% of the viewport height
vw	Configures a value relative to 1% of the viewport width

FIGURE 6.1 *The web page is set to 80% width.*

The width Property

The **width property** configures the width of an element's content in the browser viewport with either a numeric value unit (such as 100px or 20em), percentage (such as 80%, as shown in Figure 6.1) of the parent element, or viewport width value (such as 50vw, which is 50% of the viewport width). The actual width of an element displayed in the browser viewport includes the width of the element's content, padding, border, and margin—it is not the same as the value of the width property, which only configures the width of the element's content.

The min-width Property

The **min-width property** sets the minimum width of an element's content in the browser viewport. This minimum width value can prevent content from jumping around when a browser is resized. Scrollbars appear if the browser viewport is resized below the minimum width (see Figures 6.2 and 6.3).

FIGURE 6.2 *As the browser is resized, the "Coffee House" and navigation text wrap.*

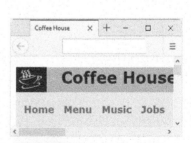

FIGURE 6.3 *The* min-width *property avoids display issues.*

The `max-width` Property

The **max-width property** sets the maximum width of an element's content in the browser viewport. This maximum width value can reduce the possibility of text stretching across large expanses of the screen by a high-resolution monitor.

The `height` Property

The **height property** configures the height of an element's content in the browser viewport with either a numeric value unit (such as `900px`), percentage (such as `60%`) of the parent element, or viewport height value (such as 50vw), which is 50% of the viewport height. Figure 6.4 shows a web page with an h1 area without a `height` or `line-height` property configured. Notice how part of the background image is truncated and is hidden from view. In Figure 6.5, the h1 area is configured with the `height` property. Notice the improved display of the background image.

FIGURE 6.4 *The background image is truncated.*

FIGURE 6.5 *The* height *property value corresponds to the height of the background image.*

 Hands-On Practice 6.1

You'll work with the height and width properties in this Hands-On Practice.

Create a new folder called coffeech6. Copy the coffeelogo.jpg file from the chapter6/ starters folder into your coffeech6 folder. Copy the chapter6/starter1.html file into your coffeech6 folder. Launch a text editor and open the file.

1. Edit the embedded CSS to configure the document to take up 80% of the browser window but with a minimum width of 750px. Add the following style rules to the body element selector:

   ```
   width: 80%; min-width: 750px;
   ```

2. Add style declarations to the h1 element selector to configure height as 150px (the height of the background image) and line height as 220%.

   ```
   height: 150px; line-height: 220%;
   ```

Save your file as index.html. Launch a browser and test your page. Your web page should look similar to Figure 6.1. A sample solution is in the chapter6/6.1 folder.

The Box Model

Each element in a document is considered to be a rectangular box. As shown in Figure 6.6, this box consists of a content area surrounded by padding, a border, and margins. This is known as the **box model**.[2]

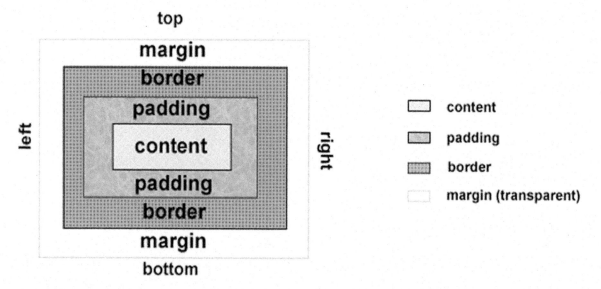

FIGURE 6.6 *The CSS box model.*

Content

The content area can consist of a combination of text and web page elements such as images, paragraphs, headings, lists, and so on. The visible width of the element on a web page is the total of the content width, the padding width, and the border width. However, the `width` property only configures the actual width of the content—not including any padding, border, or margin.

Padding

The **padding** area is between the content and the border. The default padding value is zero. When the background of an element is configured, the background is applied to both the padding and the content areas.

Border

The **border** area is between the padding and the margin. The default border has a value of 0 and does not display.

Margin

The **margin** determines the empty space between the element and any adjacent elements. The margin is always transparent—the background color of the web page or container element (such as a div) shows in this area. The solid line in Figure 6.6 that contains the margin area does not display on a web page. Browsers often have default margin values set for the web page document and for certain elements such as paragraphs, headings, forms, and so on. Use the margin property to override the default browser values.

The Box Model in Action

The web page shown in Figure 6.7 (student files chapter6/box.html) depicts the box model in action with an h1 and a div element.

▶ The h1 element is configured to have a light blue background, 20 pixels of padding (the space between the content and the border), and a black 1-pixel border.

▶ The empty space where the white web page background shows through is the margin. When two vertical margins meet (such as between the h1 element and the div element), the browser collapses the margin size to be the larger of the two margin values instead of applying both margins.

▶ The div element has a medium-blue background, the browser default padding (which is no padding), and a black 5-pixel border.

You will get more practice using the box model in this chapter. Feel free to experiment with the box model and the chapter6/box.html file.

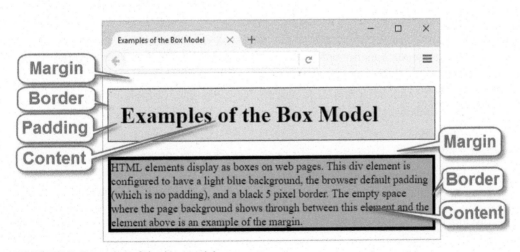

FIGURE 6.7 *Examples of the box model.*

Margin and Padding with CSS

The `margin` Property

Use the **`margin` property** to configure margins on all sides of an element. The margin determines the empty space between the element and any adjacent elements. The margin is always transparent—the background color of the web page or parent element shows in this area.

To configure the size of the margin, use a numeric value (px or em). To eliminate the margin, configure it to 0 (with no unit). Use the value `auto` to indicate that the browser should calculate the margin (more on this later in the chapter). You can also configure individual settings for `margin-top`, `margin-right`, `margin-bottom`, and `margin-left`. Table 6.2 shows CSS properties that configure margin.

TABLE 6.2   *Configuring* `margin` *with CSS*

Property	Description and Common Values
`margin`	Shorthand notation to configure the margin surrounding an element
	A numeric value (px or em) or percentage; for example, `margin: 10px;` if you set a value to 0, omit the unit
	The value `auto` is used to cause the browser to automatically calculate the margin for the element
	Two numeric values (px or em) or percentages; the first value configures the top margin and bottom margin; the second value configures the left margin and right margin; for example, `margin: 20px 10px;`
	Three numeric values (px or em) or percentage; the first value configures the top margin, the second value configures the left margin and right margin, and the third value configures the bottom margin
	Four numeric values (px or em) or percentages; the values configure the margins in the following order: `margin-top, margin-right, margin-bottom, margin-left`
`margin-bottom`	Bottom margin; a numeric value (px or em), percentage, or `auto`
`margin-left`	Left margin; a numeric value (px or em), percentage, or `auto`
`margin-right`	Right margin; a numeric value (px or em), percentage, or `auto`
`margin-top`	Top margin; a numeric value (px or em), percentage, or `auto`

The `padding` Property

The **padding property** configures empty space between the content of the HTML element (such as text) and the border. By default, the padding is set to 0. If you configure a background color or background image for an element, it is applied to both the padding and the content areas. See Table 6.3 for CSS properties that configure padding.

TABLE 6.3   *Configuring* `padding` *with CSS*

Property	Description and Common Values
`padding`	Shorthand notation to configure the amount of padding—the empty space between the element's content and border A numeric value (px or em) or percentage; for example, `padding: 10px;` if you set a value to 0, omit the unit Two numeric values (px or em) or percentages; the first value configures the top padding and bottom padding; the second value configures the left padding and right padding; for example, `padding: 20px 10px;` Three numeric values (px or em) or percentage; the first value configures the top padding, the second value configures the left padding and right padding, and the third value configures the bottom padding Four numeric values (px or em) or percentages; the values configure the padding in the following order: `padding-top, padding-right, padding-bottom, padding-left`
`padding-bottom`	Empty space between the content and bottom border; a numeric value (px or em) or percentage
`padding-left`	Empty space between the content and left border; a numeric value (px or em) or percentage
`padding-right`	Empty space between the content and right border; a numeric value (px or em) or percentage
`padding-top`	Empty space between the content and top border; a numeric value (px or em) or percentage

The web page shown in Figure 6.8 demonstrates use of the margin and padding properties. The example is in the student files at chapter6/box2.html.

The CSS is shown below:

```
body { background-color: #FFFFFF; }
h1   { background-color: #D1ECFF;
       padding-left: 60px; }
#box { background-color: #74C0FF;
       margin-left: 60px;
       padding: 5px 10px; }
```

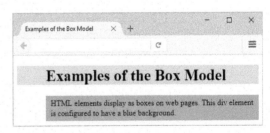

FIGURE 6.8 *Margin and padding have been configured.*

Borders with CSS

The **border property** configures the border, or boundary, around an element. By default, the border has a width set to 0 and does not display. See Table 6.4 for commonly used CSS properties that configure border.

TABLE 6.4 *Configuring* border *with CSS*

Property	Description and Common Values
border	Shorthand notation to configure the border-width, border-style, and border-color of an element; the values for border-width, border-style, and border-color separated by spaces; for example, border: 1px solid #000000;
border-bottom	Bottom border; the values for border-width, border-style, and border-color separated by spaces
border-left	Left border; the values for border-width, border-style, and border-color separated by spaces
border-right	Right border; the values for border-width, border-style, and border-color separated by spaces
border-top	Top border; the values for border-width, border-style, and border-color separated by spaces
border-width	Width of the border; a numeric pixel value (such as 1px) or the values thin, medium, thick
border-style	Style of the border; none, inset, outset, double, groove, ridge, solid, dashed, dotted
border-color	Color of the border; a valid color value

The **border-style property** offers a variety of formatting options. Be aware that these property values are not all uniformly applied by browsers. Figure 6.9 shows how a recent version of Firefox renders various border-style values.

The CSS to configure the borders shown in Figure 6.9 uses a border-width of 3 pixels, border-color of #000033, and the value indicated for the border-style property. For example, the style rule to configure the dashed border follows:

```
.dashedborder { border-width: 3px;
                border-style: dashed;
                border-color: #000033; }
```

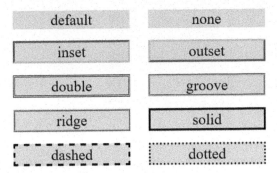

FIGURE 6.9 *Examples of the various* `border-style` *values displayed in Firefox.*

A shorthand notation allows you to configure all the border properties in one style rule by listing the values of `border-width`, `border-style`, and `border-color`. For example:

```
.dashedborder { border: 3px dashed #000033; }
```

 Hands-On Practice 6.2

You'll work with the `border` property in this Hands-On Practice. When complete, your web page will look similar to the one shown in Figure 6.10. You will use the box2.html file in the chapter6 folder of the student files as a starter file. Launch a text editor and open the box2.html file. Configure the embedded CSS as follows:

1. Configure the h1 to display a 3-pixel ridged bottom border in a dark gray color. Add the following style rule to the h1 element selector:

   ```
   border-bottom: 3px ridge #330000;
   ```

2. Configure the `box` id to display a 1-pixel solid black border. Add the following style rule to the `#box` selector:

   ```
   border: 1px solid #000000;
   ```

FIGURE 6.10 *The* `border` *property has been configured.*

3. Save your page as boxborder.html. Launch a browser and test your page. Compare your work with the sample solution at chapter6/6.2/index.html.

CSS Rounded Corners

VideoNote

CSS Rounded Corners

Now that you have worked with borders and the box model, you may have begun to notice a lot of rectangles on your web pages! The CSS **border-radius property** is used to create rounded corners and soften up those rectangles.[3]

Valid values for the `border-radius` property include one to four numeric values (using pixel or em units) or percentages that configure the radius of the corner. If a single value is provided, it configures all four corners. If four values are provided, the corners are configured in order of top left, top right, bottom right, and bottom left. You can configure corners individually with the `border-bottom-left-radius`, `border-bottom-right-radius`, `border-top-left-radius`, and `border-top-right-radius` properties.

CSS declarations to set a border with rounded corners are shown below. If you would like a visible border to display, configure the border property. Then set the value of the `border-radius` property to a value below 20px for best results.

```
border: 1px solid #000000;
border-radius: 15px;
```

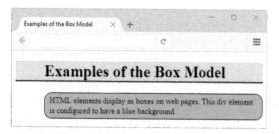

FIGURE 6.11 *Rounded corners were configured with CSS.*

See Figure 6.11 (chapter6/box3.html in the student files) for an example of this code in action.

Figure 6.12 (see chapter6/box4.html) shows a div element with only the top and bottom left corners rounded. The `border-top-left-radius` and `border-bottom-left-radius` properties were used. The code follows.

```
#box { background-color: #74C0FF;
       margin-left: 60px;
       padding: 5px 20px;
       border-top-left-radius: 90px;
       border-bottom-left-radius: 90px; }
```

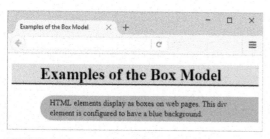

FIGURE 6.12 *Top and bottom left corners are rounded.*

You can use your creativity to configure one, two, three, or four corners of an element when using `border-radius`. With progressive enhancement in mind, note that visitors to your site who are using a browser that does not support this property will see only right-angle rather than rounded corners. However, the functionality and usability of the web page will not be affected. Keep in mind that another approach to getting a rounded look is to create a rounded rectangle background image with a graphics application.

Hands-On Practice 6.3 ────────────────────────

You'll configure a logo header area that uses a background image and rounded borders in this Hands-On Practice.

1. Create a new folder called bistroch6. Copy the lighthouselogo.jpg and background.jpg files in the chapter6/starters folder to your bistroch6 folder. A starter file is ready for you in the student files. Copy the chapter6/starter2.html file into your bistroch6 folder. Launch a browser to display the starter2.html web page shown in Figure 6.13.

2. Launch a text editor and open the starter2.html file. Save the file as index.html. Edit the embedded CSS and code an h1 element selector with style declarations that will configure the lighthouselogo.jpg image as a background image that does not repeat: height of 100px, width of 700px, font size of 3em, 150px of left padding, 30px of top padding, and a border radius of 15px. The style declarations follow:

```
h1 { background-image: url(lighthouselogo.jpg);
     background-repeat: no-repeat;
     height: 100px; width: 700px; font-size: 3em;
     padding-left: 150px; padding-top: 30px;
     border-radius: 15px; }
```

FIGURE 6.13 *The starter2.html file.*

3. Save the file. When you test your index.html file in a browser, it should look similar to the one shown in Figure 6.14 if you are using a browser that supports rounded corners. Otherwise, the logo will have right-angled corners, but the web page will still be usable. Compare your work with the solution in the student files (chapter6/6.3/index.html).

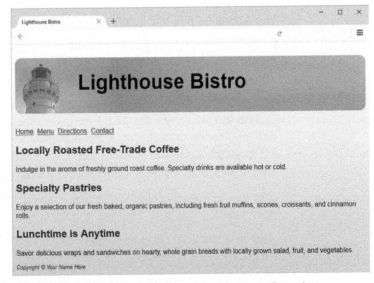

FIGURE 6.14 *The web page with the logo area configured.*

Center Page Content with CSS

You learned how to center text within a div or other block display element in Chapter 5—but what about centering the entire web page itself within the browser viewport? A popular page layout design that is easy to accomplish with just a few lines of CSS is to center the entire content of a web page within a browser viewport. The key is to configure a div element that contains or "wraps" the entire page content. The HTML follows:

```
<body>
<div id="wrapper">
... page content goes here ...
</div>
</body>
```

Next, configure CSS style rules for this container. Set the `width` property to an appropriate value. Set the `margin-left` and `margin-right` CSS properties to the value `auto`. This tells the browser to automatically divide the amount of space available for the left and right margins. The CSS follows:

```
#wrapper { width: 750px;
           margin-left: auto;
           margin-right: auto; }
```

You'll practice this technique in the next Hands-On Practice.

 Hands-On Practice 6.4

You will practice modifying and centering a web page in this Hands-On Practice as you update the index.html file from Hands-On Practice 6.3 (shown in Figure 6.14). A common design practice is to configure the background color of the wrapper or container to be a light, neutral color that provides good contrast with text. When complete, your web page will be similar to the one shown in Figure 6.15.

Create a new folder called centerch6. Copy the index.html, background.jpg, and lighthouselogo.jpg files from the chapter6/6.3 folder.

Launch a text editor and open the index.html file.

1. Edit the embedded CSS and configure the h1 selector. Remove the width style declaration. Configure a medium blue (#9DB3DC) background color and a 0 top margin (use the `margin-top` property).

2. Edit the embedded CSS and configure a new selector, an id named `container`. Add style declarations for the `background-color`, `padding`, `width`, `min-width`, `margin-left`, and `margin-right` properties as follows:

```
#container { background-color: #FFFFFF;
             padding: 2em;
             margin-left: auto; margin-right: auto;
             width: 80%;
             min-width: 800px; }
```

3. Edit the HTML. Configure a div element assigned to the id `container` that "wraps" or contains the code within the body section. Code an opening div tag on a new line after the opening body tag. Assign the div to the id named `container`. Code the closing div tag on a new line before the closing body tag.

4. Save the file. When you test your index.html file in a browser, it should look similar to Figure 6.15. The student files contain a sample solution in the chapter6/6.4 folder. A common design practice is to configure the background color of the wrapper or container to be a light, neutral color that provides good contrast with text.

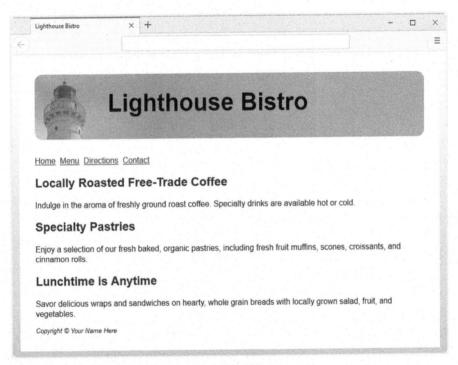

FIGURE 6.15 *The web page is centered with CSS.*

CSS Box Shadow and Text Shadow

The CSS shadow properties **box-shadow** and **text-shadow** add depth and dimension to the visual display of a web page, as shown in Figure 6.16.

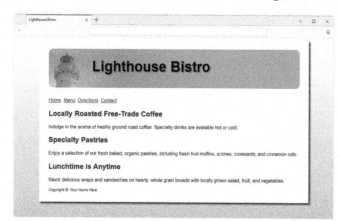

FIGURE 6.16 *Shadow properties add dimension.*

CSS box-shadow Property

The **box-shadow property** is used to create a shadow effect on the box model.[4] Configure a box shadow by coding values for the shadow's horizontal offset, vertical offset, blur radius (optional), spread distance (optional), and color:

▶ **Horizontal offset.** Use a numeric pixel value. Positive value configures a shadow on the right. Negative value configures a shadow on the left.

▶ **Vertical offset.** Use a numeric pixel value. Positive value configures a shadow below. Negative value configures a shadow above.

▶ **Blur radius (optional).** Configure a numeric pixel value. If omitted, defaults to the value 0, which configures a sharp shadow. Higher values configure more blur.

▶ **Spread distance (optional).** Configure a numeric pixel value. If omitted, defaults to the value 0. Positive values configure the shadow to expand. Negative values configure the shadow to contract.

▶ **Color value.** Configure a valid color value for the shadow.

Here's an example that configures a dark gray shadow with 5px horizontal offset, 5px vertical offset, 5px blur radius, and default spread distance:

```
box-shadow: 5px 5px 5px #828282;
```

Inner Shadow Effect. To configure an inner shadow, include the optional `inset` value. For example:

```
box-shadow: inset 5px 5px 5px #828282;
```

CSS `text-shadow` Property

The **text-shadow property** configures a text shadow by coding values for the shadow's horizontal offset, vertical offset, blur radius (optional), and color:[5]

- ▶ **Horizontal offset.** Use a numeric pixel value. Positive value configures a shadow on the right. Negative value configures a shadow on the left.

- ▶ **Vertical offset.** Use a numeric pixel value. Positive value configures a shadow below. Negative value configures a shadow above.

- ▶ **Blur radius (optional).** Configure a numeric pixel value. If omitted, defaults to the value 0, which configures a sharp shadow. Higher values configure more blur.

- ▶ **Color value.** Configure a valid color value for the shadow.

Here's an example that configures a dark gray shadow with 3px horizontal offset, 3px vertical offset, and 5px blur radius:

```
text-shadow: 3px 3px 5px #676767;
```

 Hands-On Practice 6.5

You'll configure `text-shadow` and `box-shadow` in this Hands-On Practice. When complete, your web page will look similar to the one shown in Figure 6.16. Create a new folder called shadowch6. Copy the index.html, lighthouselogo.jpg, and the background.jpg files from the chapter6/6.4 folder to your shadowch6 folder. Launch a text editor and open the index.html file.

1. Edit the embedded CSS and add the following style declarations to the `#container` selector to configure a box shadow:

   ```
   box-shadow: 5px 5px 5px #1E1E1E;
   ```

2. Add the following style declaration to the h1 element selector to configure a dark gray text shadow:

   ```
   text-shadow: 3px 3px 3px #676767;
   ```

3. Add the following style declaration to the h2 element selector to configure a light gray text shadow with no blur: `text-shadow: 1px 1px 0 #CCC;`

4. Save the file. When you test your index.html file in a browser, it should look similar to the one shown in Figure 6.16 if you are using a browser that supports the `box-shadow` and `text-shadow` properties. Otherwise, the shadows will not display, but the web page will still be usable. See the student files for a solution (chapter6/6.5/index.html).

CSS Background Clip and Origin

You're already familiar with how to configure a background image on a web page. This section introduces two CSS properties related to background images that provide you with options for clipping and sizing background images: `background-clip` and `background-origin`. As you work with these properties, keep in mind that block display elements such as div, header, and paragraph are rendered by the browser using the box model (refer to Figure 6.6), which surrounds the content of an element with padding, border, and margin.

CSS `background-clip` Property

The **background-clip property** confines the display of the background image with the following values:

- ◗ `content-box` clips off the image's display to fit the area behind the content
- ◗ `padding-box` clips off the image's display to fit the area behind the content and padding
- ◗ `border-box` (default) clips off the image's display to fit the area behind the content, padding, and border

Figure 6.17 shows div elements configured with different values of the `background-clip` property. Note that the dashed border is intentionally large in these examples.

FIGURE 6.17 *The CSS* `background-clip` *property.*

The sample page is located in the student files (chapter6/clip folder). The CSS for the first div follows:

```
.test { background-image: url(myislandback.jpg);
      background-clip: content-box;
      width: 400px; padding: 20px; margin-bottom: 10px;
      border: 10px dashed #000; }
```

CSS `background-origin` Property

The **background-origin** property positions the background image using the following values:

- ▶ `content-box` positions relative to the content area
- ▶ `padding-box` (default) positions relative to the padding area
- ▶ `border-box` positions relative to the border area

Figure 6.18 shows div elements configured with different values of the `background-origin` property. The sample page is located in the student files (chapter6/origin folder). The CSS for the first div follows:

```
.test { background-image: url(trilliumsolo.jpg);
      background-origin: content-box;
      background-repeat: no-repeat; background-position: right-top;
      width: 200px; padding: 20px; margin-bottom: 10px;
      border: 1px solid #000; }
```

You may have noticed that it's common to use several CSS properties when configuring background images. These properties typically work together. However, be aware that the `background-origin` property has no effect if the `background-attachment` property is set to the value `fixed`.

FIGURE 6.18 *The CSS* `background-origin` *property.*

CSS Background Resize and Scale

The CSS **background-size** property can be used to resize or scale the background image. Valid values for the background-size property can be:

▶ a pair of percentage values (width, height)

  If only one percentage value is provided, the second value defaults to auto and is determined by the browser.

▶ a pair of pixel values (width, height)

  If only one numeric value is provided, the second value defaults to auto and is determined by the browser.

▶ cover

  The value cover will preserve the aspect ratio of the image as it scales the background image to the *smallest* size for which both the height and width of the image can completely cover the area.

▶ contain

  The value contain will preserve the aspect ratio of the image as it scales the background image to the *largest* size for which both the height and width of the image will fit within the area.

Figure 6.19 shows two div elements that are each configured with the same background image to display without repeating.

FIGURE 6.19 *The CSS* background-size *property set to 100% 100%.*

The background image of the first div element is not configured with the `background-size` property, and the image only partially fills the space. The CSS for the second div configures the `background-size` to be 100% 100%, so the browser scales and resizes the background image to fill the space. The sample page is located in the student files (chapter6/size/sedona.html). The CSS for the second div follows:

```
#test1 { background-image: url(sedonabackground.jpg);
     background-repeat: no-repeat;
     background-size: 100% 100%; }
```

Figure 6.20 demonstrates use of the `cover` and `contain` values to configure the display of a 500×500 background image within a 200 pixel wide area on a web page. The web page on the left uses `background-size: cover;` to scale and resize the image to completely cover the area while keeping the aspect ratio of the image intact. The web page on the right uses `background-size: contain;` to scale and resize the image so that both the height and width of the image will fit within the area. Review the sample pages in the student files (chapter6/size/cover.html and chapter6/size/contain.html).

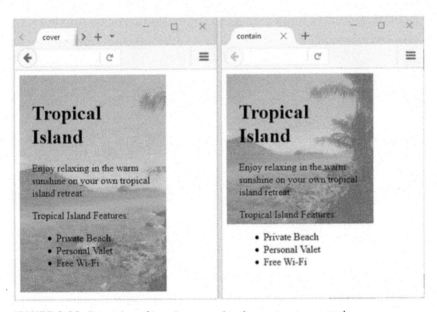

FIGURE 6.20 *Examples of* `background-size: cover;` *and* `background-size: contain;`

Practice with CSS Properties

 Hands-On Practice 6.6 ————————————————————————

In this Hands-On Practice, you will configure a web page with centered content and practice configuring CSS properties. When complete, your web page will look similar to the one shown in Figure 6.21.

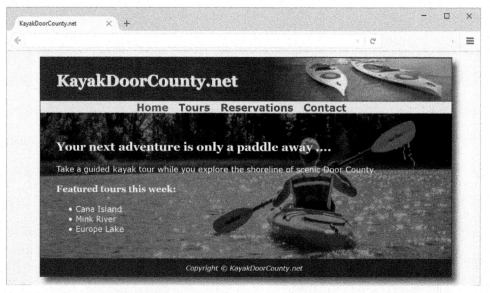

FIGURE 6.21 *New Home page.*

Create a new folder called kayakch6. Copy the headerbackblue.jpg and heroback2.jpg files from the chapter6/starters folder to your kayakch6 folder. Launch a text editor and open the chapter6/starter3.html file. Save the file in your kayakch6 folder with the name index.html. Modify the file as follows:

1. Center the page content by applying the coding technique from Hands-On Practice 6.4.

   a. Edit the embedded CSS and configure a new selector, an id named `container` with style declarations for the `width`, `margin-left` and `margin-right` properties as follows:

   ```
   #container { margin-left: auto; margin-right: auto;
                width: 80%; }
   ```

   b. Edit the HTML. Configure a div element assigned to the id `container` that "wraps" or contains the code within the body section. Code an opening div tag on a new line after the opening body tag. Assign the div to the id named `container`.

2. Edit the embedded CSS.

a. The container id selector.   Add declarations to configure a white background color, 650px minimum width, 1280px maximum width, a box shadow with a 10px offset and blur in the color #333333, and a 1px solid dark blue (#000033) border.

```
#container { margin-left: auto; margin-right: auto;
            width: 80%;
            background-color: #FFFFFF;
            min-width: 650px; max-width: 1280px;
            box-shadow: 10px 10px 10px #333333;
            border: 1px solid #000033; }
```

b. The body element selector.   Edit the styles to change the background color to #FFFFE8.

c. The header element selector.   Add declarations to configure 80px height, 5px top padding, 2em left padding, and a text shadow in the color #FFF with a 1px offset.

```
header { background-color: #000033; color: #FFFFB9;
        background-image: url(headerbackblue.jpg);
        background-position: right;
        background-repeat: no-repeat;
        height: 80px;
        padding-top: 5px; padding-left: 2em;
        text-shadow: 1px 1px 1px #FFF; }
```

d. The h1 element selector.   Code styles to configure a zero bottom margin.

```
h1 { margin-bottom: 0; }
```

e. The nav element selector.   Add a declaration to configure centered text with the text-align property.

```
nav { font-weight: bold; font-size: 1.25em;
      background-color: #FFFFCC;
      text-align: center; }
```

f. The main element selector.   Add declarations to configure heroback2.jpg as the background image and configure `background-size: 100% 100%;`. Also configure white text (use #FFFFFF) and 2em of padding.

```
main { background-color: #004D99;
        background-image: url(heroback2.jpg);
        background-size: 100% 100%;
        color: #FFFFFF; padding: 2em; }
```

g. The footer element selector.   Add a declaration for 0.5em of padding.

```
footer { font-style: italic; background-color: #000033;
        color: #FFFFCC; text-align: center; padding: 0.5em;}
```

3. Save the file. When you test your index.html file in a modern browser such as Firefox or Chrome, it should look similar to the one shown in Figure 6.21. Compare your work with the solution in the student files (chapter6/6.6/index.html). Note that if you display the page in a browser (such as Internet Explorer 11) that does not support the HTML5 main element, the display will not look as you expect. At the time this text was written, Internet Explorer did not support default styles for the HTML5 main element. You may need to nudge this browser to comply by adding the `display: block;` declaration (see Chapter 7) to the styles for the main element selector. An example solution is in the student files (chapter6/6.6/iefix.html).

CSS Opacity

The CSS **opacity property** configures the transparency of an element.[6] Opacity values range from 0 (which is completely transparent) to 1 (which is completely opaque and has no transparency). An important consideration when using the opacity property is that this property applies to both the text and the background. If you configure a semitransparent opacity value for an element with the opacity property, both the background and the text displayed will be semitransparent. See Figure 6.22 for an example of using the opacity property to configure an h1 element that is 60% opaque.

If you look very closely at Figure 6.22 or view the actual web page (student files chapter6/6.7/index.html), you'll see that both the white background and the black text in the h1 element are semitransparent. The opacity property was applied to both the background color and to the text color.

FIGURE 6.22 *The background and text of the h1 area is transparent.*

Hands-On Practice 6.7

In this Hands-On Practice, you'll work with the opacity property as you configure the web page shown in Figure 6.22.

1. Create a new folder called opacitych6. Copy fall.jpg file from the chapter6/starters folder to your opacitych6 folder. Open the chapter6/template.html file in a text editor. Save it in your opacitych6 folder with the name index.html. Change the page title to "Fall Nature Hikes".

2. Let's create the structure of the web page with a div that contains an h1 element. Add the following code to your web page in the body section:

```
<div id="content">
<h1> Fall Nature Hikes</h1>
</div>
```

3. Now, add style tags to the head section and configure the embedded CSS. You'll create an id named content to display the fall.jpg as a background image that does not repeat. The content id also has a width of 640 pixels, a height of 480 pixels, auto margins (which will center the object horizontally in the browser viewport), and 20 pixels of top padding. The code follows:

```
#content { background-image: url(fall.jpg);
           background-repeat: no-repeat;
           margin: auto;
           width: 640px;
           height: 480px;
           padding-top: 20px; }
```

4. Now configure the h1 selector to have a white background color, opacity set to 0.6, font size set to 4em, 10 pixels of padding, and a 40-pixel left margin. Sample code follows.

```
h1 { background-color: #FFFFFF;
     opacity: 0.6;
     font-size: 4em;
     padding: 10px;
     margin-left: 40px; }
```

5. Save the file. When you test your index.html file in a browser that supports opacity, the display should look similar to the page shown in Figure 6.22. See the student files for a solution (chapter6/6.7/index.html).

CSS RGBA Color

CSS supports syntax for the color property that configures transparent color, called **RGBA color**. Four values are required: red, green, blue, and alpha (transparency). RGBA color does not use hexadecimal color values. Instead, decimal color values are configured—see the partial color chart in Figure 6.23 and the Web-Safe Color Palette Appendix for examples.

#FFFFFF
rgb (255, 255, 255)
	#FFFFCC
	rgb(255, 255, 204)
		#FFFF99
		rgb(255,255,153)
			#FFFF66
			rgb(255,255,102)

#FFFF33
rgb(255,255,51)
	#FFFF00
	rgb(255,255,0)
		#FFCCFF
		rgb(255, 204, 255)
			#FFCCCC
			rgb(255,204,204)

#FFCC99
rgb(255,204,153)
	#FFCC66
	rgb(255,204,102)
		#FFCC33
		rgb(255,204,51)
			#FFCC00
			rgb(255,204,0)

#FF99FF
rgb(255,153,255)
	#FF99CC
	rgb(255,153,204)
		#FF9999
		rgb(255,153,153)
			#FF9966
			rgb(255,153,102)

FIGURE 6.23 *Hexadecimal and RGB decimal color values.*

To configure RGBA color, the values for red, green, and blue must be decimal values from 0 to 255. The alpha value must be a number from 0 (transparent) to 1 (opaque). Figure 6.24 shows a web page with the text configured to be slightly transparent with RGBA color syntax.

> **? FAQ**
>
> How is using RGBA color different from using the opacity property?
>
> The `opacity` property applies to both the background and the text within an element. If you'd like to specifically configure a semitransparent background color, code the `background-color` property with RGBA color or HSLA color (described in the next section) values. If you'd like to specifically configure semitransparent text, code the `color` property with RGBA color or HSLA color values.

In this Hands-On Practice, you'll configure white text with transparency as you configure the web page shown in Figure 6.24.

FIGURE 6.24 CSS RGBA color configures the transparent text.

1. Launch a text editor and open the file you created in the previous Hands-On Practice (also located in the student files, chapter6/6.7/index.html). Save the file with the name rgba.html.

2. Delete the current style declarations for the h1 selector. You will create new style rules for the h1 selector to configure 10 pixels of right padding and right-aligned sans-serif white text that is 80% opaque with a font size of 5em. Since not all browsers support RBGA color, you'll configure the color property twice. The first instance will be the standard color value that is supported by all browsers; the second instance will configure the RGBA color. Older browsers will not understand the RGBA color and will ignore it. Newer browsers will "see" both of the color style declarations and will apply them in the order they are coded, so the result will be transparent color. The CSS for the h1 selector follows.

```
h1 { color: #FFFFFF;
     color: rgba(255, 255, 255, 0.8);
     font-family: Verdana, Helvetica, sans-serif;
     font-size: 5em;
     padding-right: 10px;
     text-align: right; }
```

3. Save the file. When you test your rgba.html file in a browser that supports RGBA color, the display should look similar to the page shown in Figure 6.24. See the student files for a solution (chapter6/6.8/rgba.html. If you are using a nonsupporting browser, you'll see solid text instead of transparent text.

CSS HSLA Color

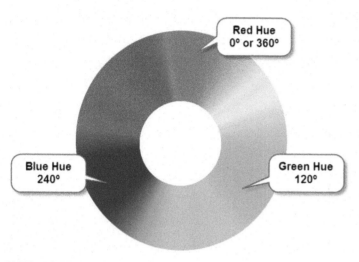

FIGURE 6.25 *A color wheel.*

For many years, web designers have configured RGB color using either hexadecimal or decimal values on web pages. Recall that RGB color is based on hardware—the red, green, and blue light that is emitted by computer monitors. A new color notation system called **HSLA color** is based on a color wheel model, which stands for hue, saturation, lightness, and alpha.

Hue, Saturation, Lightness, and Alpha

When you work with HSLA color, think of a color wheel—a circle of color—with the color red at the top of the wheel as shown in Figure 6.25. **Hue** is the actual color which is represented by numeric values ranging from 0 to 360 (like the 360 degrees in a circle). For example, red is represented by both the values 0 and 360, green is represented by 120, and blue is represented by 240. Set hue to 0 when configuring black, gray, and white. **Saturation** configures the intensity of the color and is indicated by a percentage value (full color saturation= 100% and gray=0%). **Lightness** determines the brightness or darkness of the color and is indicated by a percentage value (normal color=50% and white=100% and black=0%). **Alpha** represents the transparency of the color and has a value from 0 (transparent) to 1 (opaque). Note that you can omit the alpha value and use the `hsl` keyword instead of the `hsla` keyword.

HSLA Color Examples

Red
hsla(360, 100%, 50%, 1.0);

Green
hsla(120, 100%, 50%, 1.0);

Blue
hsla(240, 100%, 50%, 1.0);

Black
hsla(0, 0%, 0%, 1.0);

Gray
hsla(0, 0%, 50%, 1.0);

White
hsla(0, 0%, 100%, 1.0);

FIGURE 6.26 *HSLA color examples.*

Configure HSLA color as shown in Figure 6.26. with the following syntax:

`hsla (hue value, saturation value, lightness value, alpha value);`

- Red: `hsla(360, 100%, 50%, 1.0);`
- Green: `hsla(120, 100%, 50%, 1.0);`
- Blue: `hsla(240, 100%, 50%, 1.0);`
- Black: `hsla(0, 0%, 0%, 1.0);`
- Gray: `hsla(0, 0%, 50%, 1.0);`
- White: `hsla(0, 0%, 100%, 1.0);`

According to the W3C, an advantage to using HSLA color is that it is more intuitive to work with than the hardware-oriented RGB color. You can use a color wheel model (remember your art classes in grade school) to choose colors and generate the hue value from the degree placement on the circle. If you'd like to use a tone of a color, which is a color with gray added, vary the saturation value. If you'd like to use a shade or tint of a color, use the same hue value, but vary the lightness value to meet your needs. Figure 6.27 shows three shades of cyan blue configured using three different values for lightness: 25% (dark cyan blue), 50% (cyan blue), and 75% (light cyan blue).

FIGURE 6.27 *Shades of cyan blue.*

▶ Dark Cyan Blue:

`hsla(210, 100%, 25%, 1.0);`

▶ Cyan Blue:

`hsla(210, 100%, 50%, 1.0);`

▶ Light Cyan Blue:

`hsla(210, 100%, 75%, 1.0);`

 Hands-On Practice 6.9 —————

In this Hands-On Practice, you'll configure light yellow transparent text as you configure the web page shown in Figure 6.28.

1. Launch a text editor and open the file you created in the previous Hands-On Practice (see chapter6/6.8/rgba.html in the student files). Save the file with the name hsla.html.

2. Delete the style declarations for the h1 selector. You will create new style rules for the h1 selector to configure 20 pixels of padding and serif light yellow text with a 0.8 alpha value

FIGURE 6.28 *HSLA color.*

and a font size of 6em. Since not all browsers support HSLA color, you'll configure the color property twice. The first instance will be the standard color value that is supported by all browsers; the second instance will configure the HSLA color. Older browsers will not understand the HSLA color and will ignore it. Newer browsers will "see" both of the color style declarations and will apply them in the order they are coded, so the result will be transparent color. The CSS for the h1 selector follows:

```
h1 { color: #FFCCCC;
     color: hsla(60, 100%, 90%, 0.8);
     font-family: Georgia, "Times New Roman", serif;
     font-size: 6em;
     padding: 20px; }
```

3. Save the file. When you test your hsla.html file in a browser that supports HSLA color, it should look similar to the page shown in Figure 6.28. See the student files for a solution (chapter6/6.9/hsla.html). If you are using a nonsupporting browser, you'll see solid text instead of transparent text.

CSS Gradients

CSS provides a method to configure color as a **gradient**, which is a smooth blending of shades from one color to another color.[7] A CSS gradient background color is defined purely with CSS—no image file is needed! This provides flexibility for web designers along with savings in the bandwidth required to transfer gradient background image files.

Figure 6.16 displays a web page with a JPG gradient background image that was configured in a graphics application. The web page shown in Figure 6.29 (available at chapter6/lighthouse/gradient.html in the student files) does not use a JPG for the background—CSS gradient properties recreated the look of the linear gradient image.

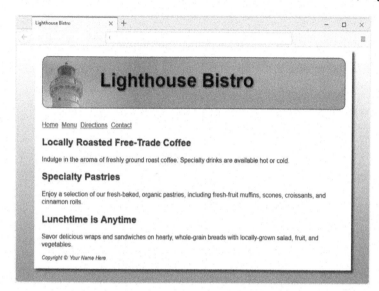

FIGURE 6.29 *The gradient in the background was configured with CSS without an image file.*

Linear Gradient Syntax

A **linear gradient** is a smooth blending of color in a single direction such as from top to bottom or from left to right. To configure a basic linear gradient, code the `linear-gradient` function as the value of the `background-image` property. Indicate the direction of the gradient by coding the keyword phrase "`to bottom`", "`to top`", "`to left`", or "`to right`". Next, list the starting color and the ending color. The basic format for a two-color linear gradient that blends from white to green follows:

```
background-image: linear-gradient(to bottom, #FFFFFF, #00FF00);
```

Radial Gradient Syntax

A **radial gradient** is a smooth blending of color emanating outward from a single point. Code the radial gradient function as the value of the `background-image` property to configure a radial gradient. List two colors as the values of the function. The first color will be displayed by default in the center of the element and gradually blend outward until the second color is displayed. The basic format for a two-color radial gradient that blends from white to blue follows:

```
background-image: radial-gradient(#FFFFFF, #0000FF);
```

CSS Gradients and Progressive Enhancement

It's very important to keep progressive enhancement in mind when using CSS gradients. Configure a "fallback" `background-color` property or `background-image` property, which will be rendered by browsers that do not support CSS gradients. The background color in Figure 6.29 was configured to be the same value as the ending gradient color.

Hands-On Practice 6.10

You'll work with CSS gradient backgrounds in this Hands-On Practice. Create a new folder called gradientch6. Copy the chapter6/starter4.html file into your gradientch6 folder. Rename the file index.html. Launch a text editor and open the file.

1. First, you will configure a linear gradient. Code embedded CSS in the head section. Configure the body of the web page to display a fallback orchid background color of #DA70D6, and a linear gradient background that blends white to orchid from top to bottom without repeating:

```
body { background-color: #DA70D6;
       background-image: linear-gradient(to
       bottom, #FFFFFF, #DA70D6);
       background-repeat: no-repeat; }
```

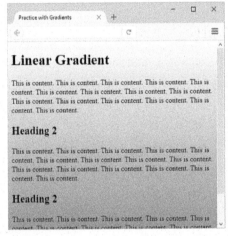

FIGURE 6.30 *Linear gradient background.*

2. Save your file and test it in a modern browser. The display should be similar to the results shown in Figure 6.30. The background gradient displays behind the page content, so scroll down the page to see the full gradient. Compare your work with the solution in the student files (chapter6/6.10/linear.html).

3. Next, you will configure a radial gradient. Edit the body section of the web page and code change the text within the h1 element to: Radial Gradient.

4. Edit the CSS and modify the value of the `background-image` property to configure a radial gradient linear gradient that blends white to orchid from center outward without repeating:

```
body { background-color: #DA70D6;
       background-image:
       radial-gradient(#FFFFFF, #DA70D6);
       background-repeat: no-repeat; }
```

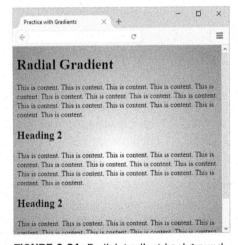

FIGURE 6.31 *Radial gradient background.*

5. Save your file and test it in a modern browser. The display should be similar to the results shown in Figure 6.31. Scroll down the page to see the full gradient. Compare your work with the solution in the student files (chapter6/6.10/radial.html).

Visit https://css-tricks.com/css3-gradients to delve deeper into CSS gradients. Experiment with generating CSS gradient code at https://www.colorzilla.com/gradient-editor and http://www.css3factory.com/linear-gradients.

Review and Apply

Review Questions

1. Which of the following is the CSS property that configures a drop shadow effect on text?

   a. `box-shadow`   **b.** `text-shadow`

   c. `drop-shadow`   **d.** `shadow`

2. Which CSS property configures the area between the content and the border?

   a. `border`   **b.** `box-shadow`

   c. `padding`   **d.** `opacity`

3. Which CSS property will configure rounded corners?

   a. `border`   **b.** `border-radius`

   c. `radial`   **d.** `margin`

4. Which CSS property can be used to resize or scale a background image?

   a. `background-repeat`

   b. `background-size`

   c. `background-clip`

   d. `background-origin`

5. Which of the following, from outermost to inner-most, are components of the box model?

   a. margin, border, padding, content

   b. content, padding, border, margin

   c. content, margin, padding, border

   d. margin, padding, border, content

6. Which CSS property configures the empty space between adjacent elements?

   a. `padding`   **b.** `border`

   c. `margin`   **d.** `letter-spacing`

7. Which of the following will configure padding that is 15 pixels on the top, 0 pixels on the left and right, and 5 pixels on the bottom?

   a. `padding: 0px 5px 0px 15px;`

   b. `padding: top-15, left-0, right-0, bottom-5;`

   c. `padding: 15px 0 5px 0;`

   d. `padding: 0 0 15px 5px;`

8. Which of the following is used along with the `width` property to configure centered page content?

   a. `margin-left: auto;`
   `margin-right: auto;`

   b. `margin: top-15, left-0, right-0, bottom-5;`

   c. `margin: 15px 0 5px 0;`

   d. `margin: 20px;`

9. Which CSS property will configure a gradient background?

   a. `background-gradient`

   b. `background-image`

   c. `background-clip`

   d. `linear-gradient`

10. Which of the following will configure a border that is 5 pixels wide, the color #330000, and a solid line?

    a. `border: 5px solid #330000;`

    b. `border-style: solid 5px;`

    c. `border: 5px, solid, #330000;`

    d. `border: 5px line #330000;`

Review Answers

1. b   2. c   3. b   4. b   5. a   6. c   7. c   8. a   9. b   10. a

Hands-On Exercises

1. Write the CSS for a class named footer with the following characteristics: a light-blue background color, Arial or sans-serif font, dark-blue text color, 10 pixels of padding, and a narrow, dashed border in a dark-blue color.

2. Write the CSS for an id named notice that is configured with width set to 80% and centered.

3. Write the CSS to configure a class that will produce a headline with a dotted line underneath it. Choose a color that you like for the text and dotted line.

4. Write the CSS to configure an h1 selector with drop shadow text, a 50% transparent background color, and sans-serif font that is 4em in size.

5. Write the CSS to configure an id named blurb with small, red, Arial font, a white background, a width of 80%, and a drop shadow.

6. Write the CSS to configure the body element with a linear gradient background that blends from black to medium blue.

7. Write the CSS to configure an id named content with 70% opacity.

Focus on Web Design

This chapter expanded your capabilities to use CSS to configure web pages. Use a search engine to search for CSS resources. The following resources can help you get started:

- https://www.w3.org/Style/CSS/learning
- https://www.noupe.com/design/40-css-reference-websites-and-resources.html
- https://developer.mozilla.org/en-US/docs/Web/CSS/Reference

Create a web page that provides a list of at least five CSS resources on the Web. For each CSS resource provide the URL, website name, and a brief description. Your web page content should take up 80% of the browser viewport and be centered. Use at least five CSS properties from this chapter to configure the color and text. Place your name in the e-mail address at the bottom of the web page elements.

Case Study

You will continue the case studies from Chapter 5 as you configure a new layout and apply your new CSS skills.

Pacific Trails Resort Case Study

In this chapter's case study, you will use the existing Pacific Trails Resort (Chapter 5) website as a starting point to create a new version of the website. The new design is a centered page layout that takes up 80% of the browser viewport with a featured hero image on each page. You'll use CSS to configure the new page layout, a background gradient, hero image, and other styles, including margin and padding. Figure 6.32 displays a wireframe with the `wrapper` div, which contains the other web page elements.

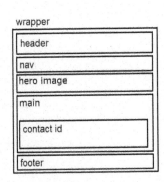

FIGURE 6.32 *New wireframe.*

You have five tasks in this case study:

1. Create a new folder for the Pacific Trails Resort website.
2. Edit the pacific.css external style sheet.
3. Update the Home page: index.html.
4. Update the Activities page: activities.html.
5. Update the Yurts page: yurts.html.

Task 1: Create a folder called ch6pacific to contain your Pacific Trails Resort website files. Copy the files from the Chapter 5 Case Study ch5pacific folder.

Task 2: The External Stylesheet. Launch a text editor and open the pacific.css external style sheet file.

- **The body element Selector.** Change the background color to light blue (#90C7E3). Add style declarations to display a linear gradient that blends from white (#FFFFFF) to light blue (#90C7E3) and does not repeat.

- **The wrapper id Selector.** Add a new selector for an id named `wrapper`. Configure the `wrapper` id to be centered (see Hands-On Practice 6.4) with a width of 80%, white background color (#FFFFFF), a minimum width of 960 pixels, a maximum width of 2048px, and a 3px offset dark (#333333) box shadow.

- **The header element Selector.** Remove declarations for line height and indented text. Add declarations to configure 60px height, centered text, and 15px top padding.

- **The h1 element Selector.** Add a style declaration to configure zero top margin.

- **The nav element Selector.** Change the background color to white (#FFFFFF). Add declarations to configure centered text and 1em padding.

- **The main element Selector.** Add a new selector for the main element. Code declarations to configure 1px top padding, 20px right padding, 20px bottom padding, and 20px left padding. Internet Explorer does not support default styles for the HTML5 main element, so add the following style declaration using the display property (see Chapter 7) to nudge this browser to display as expected: `display: block;`

- **The h2 element Selector.** Add a style declaration to configure 1px offset gray (#CCCCCC) text shadow.

- **The footer element Selector.** Add a declaration to configure 2em of padding.

- **The homehero id Selector.** Add a new selector for an id named `homehero`. Code declarations to configure 300px height and to display the coast.jpg background image to fill the space (use `background-size: 100% 100%;`) without repeating.

- **The yurthero id Selector.** Add a new selector for an id named `yurthero`. Code declarations to configure 300px height and to display the yurt.jpg background image to fill the space (use `background-size: 100% 100%;`) without repeating.

- **The trailhero id Selector.** Add a new selector for an id named `trailhero`. Code declarations to configure 300px height and to display the trail.jpg background image to fill the space (use `background-size: 100% 100%;`) without repeating.

Save the pacific.css file. Use the CSS validator (http://jigsaw.w3.org/css-validator) to check your syntax. Correct and retest, if necessary.

Task 3: The Home Page. Launch a text editor and open the index.html file.

- Code div tags to add a `wrapper` div that contains the content of the web page. Use Hands-On Practice 6.4 as a guide.

- Refer to the Figure 6.22 wireframe and note the hero image area between the nav and main elements. Locate the div element for the hero image that displays the coast.jpg file. Assign the div to the id named `homehero`. There is no HTML or text content for this div. The purpose of this div is to be a placeholder for the CSS that will display the hero image. Remove the img tag for the coast.jpg photo.

FIGURE 6.33 *Pacific Trails Home page.*

Save and test your page in a browser. It should look similar to Figure 6.33.

Task 4: The Yurts Page. Launch a text editor and open the yurts.html file.

- Code div tags to add a `wrapper` div that contains the content of the web page. Use Hands-On Practice 6.4 as a guide.

- Locate div element for the hero image area shown in Figure 6.22 that displays the yurt.jpg file. Assign the div to the id named `yurthero`. There is no HTML or text content for this div. Remove the img tag for the yurt.jpg photo.

Save and test your page in a browser. It should look similar to Figure 6.34.

FIGURE 6.34 *Pacific Trails Yurts page.*

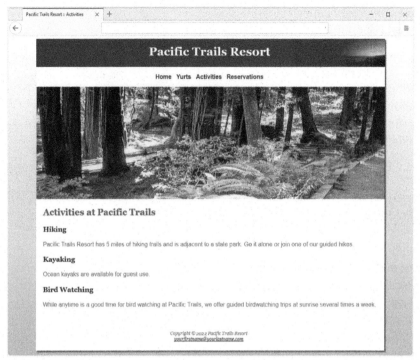

FIGURE 6.35 *Pacific Trails Activities page.*

Task 5: The Activities Page. Launch a text editor and open the activities.html file.

- Code div tags to add a `wrapper` div that contains the content of the web page. Use Hands-On Practice 6.4 as a guide.

- Locate the div element for the hero image area shown in Figure 6.22 that displays the trail.jpg file. Assign the div to the id named `trailhero`. There is no HTML or text content for this div. Remove the img tag for the trail.jpg photo.

Save and test your page in a browser. It should look similar to Figure 6.35.

Path of Light Yoga Studio Case Study

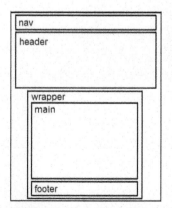

FIGURE 6.36 *New wireframe.*

In this chapter's case study, you will use the existing Path of Light Yoga Studio (Chapter 5) website as a starting point to create a new version of the website. The new design has a full width nav element, full width header element, and an 80% width centered div that contains the main element and footer element. The header area for the home page will be larger than the header area for the content pages. You'll use CSS to configure the new page layout, a background image, hero image on the content pages, and other styles, including margin and padding. Figure 6.36 displays a wireframe with the `wrapper` div, which contains the main and footer web page elements.

You have five tasks in this case study:

1. Create a new folder for Path of Light Yoga Studio website.

2. Edit the yoga.css external style sheet.

3. Update the Home page: index.html.

4. Update the Classes page: classes.html.

5. Update the Schedule page: schedule.html.

Task 1: Create a folder called ch6yoga to contain your Path of Light Yoga Studio website files. Copy the files from the Chapter 5 Case Study ch5yoga folder. This new version of the website uses a sunrise image to emphasize the "Path of Light" theme of the yoga studio. Copy the sunrise.jpg file chapter6/starters folder in the student files.

Task 2: The External Stylesheet. Launch a text editor and open the yoga.css external style sheet file.

- **The body element Selector.** Change the value of the `background-color` property to #40407A. Add declarations to configure 1600px maximum width, 900px minimum width, and zero margin (hint: `margin: 0;`).

- **The wrapper id Selector.** Add a new selector for an id named `wrapper`. Configure the `wrapper` id to be centered (see Hands-On Practice 6.4) with a width of 80%, light background color (#F5F5F5), and padding set to 2em.

- **The header element Selector.** Remove the properties that configure the background repeat and position. Set the background color to #40407A. Set text color to #FFFFFF. Configure sunrise.jpg as the background image and set 100% background size (hint: `background-size: 100% 100%;`).

- **The home class Selector.** Add a new selector for a class named `home`. This will configure the display for the home page's header area. Code a style declaration to set the height at 40% of the viewport height (hint: `height: 40vh;`). Also set top padding to 6em, left padding to 8em, 120% font size, and 300px minimum height.

- **The content class Selector.** Add a new selector for a class named `content`. This will configure the display for the content pages' header area. Set height to 200px; top padding to 2em, left padding to 8em, and bottom padding to 2em.

- **The h1 element Selector.** Remove this selector and all style declarations.

- **The nav element Selector.** Change the text alignment to right. Also set a white background color, zero margin, 0.5em top padding, 1em bottom padding, and 1em right padding.

- **The mathero id Selector.** Add a new selector for an id named `mathero`. Code declarations to configure 300px height and to display the yogamat.jpg background image to fill the space (use `background-size: 100% 100%;`) without repeating.

- **The loungehero id Selector.** Add a new selector for an id named `loungehero`. Code declarations to configure 300px height and to display the yogalounge.jpg background image to fill the space (use `background-size: 100% 100%;`) without repeating.

- **The h2 element Selector.** Add a new h2 element selector. Set the margin to 0.

- **The footer element Selector.** Remove the declaration that sets the background color.

Save the yoga.css file. Use the CSS validator (http://jigsaw.w3.org/css-validator) to check your syntax. Correct and retest if necessary.

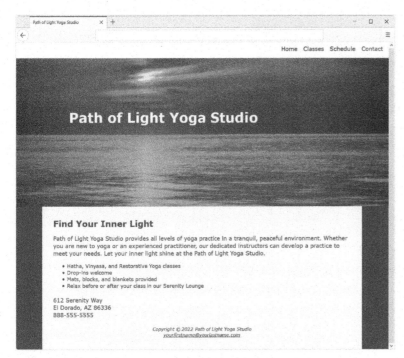

Task 3: The Home Page. You will edit the HTML to correspond to the wireframe in Figure 6.36. Launch a text editor and open the index.html file.

- Move the nav element area above the header element.

- Assign the header element to the class named `home`.

- Remove the img element.

- Configure a `wrapper` div that contains the main element and footer element.

Save and test your page in a browser. It should look similar to Figure 6.37.

FIGURE 6.37 *Path of Light Yoga Studio Home page.*

FIGURE 6.38 *Wireframe for content pages.*

Task 4: The Classes Page. You will edit the HTML to correspond to the wireframe in Figure 6.38. Launch a text editor and open the classes.html file.

- Configure a `wrapper` div that contains main element and footer element.

- Move the nav element area above the header element.

- Assign the header element to the class named `content`.

- Move the div that contains the yogamat.jpg image below the description list as in Figure 6.38. Assign the div to the id named `mathero`. Remove the line break and img tag from this div. The purpose of this div is to be a placeholder for the CSS that will display the yogamat.jpg hero image.

Save and test your page in a browser. It should look similar to Figure 6.39.

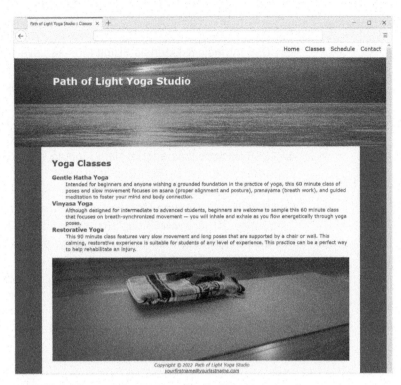

FIGURE 6.39 *Path of Light Yoga Studio Classes page.*

Task 5: The Schedule Page.

You will edit the HTML to correspond to the wireframe in Figure 6.38. Launch a text editor and open the schedule.html file.

- Code div tags to add a `wrapper` div that contains the main element and footer element.

- Move the nav element area above the header element.

- Assign the header element to the class named `content`.

- Move the div that contains the yogalounge.jpg image below the other elements within the main element. Assign the div to the id named `loungehero`. Remove the line break and img tags from this div. The purpose of this div is to be a placeholder for the CSS that will display the yogalounge.jpg hero image.

Save and test your page in a browser. It should look similar to Figure 6.40.

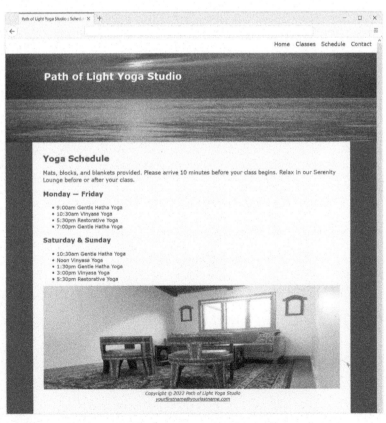

FIGURE 6.40 *Path of Light Yoga Studio Schedule page.*

Endnotes

1. "CSS Intrinsic & Extrinsic Sizing Module Level 3." Edited by Tab Atkins et al., *W3C*, 22 May 2019, www.w3.org/TR/css-sizing-3/.
2. "CSS Box Model Module Level 3." Edited by Elika J. Etemad, *W3C*, 18 Dec. 2018, www.w3.org/TR/css-box-3/.
3. "CSS Backgrounds and Borders Module Level 3: 5. Rounded Corners." Edited by Bert Bos et al., *W3C*, 17 Oct. 2017, www.w3.org/TR/css-backgrounds-3/#corners.
4. "CSS Backgrounds and Borders Module Level 3: 7.1. Drop Shadows: the Box-Shadow Property." Edited by Bert Bos et al., *W3C*, W3C, 17 Oct. 2017, www.w3.org/TR/css-backgrounds-3/#the-box-shadow.
5. "CSS Text Decoration Module Level 3." Edited by Elika J. Etemad and Koji Ishii, *W3C*, 13 Aug. 2019, www.w3.org/TR/css-text-decor-3/.
6. "CSS Color Module Level 3." Edited by Tantek Çelik et al., *W3C*, W3C, 19 June 2018, www.w3.org/TR/css-color-3/.
7. "CSS Images Module Level 3." Edited by Tab Atkins et al., *W3C*, 10 Oct. 2019, www.w3.org/TR/css-images-3/.

CHAPTER 7

Page Layout Basics

You've already configured centered page layout with Cascading Style Sheets (CSS). We'll add to your toolbox of CSS page layout techniques in this chapter. You'll explore floating and positioning elements with CSS. You'll be introduced to using CSS to add interactivity to hyperlinks with pseudo-classes and use CSS to style navigation in unordered lists. You will configure a single-page website using hyperlinks to named fragments. You will build many page layout skills in this chapter.

You'll learn how to...

- Configure float with CSS
- Create two-column page layouts with CSS
- Configure navigation in unordered lists and style with CSS
- Add interactivity to hyperlinks with CSS pseudo-classes
- Configure printed pages with CSS

- Configure fixed, relative, absolute, and sticky positioning with CSS
- Configure stacking order with CSS
- Configure CSS sprites
- Configure a hyperlink to a named fragment internal to a web page
- Configure a single page website

Normal Flow

Browsers render your web page code line by line in the order it appears in the .html document. This processing is called normal flow. **Normal flow** displays the elements on the page in the order they appear in the web page source code.

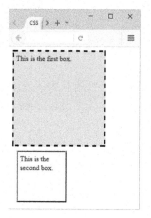

FIGURE 7.1 *The div elements.*

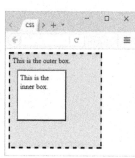

FIGURE 7.2 *Nested elements.*

Figures 7.1 and 7.2 each display two div elements that contain text content. Let's take a closer look. Figure 7.1 shows a screenshot of two div elements placed one after another on a web page. In Figure 7.2, the boxes are nested inside each other. In both cases, the browser used normal flow (the default) and displayed the elements in the order in which they appeared in the source code. As you've worked through the exercises in the previous chapters, you created web pages that the browser rendered using normal flow.

You'll practice normal flow a bit more in the next Hands-On Practice. Then, later in the chapter, you'll experiment with CSS positioning and float to configure the flow, or placement, of elements on a web page.

 Hands-On Practice 7.1 ————————————————————

You will explore the box model and normal flow in this Hands-On Practice as you work with the web pages shown in Figures 7.1 and 7.2.

Practice with Normal Flow

Launch a text editor and open the chapter7/starter1.html file from the student files. Save the file with the name box1.html. Add the following code in the body of the web page to configure the two div elements:

```
<div class="div1">
This is the first box.
</div>
<div class="div2">
This is the second box.
</div>
```

Now let's add embedded CSS in the head section to configure the "boxes." Add a new style rule for a class named `div1` to configure a light blue background, dashed border, width of 200 pixels, height of 200 pixels, and 5 pixels of padding. The code follows:

```
.div1 { width: 200px;
        height: 200px;
        background-color: #D1ECFF;
        border: 3px dashed #000000;
        padding: 5px; }
```

Create a style rule for a class named `div2` to configure a width and height of 100 pixels, white background color, ridged border, 10 pixel margin, and 5 pixels of padding. The code follows:

```
.div2 { width: 100px;
        height: 100px;
        background-color: #FFFFFF;
        border: 3px ridge #000000;
        padding: 5px;
        margin: 10px; }
```

Save the file. Launch a browser and test your page. It should look similar to the one shown in Figure 7.1. The student files contain a sample solution at chapter7/7.1/box1.html.

Practice with Normal Flow and Nested Elements

Launch a text editor and open the box1.html file from the student files (chapter7/7.1/box1.html). Save the file with the name box2.html. Delete the content from the body section of the web page. Add the following code to configure two div elements—one nested inside the other:

```
<div class="div1">
This is the outer box.
  <div class="div2">
  This is the inner box.
  </div>
</div>
```

Save the file. Launch a browser and test your page. It should look similar to the one shown in Figure 7.2. Notice how the browser renders the nested div elements—the second box is nested within the first box because it is coded inside the first div element in the web page source code. This is an example of normal flow. The student files contain a sample solution at chapter7/7.1/box2.html.

A Look Ahead—CSS Layout Properties

You've seen how normal flow causes the browser to render the elements in the order that they appear in the HTML source code. When using CSS for page layout, there are situations in which you will want to specify the location of an element on the page—including the absolute pixel location, the location relative to where the element would normally display, floating on the page, flexible box layout (flexbox), and grid layout. The CSS properties that configure float and positioning are introduced in this chapter. Flexbox and grid layout are introduced in Chapter 8.

Float

The `float` Property

Elements that seem to float on the right or left side of either the browser window or another element are often configured using the CSS **float property**.[1] The browser renders these elements using normal flow and then shifts them to either the right or left as far as possible within their container (usually either the browser viewport or a div element).

- Use `float: right;` to float the element on the right side of the container.
- Use `float: left;` to float the element on the left side of the container.
- Specify a width for a floated element unless the element already has an implicit width—such as an img element.
- Other elements and web page content will flow around the floated element, so floated elements should always be coded before the elements that will display alongside them.

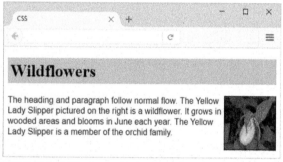

FIGURE 7.3 *The image is configured to float.*

Figure 7.3 shows a web page with an image configured with `float: right;` to float on the right side of the browser viewport (see the student files, chapter7/float.html). When floating an image, the `margin` property is useful to configure empty space between the image and text on the page.

View Figure 7.3 and notice how the image stays on the right side of the browser viewport. An id called `yls` was created that applies the `float`, `margin`, and `border` properties. The attribute `id="yls"` was placed on the image tag. The CSS follows:

```
h1 { background-color: #A8C682;
     padding: 5px;
     color: #000000; }
p {  font-family: Arial, sans-serif; }
#yls { float: right;
       margin: 0 0 5px 5px;
       border: 1px solid #000000; }
```

The HTML source code follows:

```
<h1>Wildflowers</h1>
<img  id="yls"  src="yls.jpg" alt="Yellow Lady Slipper" height="100"
width="100">
<p>The heading and paragraph follow normal flow. The Yellow Lady Slip-
per pictured on the right is a wildflower. It grows in wooded areas
and blooms in June each year. The Yellow Lady Slipper is a member of
the orchid family.</p>
```

In this Hands-On Practice, you'll practice using the CSS `float` property as you configure the web page shown in Figure 7.4.

Create a folder named ch7float. Copy the starter2.html and yls.jpg files from the chapter7 folder in the student files into your ch7float folder. Launch a text editor and open the starter2.html file. Notice the order of the image and paragraphs. Notice that there is no CSS to float the image. Display starter2.html in a browser. The browser renders the page using normal flow and displays the elements in the order they are coded.

Let's add CSS to float the image. Save the file with the name floatyls.html. Open floatyls.html in a text editor and modify the code as follows:

1. Add a style rule for a class named `float` that configures float, margin, and border properties:

```
.float { float: left;
         margin-right: 10px;
         border: 3px ridge #000000; }
```

2. Assign the image element to the class named `float` (use `class="float"`).

Save the file. Launch a browser and test your page. It should look similar to the web page shown in Figure 7.4. The student files contain a sample solution at chapter7/7.2/float.html.

The Floated Element and Normal Flow

Take a moment to examine your file in a browser (see Figure 7.4) and consider how the browser rendered the page. The div element is configured with a light background color to demonstrate how floated elements are rendered outside of normal flow. Observe that the floated image and the first paragraph are contained within the div element. The h2 element follows the div. If all the elements were rendered using normal flow, the area with the light background color would contain both the child elements of the div: the image and the first paragraph. In addition, the h2 element would be placed on its own line under the div element.

However, once the image is placed vertically on the page, it is floated *outside of normal flow*—that's why the light background color only appears behind the first paragraph and the h2 element's text begins immediately after the first paragraph and appears next to the floated image.

FIGURE 7.4 *The CSS* `float` *property left aligns the image.*

Clear a Float

The `clear` Property

The CSS **`clear` property** is often used to terminate, or "clear," a float. You can set the value of the `clear` property to `left`, `right`, or `both`—depending on the type of float you need to clear.[2]

Review Figure 7.5 and the code sample in the student files at chapter7/7.2/float.html. Notice that although the div element contains both an image and the first paragraph, the light background color of the div only displays behind the screen area occupied by the first paragraph—it stops a bit earlier than expected. Clearing the float will help take care of this display issue.

FIGURE 7.5 *The float needs to be cleared to improve the display.*

Clearing a Float with a Line Break

A common technique to clear a float within a container element is to add a line break element configured with the `clear` property. See the example in the student files at chapter7/float/clear1.html.

Observe that a CSS class is configured to clear the left float:

```
.clearleft { clear: left; }
```

Also, a line break tag assigned to the `clearleft` class is coded before the closing `</div>` tag. The code snippet for the div element follows:

```
<div>
<img class="float" src="yls.jpg" alt="Yellow Lady Slipper"
height="100" width="100">
<p>The Yellow Lady Slipper grows in wooded areas and blooms in June
each year. The flower is a member of the orchid family.</p>
<br class="clearleft">
</div>
```

Figure 7.6 displays a screenshot of this page. Note that the light background color of the div element extends farther down the page and the h2 element's text begins on its own line under the image.

FIGURE 7.6 The `clear` property is applied to a line break tag.

If you are not concerned about the light background color display, another option is to omit the line break tag and instead apply the `clearleft` class to the h2 element. This does not change the display of the light background color, but it does force the h2 element's text to begin on its own line, as shown in Figure 7.7 (see the student files at chapter7/float/clear2.html).

FIGURE 7.7 The `clear` property is applied to the h2 element.

Overflow

The `overflow` Property

The CSS **`overflow` property** is often used to clear a float, although its intended purpose is to configure how content should display if it is too large for the area allocated. See Table 7.1 for a list of commonly used values for the `overflow` property.[3]

TABLE 7.1 *The `overflow` Property*

Value	Purpose
visible	Default value; the content is displayed, and if it's too large, the content will "overflow" outside the area allocated to it
hidden	The content is clipped to fit the room allocated to the element in the browser viewport
auto	The content fills the area allocated to it and, if needed, scroll bars are displayed to allow access to the remaining content
scroll	The content is rendered in the area allocated to it and scroll bars are displayed

Clearing a Float with the `overflow` Property

Review Figure 7.8 and the code sample in the student files at chapter7/7.2/float.html. Observe the div element, which contains the floated image and first paragraph on the page. Notice that although the div element contains both an image and the first paragraph, the div element's light background color does not extend as far as expected; it is only visible in the area occupied by the first paragraph. You can use the `overflow` property assigned to the container element to resolve this display issue and clear the float. In this case, we'll apply the `overflow` and `width` properties to the div element selector. The CSS to configure the div in this manner follows:

FIGURE 7.8 *The display can be improved by clearing the float with overflow.*

```
div { background-color: #F3F1BF;
      overflow: auto;
      width: 100%; }
```

This CSS is all that is needed to be added to the code to clear the float and cause the web page to display similar to Figure 7.9 (see the student files at chapter7/float/overflow.html).

The `clear` Property Versus the `overflow` Property

Notice that Figure 7.9 (using the `overflow` property) and Figure 7.6 (applying the `clear` property to a line break tag) result in a similar web page display. You may be wondering about which CSS property (`clear` or `overflow`) is the best to use when you need to clear a float.

Although the `clear` property is widely used, in this example, it is more efficient to apply the `overflow` property to the container element such as a div element. This will clear the float, avoid adding an extra line break tag, and ensure that the container element expands to enclose the entire floated element. You'll get more practice with the `float`, `clear`, and `overflow` properties as you continue working through the book. Floating elements can be useful when designing multicolumn page layouts with CSS.

FIGURE 7.9 *The* `overflow` *property is applied to the div element selector.*

Configuring Scrollbars with the `overflow` Property

The web page in Figure 7.10 demonstrates the use of `overflow: auto;` to automatically display scroll bars if the content exceeds the space allocated to it. In this case, the div that contains the paragraph and the floated image was configured with a width of 300px and a height of 100px. See the example web page in the student files at chapter7/float/scroll.html. The CSS for the div is shown below:

```
div { background-color: #F3F1BF;
      overflow: scroll;
      width: 300px;
      height: 100px;
}
```

FIGURE 7.10 *The browser displays scrollbars.*

? FAQ Why aren't we using external styles?

Since we are only creating sample pages to practice new coding techniques, it is practical to work with a single file. However, if this were an actual website, you would be using an external style sheet for maximum productivity and efficiency.

CSS Box Sizing

When you view an element on a web page, it's intuitive to expect that the width of an element on a page includes the size of the element's padding and border. However, this isn't the default behavior of browsers.

Recall from the box model introduction in Chapter 6 that the `width` property by default only includes the actual width of the content itself within the element and does not also include the width of any padding or border that may exist for the element. This can sometimes be confusing when designing page layout with CSS. The purpose of the `box-sizing` property is to alleviate this issue.

The CSS **box-sizing property** causes the browser calculation of the width or height to include the content's actual width or height in addition to the width or height of any padding and border that may exist.

Valid `box-sizing` property values include `content-box` (the default) and `border-box`. Use the `box-sizing: border-box;` declaration to configure the browser to also include the values of the border and padding when calculating the width and height properties of an element.[4]

Figures 7.11 and 7.12 show web pages (chapter7/boxsizing1.html and chapter7/boxsizing2.html in the student files) that each have floated elements configured

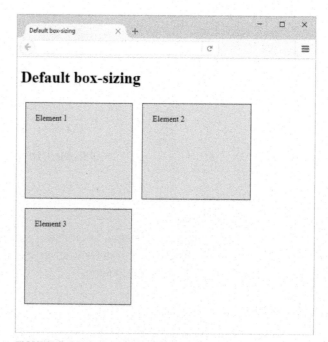

FIGURE 7.11 *Default box-sizing.*

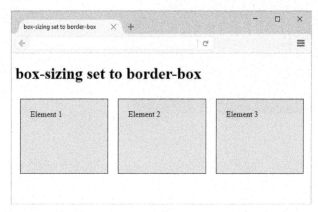

FIGURE 7.12 *The box-sizing property set to border-box.*

with 30% width, 150px height, 20px padding, and 10px margin. The page in Figure 7.11 uses default box-sizing. The page in Figure 7.12 uses box-sizing set to `border-box`. The size of the elements and the placement of the elements on the pages differ.

You may notice at first glance that the elements look larger in Figure 7.11. The larger display is because the browser sets the content to 30% width before adding the 20 pixels of padding on each side. The elements are smaller in Figure 7.12. The smaller display is because the browser applies the 30% width to the combination of the padding and the content.

Let's take a closer look at the placement of the three floated elements on the pages.

Figure 7.11 does not display all three elements side by side. This web page uses default `box-sizing` so the browser assigned the 30% width to each element's content only and then added 20 pixels of padding to each side of each element. Due to these calculations, the browser determined there was not enough room in the browser viewport to display all three elements next to each other and the browser dropped the third floated element to the next line.

The web page in Figure 7.12 is coded with `box-sizing` set to `border-box,` which configures the three floated elements to be displayed side by side because the browser assigned the 30% width to the combined content and padding areas (including 20 pixels of padding on each side).

It is common practice for web developers to apply border-box box-sizing when they plan to use floated elements or multicolumn layouts. It's also common practice to apply `box-sizing` by configuring the * **universal selector**, which will target all HTML elements. The CSS style rule to apply border-box box-sizing to all elements with the universal selector follows:

`* { box-sizing: border-box; }`

Feel free to experiment with the box-sizing property and the examples (chapter7/boxsizing1.html and chapter7/boxsizing2.html in the student files). You will use the box-sizing property as you explore page layout in this chapter.

Basic Two-Column Layout

A common design for a web page is a two-column layout. This is often accomplished with CSS by configuring one of the columns to float on the web page. Coding HTML is a skill and skills are best learned by practice. This Hands-On Practice guides you as you convert a single-column page layout (Figure 7.13) into your first two-column layout (Figure 7.14).

FIGURE 7.13 *Single-column layout.*

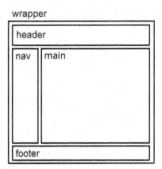

FIGURE 7.14 *Two-column layout.*

 Hands-On Practice 7.3

FIGURE 7.15 *Web page with single-column layout.*

A. Review single-column layout. Launch a text editor and open the singlecol.html file from the chapter7 folder in the student files. Take a moment to look over the code. Notice the structure of the HTML tags corresponds to the wireframe in Figure 7.13.

```
<body>
<div id="wrapper">

  <header> </header>

  <nav> </nav>

  <main> </main>

  <footer> </footer>

</div>
</body>
```

Save the file with the name index.html. When you display index.html in a browser, your display should be similar to Figure 7.15.

B. Configure a two-column layout. Launch a text editor and open the index.html file. You will edit the HTML and CSS to configure a two-column layout as shown in Figure 7.14 wireframe.

1. **Edit the HTML.** The single-column navigation is horizontal, but the two-column navigation will be displayed in a vertical orientation. Later in this chapter, you'll learn how to configure navigation hyperlinks within an unordered list but for now, a quick adjustment is to code a line break tag after each of the first two hyperlinks in the nav area.

2. **Configure the float with CSS.** Locate the style tags in the head section of the document and code the following style rule as embedded CSS to configure a nav element with a width of 150px that floats to the left.

   ```
   nav { float: left;
         width: 150px; }
   ```

   Save the file and test it in the Firefox or Chrome browser. Your display will be similar to Figure 7.16. Notice that the content in the main area wraps around the floated nav element.

3. **Configure two columns with CSS.** You just configured the nav element to float on the left. The main element will be in the right-side column and will be configured with a left margin (the same side as the float). To get a two-column look, the value of the margin should be greater than the width of the floated element. Open the index.html file in a text editor and code the following style rule to configure a 160px left margin for the main element.

   ```
   main { margin-left: 160px; }
   ```

   Save the file and test it in the Firefox or Chrome browser. Your display will be similar to Figure 7.17 with a two-column layout.

FIGURE 7.16 *The nav is floating on the left.*

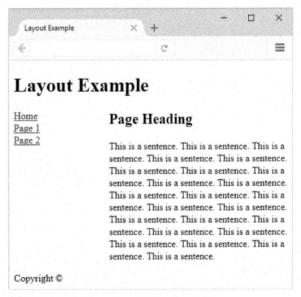

FIGURE 7.17 *Two-column layout.*

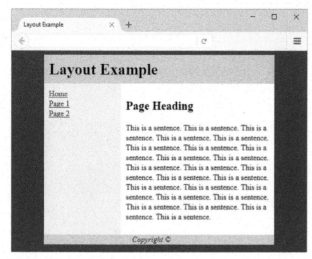

FIGURE 7.18 *Final two-column layout.*

4. **Enhance the page with CSS.** Code the following style rules as embedded CSS to create a more appealing web page. When you have completed this step, your page should be similar to Figure 7.18.

   a. **The body element selector.** Configure a dark background color.

   ```
   body { background-color: #000066; }
   ```

   b. **The wrapper id selector.** Configure 80% width, centered on the page, and a light (#EAEAEA) background color. This

background color will display behind child elements (such as the nav element) that do not have a background color configured.

```
#wrapper { width: 80%;
           margin-left: auto;
           margin-right: auto;
           background-color: #EAEAEA; }
```

   c. **The header element selector.** Configure #CCCCFF background color.

   ```
   header { background-color: #CCCCFF; }
   ```

   d. **The h1 element selector.** Configure 0 margin and 10px of padding.

   ```
   h1 { margin: 0;
        padding: 10px; }
   ```

   e. **The nav element selector.** Edit the style rule and add a declaration for 10 pixels of padding.

   ```
   nav { float: left;
         width: 150px;
         padding: 10px;   }
   ```

   f. **The main element selector.** Edit the style rule and add a declaration for 10 pixels of padding and #FFFFFF background color.

   ```
   main { margin-left: 160px;
          padding: 10px;
          background-color: #FFFFFF;   }
   ```

   g. **The footer element selector.** Configure centered, italic text, and a #CCCCFF background color. Also configure the footer to clear all floats.

   ```
   footer { text-align: center;
            font-style: italic;
            background-color: #CCCCFF;
            clear: both; }
   ```

Save your file and test it in the Firefox or Chrome browser. Your display should be similar to Figure 7.18. You can compare your work to the sample in the student files (chapter7/7.3/ index.html). Internet Explorer does not support default styles like the HTML5 main element. You may need to nudge this browser to comply by adding the `display: block;` declaration (introduced later in this chapter) to the styles for the main element selector. An example solution is in the student files (chapter7/7.3/iefix.html).

Two-Column Layout Example

The web page you coded in Hands-On Practice 7.3 is just one example of a two-column layout design. Let's explore coding the two-column layout with a footer in the right column as shown in Figure 7.19 wireframe. The HTML template for the page layout follows:

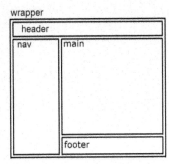

FIGURE 7.19 *Alternate wireframe.*

```
<div id="wrapper">
  <header>
  </header>
  <nav>
  </nav>
  <main>
  </main>
  <footer>
  </footer>
</div>
```

The key CSS configures a floating nav element, a main element with a left margin, and a footer with a left margin.

```
nav { float: left; width: 150px; }
main { margin-left: 165px; }
footer { margin-left: 165px; }
```

The web page shown in Figure 7.20 implements this layout. An example is in the student files, chapter7/float/twocolumn.html.

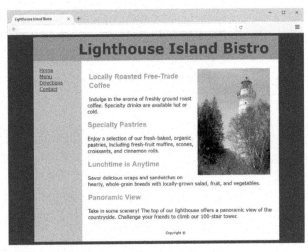

FIGURE 7.20 *Page with alternate layout.*

Do I have to use a wrapper?

No, you are not required to use a wrapper or container for a web page layout. However, it does make it easier to get the two-column look because the background color of the wrapper div will display behind any of its child elements that do not have their own background color configured.

Quick TIP

A key to coding successful layouts with float is in the HTML—place the element that needs to float BEFORE its companion elements. The browser will shift the floated element over to the side of the browser viewport and display the elements that follow alongside the floated element.

Vertical Navigation with an Unordered List

One of the advantages of using CSS for page layout involves the use of semantically correct code. Writing semantically correct code means using the markup tag that most accurately reflects the purpose of the content. Using the various levels of heading tags for content headings and subheadings or placing paragraphs of text within paragraph tags (rather than using line breaks) are examples of writing semantically correct code. This type of coding is a step in the direction to support the Semantic Web.

Leading Web developers such as Eric Meyer and Jeffrey Zeldman have promoted the idea of using unordered lists to configure navigation menus.[5,6] After all, a navigation menu is a list of hyperlinks. Recall from Chapter 5 that you can configure an unordered list to omit the display of the list markers or even display an image instead of a standard list marker.

Configuring navigation with a list also helps to provide for accessibility. Screen reader applications offer easy keyboard access and verbal cues for information organized in lists, such as the number of items in the list.

Figure 7.21 shows the navigation area of a web page (found in the student files chapter7/twocolumn3.html) that uses an unordered list to organize the navigation links. The HTML code follows:

```
<ul>
<li><a href="index.html">Home</a></li>
<li><a href="menu.html">Menu</a></li>
<li><a href="directions.html">Directions</a></li>
<li><a href="contact.html">Contact</a></li>
</ul>
```

FIGURE 7.21 *Navigation in an unordered list.*

FIGURE 7.22 *The list markers have been eliminated with CSS.*

Configure an Unordered List with CSS

OK, so now that we're semantically correct, how about improving the visual aesthetic? Let's use CSS to eliminate the list marker (refer back to Chapter 5). We also need to make sure that our special styles only apply to the unordered lists in the navigation area (within the nav element) so we'll use a descendant selector. The CSS to configure the list in Figure 7.22 follows:

```
nav ul { list-style-type: none; }
```

Remove the Underline with the CSS `text-decoration` Property

The **text-decoration property** modifies the display of text in the browser and is most often used to eliminate the underline from the hyperlinks. As shown in Figure 7.23, the navigation hyperlinks are configured without an underline by coding:

```
text-decoration: none;
```

FIGURE 7.23 The CSS `text-decoration` property has been applied.

 ### Hands-On Practice 7.4 ───────────────────

You will configure vertical navigation with an unordered list in this Hands-On Practice. Create a folder named ch7vert. Copy the files lighthouseisland.jpg, lighthouselogo.jpg, and starter3.html from the chapter7 folder in the student files into your ch7vert folder. Display the web page in a browser. It should look similar to Figure 7.24—notice that the navigation area needs to be configured.

Launch a text editor and open the starter3.html file. Save the file as index.html in your ch7vert folder.

FIGURE 7.24 Notice that the navigation area needs to be configured.

1. Review the code for this page, which uses a two-column layout. Examine the nav element and modify the code surrounding the hyperlinks to configure the navigation in an unordered list.

```
<nav>
  <ul>
    <li><a href="index.html">Home</a></li>
    <li><a href="menu.html">Menu</a></li>
    <li><a href="directions.html">Directions</a></li>
    <li><a href="contact.html">Contact</a></li>
  </ul>
</nav>
```

2. Let's add CSS to the embedded styles to configure the unordered list elements only within the nav element: eliminate the list marker and set the padding to 10 pixels.

```
nav ul { list-style-type: none;
         padding: 10px; }
```

3. Next, configure the anchor tags within the nav element to have 10 pixels of padding, use bold font, and display no underline.

```
nav a { text-decoration: none;
        padding: 10px;
        font-weight: bold; }
```

Save your page and test it in a browser. Your page should look similar to Figure 7.25. A sample is found in the student files (chapter7/7.4/index.html).

FIGURE 7.25 Two-column layout with vertical navigation.

Horizontal Navigation with an Unordered List

You may be wondering how to use an unordered list for a horizontal navigation menu. The answer is CSS! List item elements are block display elements. They need to be configured as inline display to appear in a horizontal line. The CSS `display` property makes this possible.

CSS `display` Property

The CSS **display property** configures the way that browsers render elements. See Table 7.2 for a list of commonly used values.[7]

TABLE 7.2 *The* `display` *Property*

Value	Purpose
`none`	The element will not display
`inline`	The element will display as an inline display element without line breaks above and below
`inline-block`	The element will display as an inline display element adjacent to other inline display elements but also can be configured with properties of block display elements including width and height
`block`	The element will display as a block display element with line breaks above and below
`flex`	The element will display as a block-level flex container (see Chapter 8)
`grid`	The element will display as a block-level grid container (see Chapter 8)

Home   Menu   Directions   Contact

FIGURE 7.26 *Navigation in an unordered list.*

Figure 7.26 shows the navigation area of a web page (student files chapter7/navigation.html) with a horizontal navigation area organized by an unordered list. The HTML follows:

```
<nav>
  <ul>
    <li><a href="index.html">Home</a></li>
    <li><a href="menu.html">Menu</a></li>
    <li><a href="directions.html">Directions</a></li>
    <li><a href="contact.html">Contact</a></li>
  </ul>
</nav>
```

Configure with CSS

The following CSS was applied in the code sample for Figure 7.26:

▶ To eliminate the list marker from unordered lists within the nav element, apply
`list-style-type: none;` to the `nav ul` selector:

```
nav ul { list-style-type: none; }
```

▶ To render the list items within the nav element horizontally instead of vertically, apply
`display: inline;` to the `nav li` selector:

```
nav li { display: inline; }
```

▶ To eliminate the underline from the hyperlinks within the nav element, apply
`text-decoration: none;` to the `nav a` selector. Also, configure right padding to
add some space between the hyperlinks:

```
nav a { text-decoration: none; padding-right: 10px; }
```

 Hands-On Practice 7.5 ─────────────────────────

You will configure horizontal navigation with an unordered list in this Hands-On Practice. Create a folder named ch7hort. Copy the files lighthouseisland.jpg, lighthouselogo.jpg, and starter4.html from the chapter7 folder in the student files into your ch7hort folder. Display the web page in a browser. It should look similar to Figure 7.27—notice that the navigation area needs to be configured to display in a single line. Launch a text editor and open the starter4.html file. Save the file as index.html in your ch7hort folder.

FIGURE 7.27 *Notice that the navigation area needs to be configured.*

1. Examine the nav element and notice that it contains an unordered list with navigation hyperlinks. Let's add CSS to the embedded styles to configure the unordered list element within the nav element: eliminate the list marker, center the text, set the font size to 1.5em, and set the margin to 5 pixels.

```
nav ul { list-style-type: none;
        text-align: center;
        font-size: 1.5em;
        margin: 5px; }
```

2. Configure the li elements within the nav element to display as inline elements.

```
nav li { display: inline; }
```

3. Configure the anchor elements within the nav element to display no underline. Also set the left and right padding to 10 pixels.

```
nav a { text-decoration: none;
        padding-left: 10px;
        padding-right: 10px; }
```

Save your page and test it in a browser. Your page should look similar to Figure 7.28. A sample is found in the student files (chapter7/7.5/index.html).

FIGURE 7.28 *Horizontal navigation within an unordered list.*

CSS Interactivity with Pseudo-Classes

VideoNote

Interactivity with CSS Pseudo-Classes

Have you ever visited a website and found that the text hyperlinks changed color when you moved the mouse pointer over them? Often, this is accomplished using a CSS **pseudo-class**, which can be used to apply a special effect to a selector. The five pseudo-classes that can be applied to the anchor element are shown in Table 7.3.[8]

TABLE 7.3 *Commonly Used CSS Pseudo-Classes*

Pseudo-Class	When Applied
:link	Default state for a hyperlink that has not been clicked (visited)
:visited	Default state for a visited hyperlink
:focus	Triggered when the hyperlink has keyboard focus
:hover	Triggered when the mouse pointer moves over the hyperlink
:active	Triggered when the hyperlink is actually clicked

Notice the order in which the pseudo-classes are listed in Table 7.3. Anchor element pseudo-classes *must be coded in this order* (although it's OK to omit one or more of those listed). If you code the pseudo-classes in a different order, the styles will not be reliably applied. It's common practice to configure the :focus and :active pseudo-classes with the same styles.

To apply a pseudo-class, write it after the selector. The following code sample will configure text hyperlinks to be red initially. The sample also uses the :hover pseudo-class to configure the hyperlinks to change their appearance when the visitor places the mouse pointer over them so that the underline disappears and the color changes.

```
a:link { color: #ff0000; }
a:hover { text-decoration: none;
          color: #000066; }
```

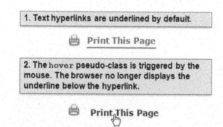

FIGURE 7.29 *Using the hover pseudo-class.*

Figure 7.29 shows part of a web page that uses a similar technique. Note the position of the mouse pointer over the "Print This Page" hyperlink—the text color has changed and has no underline. Most modern browsers support CSS pseudo-classes.

Hands-On Practice 7.6

You will use pseudo-classes to create interactive hyperlinks in this Hands-On Practice. Create a folder named ch7hover. Copy the lighthouseisland.jpg, lighthouselogo.jpg, and starter3.html files from the chapter7 folder in the student files into your ch7hover folder. Display the web page in a browser. It should look similar to Figure 7.30—notice that the navigation area needs to be configured. Launch a text editor and open the starter3.html file. Save the file as index.html in your ch7hover folder.

FIGURE 7.30 *The navigation area needs to be styled in this two-column page layout.*

1. Review the code for this page, which uses a two-column layout. Examine the nav element and modify the code surrounding the hyperlinks to configure the navigation in an unordered list.

```
<nav>
  <ul>
    <li><a href="index.html">Home</a></li>
    <li><a href="menu.html">Menu</a></li>
    <li><a href="directions.html">Directions</a></li>
    <li><a href="contact.html">Contact</a></li>
  </ul>
</nav>
```

2. Let's add CSS to the embedded styles to configure the unordered list element within the nav element: eliminate the list marker and set the padding to 10 pixels.

```
nav ul { list-style-type: none; padding: 10px; }
```

3. Next, configure basic interactivity with pseudo-classes.

▶ Configure the anchor elements within the nav element to have 10 pixels of padding, use bold font, and display no underline.

```
nav a { text-decoration: none; padding: 10px;
        font-weight: bold; }
```

▶ Use pseudo-classes to configure anchor tags within the nav element to display white (#FFFFFF) text for unvisited hyperlinks, light-gray (#EAEAEA) text for visited hyperlinks, and dark blue (#000066) text when the mouse pointer hovers over hyperlinks:

```
nav a:link { color: #FFFFFF; }
nav a:visited { color: #EAEAEA; }
nav a:hover { color: #000066; }
```

FIGURE 7.31 *CSS pseudo-classes add interactivity to the navigation.*

Save your page and test it in a browser. Move your mouse pointer over the navigation area and notice the text color change. Your page should look similar to Figure 7.31. A sample is found in the student files (chapter7/7.6/index.html).

Practice with CSS Two-Column Layout

 Hands-On Practice 7.7 ———————————————————

In this Hands-On Practice, you'll create a new version of the Lighthouse Island Bistro home page with a top header section spanning two columns, content in the left column, navigation in the right column, and a footer section below the two columns. See Figure 7.32 for the wireframe. You will configure the CSS in an external style sheet. Create a new folder named ch7bistro. Copy the starter5.html, lighthouseisland.jpg, and lighthouselogo.jpg files from the chapter7 folder in the student files into your ch7bistro folder.

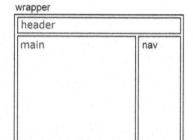

FIGURE 7.32 *The wireframe for a two-column layout with a top logo area.*

1. Launch a text editor and open the starter5.html file. Add a link element to the head section of the web page that associates this file with an external style sheet named bistro.css. A code sample follows:

   ```
   <link href="bistro.css" rel="stylesheet">
   ```

   Save the file with the name index.html.

2. Launch a text editor and create a new file named bistro.css in your ch7bistro folder. Configure the CSS as follows:

   ▶ **The universal selector:** set the box-sizing property to border-box.

   ```
   *{ box-sizing: border-box; }
   ```

   ▶ **The body element selector:** very dark blue background (#00005D) and Verdana, Arial, or the default sans-serif font typeface

   ```
   body { background-color: #00005D;
          font-family: Verdana, Arial, sans-serif; }
   ```

   ▶ **The wrapper id:** centered, take up 80% of the browser viewport, a minimum width of 940px, dark blue text (#000066), and medium-blue (#B3C7E6) background (*this color will display behind the nav section*)

   ```
   #wrapper { margin: 0 auto; width: 80%; min-width: 940px;
              background-color: #B3C7E6; color: #000066; }
   ```

   ▶ **The header element selector:** slate blue (#869DC7) background; very dark blue text (#00005D); 150% font size; 10px top, right, and bottom padding; 155px left padding; height set to 150 pixels; and the lighthouselogo.jpg background image

   ```
   header { background-color: #869DC7; color: #00005D;
            font-size: 150%; padding: 10px 10px 10px 155px;
            height: 150px;
            background-repeat: no-repeat;
            background-image: url(lighthouselogo.jpg); }
   ```

▶ **The nav element selector:** float on the right, 150px width, bold text, and 0.1em letter spacing

```
nav { float: right; width: 150px; font-weight: bold;
      letter-spacing: 0.1em; }
```

▶ **The main element selector:** white background (#FFFFFF), black text (#000000), 10 pixels top and bottom padding, and 20 pixels left and right padding, auto overflow, and block display (fixes an Internet Explorer 11 rendering issue).

```
main { background-color: #FFFFFF; color: #000000;
       padding: 10px 20px; overflow: auto; display: block; }
```

▶ **The footer element selector:** 70% font size, centered text, 10 pixels of padding, a slate blue background color (#869DC7), and `clear` set to `both`.

```
footer { font-size: 70%; text-align: center; padding: 10px;
         background-color: #869DC7; clear: both; }
```

Save the bistro.css file. Display index.html in a browser. Your page should look similar to Figure 7.33.

3. Continue editing the bistro.css file to style the h2 element selector and floating image. Configure the h2 element selector with slate blue text color (#869DC7) and Arial or sans-serif font typeface. Configure the `floatright` id to float on the right side with 10 pixels of margin.

FIGURE 7.33 *The home page with major page sections configured using CSS.*

```
h2 { color: #869DC7;
     font-family: Arial, sans-serif; }
#floatright { float: right; margin: 10px; }
```

4. Continue editing the bistro.css file and configure the vertical navigation bar.

▶ Configure the unordered list: eliminate list markers, set zero margin, and set zero padding:

```
nav ul { list-style-type: none; margin: 0; padding: 0; }
```

▶ Configure hyperlinks: no underline, 20 pixels padding, medium-blue background color (#B3C7E6), and 1 pixel solid white bottom border. Use `display: block;` to allow the web page visitor to click anywhere in the anchor "button" to activate the hyperlink.

```
nav a { text-decoration: none; padding: 20px; display: block;
        background-color: #B3C7E6;
        border-bottom: 1px solid #FFFFFF; }
```

▶ Configure the `:link`, `:visited`, and `:hover` pseudo-classes as follows:

```
nav a:link { color: #FFFFFF; }
nav a:visited { color: #EAEAEA; }
nav a:hover { color: #869DC7;
              background-color: #EAEAEA; }
```

FIGURE 7.34 *CSS pseudo-classes add interactivity to the page.*

Save your file. Display your index.html page in a browser. Move your mouse pointer over the navigation area and notice the interactivity, as shown in Figure 7.34. A sample solution is in the chapter7/7.7/index.html file.

CSS for Print

Even though the "paperless society" has been talked about for decades, the fact is that many people still love paper, and you can expect your web pages to be printed. CSS offers you some control over what gets printed and how the printouts are configured. This is easy to do using external style sheets. Create one external style sheet with the configurations for browser display and a second external style sheet with the special printing configurations. Associate both of the external style sheets to the web page using two link elements. Configure a **media attribute** on each link element. Table 7.4 describes commonly used values of the media attribute.

TABLE 7.4 *The* media *Attribute*

Value	Purpose
screen	The default value; indicates the style sheet that configures typical browser viewport display on a color computer screen
print	Indicates the style sheet that configures the printed formatting

Modern browsers will use the correct style sheet depending on whether they are rendering a screen display or preparing to print a document. Use media="screen" to configure the link element for your browser display. Use media="print" to configure the link element for your printout. Sample HTML follows:

```
<link rel="stylesheet" href="lighthouse.css" media="screen">
<link rel="stylesheet" href="lighthouseprint.css" media="print">
```

Print Styling Best Practices

You might be wondering how a print style sheet should differ from the CSS used to display the web page in a browser. Let's explore some commonly used techniques for styling printed web pages.

▶ **Hide Nonessential Content.**   It's common practice to prevent banner ads, navigation, or other extraneous areas from appearing on the printout. Use the display: none; style declaration to hide content that is not needed on a printout of the web page.

▶ **Configure Font Size and Color for Printing.**   Another common practice is to configure the font sizes on the print style sheet to use pt units. This will better control the text on the printout. You might also consider configuring the text color to black (#000000) if you envision the need for visitors to print your pages often. The default setting on most browsers prevents background colors and background images from printing, but you can also prevent background image and background color display in your print style sheet.

▶ **Control Page Breaks.** Use the CSS **page-break-before** or **page-break-after** properties to control page breaks when printing the web page.[9] Well-supported values for these properties are `always` (the page break will always occur as designated), `avoid` (if possible, the page break will not occur before or after, as designated), and `auto` (default). For example, to configure a page break at a specific point in the document (in this case, right before an element assigned to the class named `newpage`), configure the CSS as shown below:

```
.newpage { page-break-before: always; }
```

 Hands-On Practice 7.8 ──────────────────────────────────

In this Hands-On Practice, you'll rework the Lighthouse Island Bistro page from Hands-On Practice 7.7 to be configured for optimal screen display and printing. Create a new folder named ch7print. Copy the files from either your ch7bistro folder or the student files chapter7/7.7 folder into the ch7print folder.

1. Launch a text editor and open the index.html file. Examine the source code and locate the style element. Copy the CSS between the style tags and paste into a new text document named bistro.css. Save the bistro.css file in the ch7print folder.

2. Edit the index.html file and edit the link tag in the head section that associates the web page with the bistro.css file to specify screen display (use `media="screen"`).

3. Edit the index.html file and add another link tag that associates the web page with a file named bistroprint.css for printing (use `media="print"`). Save the index.html file.

4. Launch a text editor and open bistro.css. Since you want to keep most of the styles for printing, you will start by creating a new version of the external style sheet. Save bistro.css with the name of bistroprint.css in the ch7print folder. You will modify three areas on this style sheet: the header selector, the main selector, and the nav selector.

   ▶ Modify the header styles to print using black text in 20 point font size:
   ```
   header { color: #000000; font-size: 20pt; }
   ```

   ▶ Modify the main element area to print using a serif typeface in a 12 point font size:
   ```
   main { font-family: "Times New Roman", serif; font-size: 12pt; }
   ```

   ▶ Modify the navigation area to not display:

   ```
   nav { display: none; }
   ```

   Save your file.

5. Test your work. Display your index.html file in a browser. Select Print from the menu. Your display should look similar to the page shown in Figure 7.35. The header and content font sizes have been configured. The navigation does not display. The student files contain a sample solution in the chapter7/7.8 folder.

FIGURE 7.35 *The print preview display of the web page.*

CSS Sprites

When browsers display web pages, they must make a separate http request for every file used by the page, including .css files and image files such as .gif, .jpg, and .png files. Each http request takes time and resources. A **sprite** is an image file that contains multiple small graphics. The single graphics file saves download time because the browser only needs to make one http request for the combined image instead of many requests for the individual smaller images. Using CSS to configure the small graphics combined in the sprite as background images for various web page elements is called **CSS sprites**, a technique made popular by David Shea.[10]

The first image in the sprite begins at the top of the image.

The checkerboard background indicates a transparent image

The second image in the sprite begins 100 pixels down from the top.

FIGURE 7.36 *The sprite consists of two images.*

The CSS sprites technique uses the CSS `background-image`, `background-repeat`, and `background-position` properties to manipulate the placement of the background image.

Figure 7.36 shows a sprite with two lighthouse images on a transparent background. These images are configured as background images for the navigation hyperlinks with CSS as shown in Figure 7.37. You'll see this in action as you complete the next Hands-On Practice.

FIGURE 7.37 *Sprites in action.*

Hands-On Practice 7.9

You will rework the Lighthouse Island Bistro page from Hands-On Practice 7.7 to implement CSS sprites. Create a new folder named ch7sprites. Copy the files from either your ch7bistro folder or the student files chapter7/7.7 folder into the your ch7sprites folder. Copy the sprites.gif file from the student files chapter7 folder. The sprites.gif, shown in

Figure 7.36, contains two lighthouse images. The first lighthouse image starts at the top of the graphics file. The second lighthouse image begins 100 pixels down from the top of the graphics file. We'll use this information about the location of the second image within the graphics file when we configure the display of the second image. Launch a text editor and open bistro.css.

You will edit the styles to configure background images for the navigation hyperlinks.

1. Configure the background image for navigation hyperlinks. Add the following styles to the `nav a` selector to set the background image to the sprites.gif with no repeat. The value `right` in the `background-position` property configures the lighthouse image to display at the right of the navigation element. The value 0 in the `background-position` property configures the display at offset 0 from the top (at the very top) so the first lighthouse image displays.

```
nav a { text-decoration: none;
        display: block;
        padding: 20px;
        background-color: #B3C7E6;
        border-bottom: 1px solid #FFFFFF;
        background-image: url(sprites.gif);
        background-repeat: no-repeat;
        background-position: right 0;  }
```

2. Configure the second lighthouse image to display when the mouse pointer passes over the hyperlink. Add the following styles to the `nav a:hover` selector to display the second lighthouse image. The value `right` in the `background-position` property configures the lighthouse image to display at the right of the navigation element. The value `-100px` in the `background-position` property configures the display at an offset of 100 pixels down from the top so the second lighthouse image appears.

```
nav a:hover { background-color: #EAEAEA;
              color: #869DC7;
              background-position: right -100px;  }
```

Save the file and test your index.html file it in a browser. Your page should look similar to Figure 7.37. Move your mouse pointer over the navigation hyperlinks to see the background images change. Compare your work with the student files (chapter7/7.9/index.html).

FAQ

How can I create my own sprite graphics file?

Most web developers use a graphics application such as Adobe Photoshop or GIMP to edit images and save them in a single graphics file for use as a sprite. Or, you could use a web-based sprite generator such as the ones listed below:

▶ CSS Sprites Generator: https://csssprites.com
▶ CSS Sprite Generator: https://spritegen.website-performance.org
▶ SpritePad: https://spritepad.wearekiss.com/

Positioning with CSS

You've seen how normal flow causes the browser to render the elements in the order that they appear in the HTML source code. When using CSS for page layout, there are situations when you may want more control over the position of an element. The **position property** configures the type of positioning used when the browser renders an element. Table 7.5 lists `position` property values and their purpose.[11]

TABLE 7.5 *The* `position` *Property*

Value	Purpose
static	Default value; the element is rendered in normal flow
fixed	Configures the location of an element within the browser viewport; the element does not move when the page is scrolled
relative	Configures the location of an element relative to where it would otherwise render in normal flow
sticky	Combines features of relative and fixed positioning; not supported in Internet Explorer
absolute	Precisely configures the location of an element outside of normal flow

Static Positioning

Static positioning is the default and causes the browser to render an element in normal flow. As you've worked through the exercises in this book, you have created web pages that the browser rendered using normal flow.

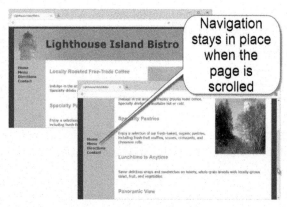

FIGURE 7.38 *Fixed positioning navigation.*

Fixed Positioning

Use **fixed positioning** to cause an element to be removed from normal flow and to remain stationary, or "fixed in place," when the web page is scrolled in the browser viewport.

Figure 7.38 shows a web page (found in the student files chapter7/fixed.html) with a navigation area configured with fixed position. The navigation stays in place even though the user has scrolled down the page.

The CSS follows:

```
nav { position: fixed; }
```

Relative Positioning

Use **relative positioning** to change the location of an element slightly, relative to where it would otherwise appear in normal flow. However, the area in normal flow is still reserved for the element, and other elements will flow around that reserved space. Configure relative positioning with the `position: relative;` property along with one or more of the following offset properties: `left`, `right`, `top`, `bottom`. Table 7.6 lists the offset properties.[12]

TABLE 7.6 *The Position Offset Properties*

Property	Value	Purpose
left	Numeric value or percentage	The position of the element offset from the left side of the container element
right	Numeric value or percentage	The position of the element offset from the right side of the container element
top	Numeric value or percentage	The position of the element offset from the top of the container element
bottom	Numeric value or percentage	The position of the element offset from the bottom of the container element

Figure 7.39 shows a web page (see the student files chapter7/relative.html) that uses relative positioning along with the `left` property to configure the placement of an element in relation to the normal flow. In this case, the container element is the body of the web page. The result is that the content of the element is rendered as being offset or shifted by 30 pixels from the left where it would normally be placed at the browser's left margin. Notice also how the `padding` and `background-color` properties configure the heading element. The CSS follows:

FIGURE 7.39 *The paragraph is configured using relative positioning.*

```
p {  position: relative;
     left: 30px;
     font-family: Arial, sans-serif; }
h1 { background-color: #cccccc;
     padding: 5px;
     color: #000000; }
```

The HTML source code follows:

```
<h1>Relative Positioning</h1>
<p>This paragraph uses CSS relative positioning to be placed 30 pixels in from the
left side.</p>
```

Sticky Positioning

Sticky positioning combines features of relative and fixed positioning.

With **sticky positioning**, the element initially displays relative to where it would be rendered in normal flow (which may even be its final sticky position, depending on where the element is located within the source code and other CSS that may be applied).

If the element is not initially rendered at its final sticky position, as soon as the browser scrolls the element to the specified position the element gets "stuck," remains there as a fixed position element, and no longer moves when the page is scrolled.

If an element is initially rendered at its sticky position, it will remain there when the browser scrolls. Explore an example in the student files (chapter7/layout/sticky.html) of using sticky positioning to force a navigation bar to move up to the top of the page when scrolled. The positioning CSS follows:

```
nav { position: sticky;
      top: 0; }
```

Absolute Positioning

FIGURE 7.40 *The paragraph is configured with absolute positioning.*

Use **absolute positioning** to precisely specify the location of an element outside of normal flow in relation to its first non-static parent element. If there is no non-static parent element, the absolute position is specified in relation to the browser viewport. Configure absolute positioning with the `position: absolute;` property along with one or more of the offset properties (`left`, `right`, `top`, `bottom`) listed in Table 7.6.

Figure 7.40 depicts a web page that configures an element with absolute positioning to display the content 200 pixels in from the left margin and 100 pixels down from the top of the web page document. An example is in the student files, chapter7/absolute.html. The CSS follows:

```
p { position: absolute;
    left: 200px;
    top: 100px;
    font-family: Arial, sans-serif;
    width: 300px; }
```

The HTML source code follows:

```
<h1> Absolute Positioning</h1>
<p>This paragraph is 300 pixels wide and uses CSS absolute positioning
to be placed 200 pixels in from the left and 100 pixels down from the
top of the browser window.</p>
```

The z-index Property

The **z-index property** provides a way to configure the stacking of positioned elements on a web
page.[13] The element must also be positioned with
either absolute, relative, fixed, or sticky positioning.
The default z-index for a positioned element is 0. To
configure a different z-index use an integer value.
Elements with higher z-index values will stack on top
of elements with lower z-index values. Given two or
more elements placed in overlapping space on a web
page, the positioned element with the highest z-index
value will stack on the top and display over the other
element(s).

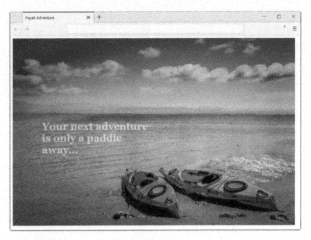

FIGURE 7.41 *Z-index in action.*

 Hands-On Practice 7.10 ────────────────────

In this Hands-On Practice you'll apply CSS positioning properties as you configure a heading
to display over an image, as shown in Figure 7.41. Create a new folder named ch7p. Copy
beached.jpg and template.html from the student files chapter7 folder into your ch7p folder.
Launch a text editor and modify the template.html as indicated:

Edit the HTML. Configure the title element with: Kayak Adventure. Code a div assigned to
the class named `hero`. This div will contain a heading and an image. Code an h1 element
within the div that contains: "Your next adventure is only a paddle away...". Code an img
element within the div to display beached.jpg.

Edit the CSS. Code a style element in the head section and add embedded CSS to the
head section.

1. Configure a class named `hero`. This class will use relative positioning instead of the
   default static positioning. This does not change the location of the `hero` div but sets
   the stage to use absolute positioning on the h1 element in relation to its container div
   instead of in relation to the entire web page document. This won't matter too much for
   our example, but it would be very helpful if the hero div were part of a more complex
   web page. Set the `hero` class to use relative positioning.

   ```
   .hero { position: relative; }
   ```

2. Configure the `.hero h1` descendant selector to use absolute positioning, 40% from
   the top, 10% from the left, 40% width, Georgia font, z-index set to 999, and white text
   that is 60% opaque.

   ```
   .hero h1 { position: absolute; top: 40%; left: 10%; width: 40%;
              font-family: Georgia, serif; z-index: 999;
              color: rgba(255, 255, 255, 0.6); }
   ```

Save your page as index.html and display it in a browser. Compare your work to Figure 7.41
and the sample in the student files (chapter7/7.10).

Fixed Position Navigation Bar

You've probably seen web pages that feature a header area or navigation bar that is fixed in place across the top of the browser window. A fixed top navigation bar that does not take up too much vertical space and is not distracting can provide convenient, site-wide navigation that is always available to website visitors. This popular page layout technique is easy to accomplish with CSS positioning and the CSS z-index property.

 Hands-On Practice 7.11

In this Hands-On Practice, you will configure a fixed navigation area across the top of a web page that remains in place when the page is vertically scrolled in a browser. Create a new folder named ch7z. Copy the files from your ch7hort folder or the student files chapter7/7.5 folder into the ch7z folder.

1. Open the index.html file in a browser. The top of the page should look similar to Figure 7.28 with a horizontal navigation bar below the header area. You will modify the layout of this page so that it looks like Figure 7.42—with a fixed top navigation bar above the header area. When you scroll the page in the browser, the fixed navigation area will remain in place as the rest of the page moves.

FIGURE 7.42 *The web page has a fixed top navigation bar.*

2. Launch a text editor and open the index.html file. You will edit the embedded CSS as follows.

▶ **The nav element selector.** Code a new style rule that sets fixed position beginning at the top left of the page, 40px height, 100% width, 40em minimum width, #B3C7E6 background color, and z-index set to a high value, such as 9999.

```
nav { position: fixed; top: 0; left: 0;
        height: 40px; width: 100%; min-width: 40em;
        background-color: #B3C7E6;
        z-index: 9999; }
```

▶ **The nav ul element selector.** Edit the style rule. Change the text alignment to right. Also set 10% padding on the right.

```
nav ul { list-style-type: none; text-align: right;
        font-size: 1.5em; margin: 5px;
        padding-right: 10%; }
```

▶ **The header element selector.** Edit the style rule. To allow room for the navigation bar at the top of the page, configure a 40px top margin.

```
header { background-color: #869DC7; color: #00005D;
        font-size: 150%; padding: 10px 10px 10px 155px;
        background-image: url(lighthouselogo.jpg);
        background-repeat: no-repeat; height: 130px;
        margin-top: 40px; }
```

3. Edit the HTML. The current page content is too short to showcase the fixed navigation bar. You will need to edit the content of the page to make it longer. Since this is a practice page, a quick way to get length is to copy and paste the h2 and paragraph code three or four times on the page within the main element.

Save the file and test in a browser. Your page should be similar to Figure 7.42. Scroll down the page and your display should be similar to Figure 7.43. The navigation bar is fixed in place even though the rest of the place scrolls up and down. A sample solution is in the student files chapter7/7.11 folder.

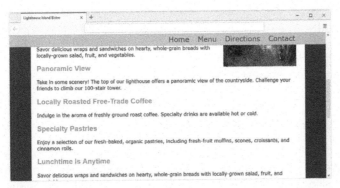

FIGURE 7.43 *The navigation bar stays in place while the content is scrolled.*

Fragment Identifiers

VideoNote

Linking to
a Named
Fragment

Browsers begin the display of a web page at the top of the document. However, there are times when you need to narrow down the target of the hyperlink and link to a specific portion of a web page. You can accomplish this by coding a hyperlink to a **fragment identifier** (sometimes called a named fragment or fragment id), which is simply an HTML element with an id attribute.

There are two components to your coding when using fragment identifiers:

1. The tag that identifies the named fragment of a web page. This tag must be assigned to an id. For example: `<div id="content">`

2. The anchor tag that links to the named fragment on a web page.

Fragment Identifiers and FQAs

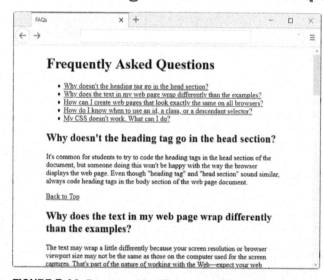

FIGURE 7.44 *Fragment identifiers are used on this page.*

Providing a hyperlink to a named fragment is often done on long web pages. For example, lists of frequently asked questions (FAQs) often use fragment identifiers to jump to a specific part of the page and display the answer to a question. You might see a "Back to Top" hyperlink that a visitor can click to cause the browser to quickly scroll the page up to the top for easy page navigation.

The web page in Figure 7.44 displays a list of FAQs that uses hyperlinks to fragment identifiers that jump to each question's answer when clicked. Explore the example page in the student files (chapter7/faqs.html).

Fragment Identifiers and Accessibility

Another use of named fragments helps to provide for accessibility. Web pages may have a fragment identifier to indicate the beginning of the actual page content. When the visitor clicks on the "Skip to content" hyperlink, the browser links to the named fragment and shifts focus to the content area of the page. The "Skip to content" or "Skip navigation" link provides a way for screen reader users to skip repetitive navigation links.

Figure 7.45 shows a web page with a "Skip to Content" hyperlink. When the user activates the "Skip to Content" link (by clicking on it with a mouse, tabbing over to it and pressing enter using a keyboard, or tapping on it using a touch device), the browser shifts the display to display the named fragment at or near the top of the browser viewport.

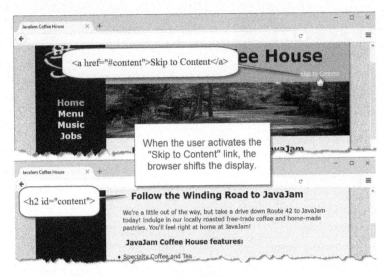

Working with a Fragment Identifier

Two separate tasks are required when configuring a hyperlink to a fragment identifier.

FIGURE 7.45 *The "skip to content" link in action.*

1. Establish the Target. Assign an HTML element to an id attribute. For example, create the "Skip to content" fragment identifier by configuring an element that begins the page content with an id, such as:

   ```
   <h2 id="content">
   ```

2. Reference the Target. At the point of the page where you want to place a hyperlink to the content, code an anchor element. Use the href attribute and place a # symbol (called a hash mark) before the name of the fragment identifier. The code for a hyperlink to the named fragment "content" follows:

   ```
   <a href="#content">Skip to Content</a>
   ```

The hash mark indicates that the browser should search for an id on the same page. If you forget to type the hash mark, the browser will not look on the same web page; it will look for an external file.

There may be times when you need to link to a named fragment on another web page. To accomplish this, place a # symbol followed by the fragment identifier id value after the file name in the anchor tag. For example, to link to a specific area, such as the Contact Info, configured with id="contact" in the index.html file from any other page on the same website, you could use the following HTML:

```
<a href="index.html#contact">Contact Info</a>
```

In the next section, you'll use fragment identifiers to configure the navigation on a single page website, which consists of a single long page with navigation hyperlinks at the top to specific areas on the page.

Single Page Website

Top Fixed Navigation
Home "page" with hero image
Tours "page"
Tours hero image
Rentals "page"
Rentals hero image
Contact "page"
Contact hero image
Bottom Fixed Footer

FIGURE 7.46 *SPW wireframe.*

Recall from Chapter 3 that a single page website (SPW) consists of one very long page (a single HTML file) with a clearly defined navigation area, typically at the top of the page. The navigation takes you to specific areas on the page because each hyperlink points to a specific element on the page indicated by a fragment identifier.

 Hands-On Practice 7.12 ──────────────

In this Hands-On Practice you'll configure an SPW with a fixed top navigation bar, fixed bottom footer, and four sections: home, tours, rentals, and contact. Each of these sections will function as a "page" in the SPW, as shown in Figure 7.46. Notice the fixed navigation is at the top of the browser viewport and the fixed footer is at the bottom of the browser viewport. The screen captures shown in Figure 7.47 are all using a single HTML file—a single page website.

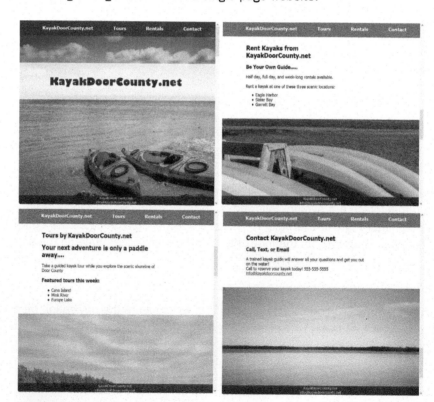

FIGURE 7.47 *Single page website.*

Create a new folder called ch7spw. Copy the files from the chapter7/spwstart folder into your ch7spw folder. The wireframe in Figure 7.46 shows the structural areas of the web page and the organization of the HTML. Observe the top fixed navigation area, Home "page" text content and hero image, Tours "page" text content, Tours hero image, Rentals "page" text content, Rentals hero image, Contact "page" text content, Contact hero image, and footer.

1. Launch a text editor and open your index.html file. Scroll down to the HTML and notice that there are comments indicating Home Page, Tours Page, Rentals Page, and Contact Page. Notice that each page area begins with a section element. Configure a named fragment for each area. Assign the Home Page section to the named fragment home by coding id="home" on the opening section tag. Assign the Tours Page section to the named fragment tours by coding id="tours" on the opening section tag. Assign the Rentals Page section to the named fragment rentals by coding id="rentals" on the opening section tag. Assign the Contact Page section to the named fragment contact by coding id="contact" on the opening section tag.

Continue editing the file and configure the navigation area. Locate the nav element below the opening body tag. Code an ordered list to configure the navigation links within the nav element. The text "KayakDoorCounty.net" will hyperlink to #home. The text "Tours" will hyperlink to #tours. The text "Rentals" will hyperlink to #rentals. The text "Contact" will hyperlink to #contact. The HTML is shown below:

```
<nav>
 <ul>
  <li><a href="#home">KayakDoorCounty.net</a></li>
  <li><a href="#tours">Tours</a></li>
  <li><a href="#rentals">Rentals</a></li>
  <li><a href="#contact">Contact</a></li>
 </ul>
</nav>
```

Save your file as index.html and display it in a browser. Your display should be similar to the first browser screen shot in Figure 7.47. Click the navigation hyperlinks view the other page sections. Scroll down and up the single page website to view all the information and view the photographs. You can compare your work to chapter7/7.12/index.html in the student files.

If you would prefer that the footer does not always appear in the browser viewport, edit the CSS and remove the style declarations for the position and bottom properties. Save your file and display in a browser. Notice how the footer only appears when you scroll all the way to the bottom of the web page. The student files (chapter7/7.12/index1.html) contains a sample solution.

If you'd like to create the effect of parallax scrolling in which the background images scroll at a different speed than the text content, edit the CSS and add the following style declaration to the style rules for the hero, tourshero, rentalshero, and contacthero class selectors: background-attachment: fixed;

Save your file and display in a browser. Notice how the images scroll at a different rate than the text content. The student files (chapter7/7.12/index2.html) contains a sample solution.

CHAPTER 7

Review and Apply

Review Questions

1. Which of the following pseudo-classes is the default state for a hyperlink that has already been clicked?
   - **a.** :hover
   - **b.** :link
   - **c.** :onclick
   - **d.** :visited

2. Which of the following is used to change the location of an element slightly in relation to where it would otherwise appear on the page?
   - **a.** relative positioning
   - **b.** static positioning
   - **c.** absolute positioning
   - **d.** fixed positioning

3. Which of the following properties can be used to clear a float?
   - **a.** float or clear
   - **b.** clear or overflow
   - **c.** position or clear
   - **d.** overflow or float

4. Which of the following causes an element not to display?
   - **a.** display: block;
   - **b.** display: 0px;
   - **c.** display: none;
   - **d.** display: inline;

5. Which of the following causes an element to display without empty space above and below?
   - **a.** display: block;
   - **b.** display: static;
   - **c.** display: none;
   - **d.** display: inline;

6. Which of the following is an image file that contains multiple small graphics?
   - **a.** thumbnail
   - **b.** snap
   - **c.** sprite
   - **d.** float

7. Which of the following configures a class called notes to float to the left?
   - **a.** .notes { left: float; }
   - **b.** .notes { float: left; }
   - **c.** .notes { float-left: 200px; }
   - **d.** .notes { position: float; }

8. Which of the following is the rendering flow used by a browser by default?
   - **a.** HTML flow
   - **b.** normal display
   - **c.** browser flow
   - **d.** normal flow

9. Which of the following is an example of using a descendant selector to configure the anchor tags within the nav element?
   - **a.** nav. a
   - **b.** a nav
   - **c.** nav a
   - **d.** #nav a

10. Which of the following is used along with the left, right, and/or top property to precisely configure the position of an element outside of normal flow?
    - **a.** position: relative;
    - **b.** position: absolute;
    - **c.** position: float;
    - **d.** absolute: position;

Review Answers

1. d 2. a 3. b 4. c   5. d   6. c   7. b   8. d   9. c   10. b

Hands-On Exercises

1. Write the CSS for an id with the following characteristics: fixed position, light gray background color, bold font weight, and 10 pixels of padding.

2. Write the CSS for an id with the following characteristics: float to the left of the page, light-beige background, Verdana or sans-serif large font, and 20 pixels of padding.

3. Write the CSS for an id that will be absolutely positioned on a page 20 pixels from the top and 40 pixels from the right. This area should have a light-gray background and a solid border.

4. Write the CSS for a class that is relatively positioned. This class should appear 15 pixels from the left. Configure the class to have a light-green background.

5. Create a web page about your favorite hobby, movie, or music group. Configure the text, color, and a two-column layout with CSS.

6. Write the HTML code to associate a web page with an external style sheet named myprint.css to configure a printed web page.

7. Write the HTML code to create a fragment identifier designated by "main" to indicate the main content area of a web page document.

8. Write the HTML code to create a hyperlink to the named fragment designated by an id named main.

9. Create a single page website about a favorite location—it could be somewhere you have gone on vacation or somewhere you would like to visit. Use the coding techniques in Hands-On Practice 7.12 as a guide. Make a list of things to do (activities) at this location. Make a list of things to see or famous places (sights) to visit at this location. The single page website should include three "pages"—Home, Activities, and Sights. Either use your own vacation photos or select relevant royalty-free photos from the Web. Use one photo for each "page." If you use photos from the Web, be sure to provide appropriate credit in the footer area. Include your name in an e-mail address in the page footer area. Save the file as location.html.

Focus on Web Design

There is still much for you to learn about CSS. A great place to learn about web technology is on the Web itself. Use a search engine to search for CSS page layout tutorials. Choose a tutorial that is easy to read. Select a section that discusses a CSS technique that was not covered in this chapter. Create a web page that uses this new technique. Consider how the suggested page layout follows (or does not follow) principles of design such as contrast, repetition, alignment, and proximity (refer back to Chapter 3). The web page should provide the URL of your tutorial, the name of the website, a description of the new technique you discovered, and a discussion of how the technique follows (or does not follow) principles of design.

Case Study

You will continue the case studies from Chapter 6 as you update the websites to configure columns and more sophisticated navigation.

wrapper
header
nav | hero image
main
contact id
footer

FIGURE 7.48 *Pacific Trails two-column page layout.*

Pacific Trails Resort Case Study

In this chapter's case study, you will use the Pacific Trails Resort existing website (Chapter 6) as a starting point to create a new version of the website that uses a two-column page layout. Other changes include configuring navigation links within an unordered list, configuring the header area text to be a hyperlink to the Home page, and reworking the text on the content pages. Figure 7.48 displays a wireframe with the new layout. The new Home page is shown in Figure 7.49.

You have five tasks in this case study:

1. Create a new folder for the Pacific Trails Resort website.
2. Edit the pacific.css external style sheet.
3. Edit the Home page (index.html).
4. Edit the Yurts page (yurts.html).
5. Edit the Activities page (activites.html).

Task 1: Create a folder called ch7pacific to contain your Pacific Trails Resort website files. Copy the files from the Chapter 6 Case Study ch6pacific folder. Copy the coast2.jpg image from the chapter7 folder.

Task 2: Configure the CSS. Launch a text editor and open the pacific.css external style sheet file. Edit the CSS as follows:

- **The universal selector.** Set the box-sizing property to border-box.

  `* { box-sizing: border-box; }`

- **The wrapper id selector.** Change the background color from white (#FFFFFF) to blue (#90C7E3). Configure a 1px solid dark blue (#000033) border. Copy the background image style declaration for the linear gradient from the body selector. This will display behind the navigation area.

- **The body element selector.** Change the background color to #EAEAEA. Remove the background-image and background-repeat style declarations.

- **The header element selector.** Remove the style declarations that configure the background image. Set height to 120px, top padding to 30px, and left padding to 3em.

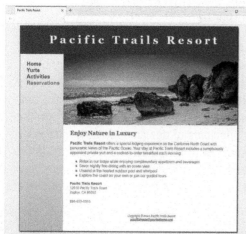

FIGURE 7.49 *The new Pacific Trails Home page with a two-column layout.*

- **The h1 element selector.** Configure 3em font size and 0.25em of letter spacing.

- **The nav element selector.** This is the area that will float on the page. Remove the background-color declaration—the nav area will pick up the background color of the `wrapper` id. Remove the text-align declaration. Change the padding to 1.5em. Set 120% font size. Configure left float and a width of 160 pixels.

- **The homehero id selector.** Configure a 190px left margin. Change the background image to coast2.jpg.

- **The yurthero id selector.** Configure a 190px left margin.

- **The trailhero id selector.** Configure a 190px left margin.

- **The main element selector.** Configure style declarations to set a white (#FFFFFF) background, 190 pixels of left margin, and change the left padding to 30px. To allow for the main element to contain floated elements, also set `overflow: auto;`

- **The section element selector.** Configure a style rule to set a left float, 33% width, 2em left padding, and 2em right padding.

- **Configure the unordered list in the main content area.** Replace the ul element selector with a descendant selector (`main ul`) to specify only ul elements within the main content.

- **The footer element selector.** Configure styles to set a 190 pixel left margin and white (#FFFFFF) background color.

- **Configure the navigation area.** Use descendant selectors to configure the unordered list and anchor elements *within the nav element*.

  - **Style the unordered list.** Configure the nav `ul` selector with no list markers, zero margin, zero left padding, and 1.2em font size.

  - **Style unvisited navigation hyperlinks.** Configure the `:link` pseudo-class with medium blue text color (#5C7FA3).

  - **Style visited navigation hyperlinks.** Configure the `:visited` pseudo-class with dark blue text color (#344873).

  - **Style interactive hyperlinks.** Configure the `:hover` pseudo-class with dark red text color (#A52A2A).

- **Configure hyperlinks in the header area.** Use descendant selectors to configure hyperlinks within the header element with no underline, white (#FFFFFF) text color for the `:link` and `:visited` pseudo-classes, and light blue (#90C7E3) text color for the `:hover` pseudo-class.

Save the pacific.css file. Check your syntax with the CSS validator (http://jigsaw.w3.org/css-validator). Correct and retest if necessary.

Task 3: Edit the Home Page. Launch a text editor and open the index.html file. Configure the navigation hyperlinks using an unordered list. Remove the ` ` special characters. Configure the "Pacific Trails Resort" text in the header area to be a hyperlink to the Home page (index.html). It should be similar to the page shown in Figure 7.49.

Task 4: Edit the Yurts Page. Launch a text editor and open yurts.html. Modify the page in a similar manner as the Home page. Examine the wireframe in Figure 7.50 and notice that there are three sections with the main element. Remove the tags that configure the description list from the page. Notice the text content is a series of questions and answers. Configure each question within an h3 element. Configure each answer within a paragraph element. Code a section element to contain each question and answer pair. Save your file and test it in a browser. It should be similar to the page shown in Figure 7.51.

Task 5: Edit the Activities Page. Launch a text editor and open activities.html. Modify the page in a similar manner as the Home page. Examine the wireframe in Figure 7.50 and notice that there are three sections with the main element. Code a section element to contain each pair of h3 and p elements. Save your file and test it in a browser. It should be similar to the overall page layout shown in Figure 7.51.

You have successfully implemented multiple columns in this case study. The Pacific Trails Resort website has a two-column page layout!

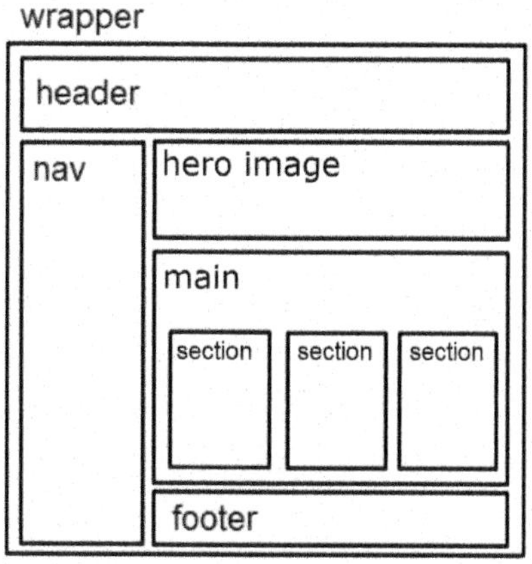

wrapper

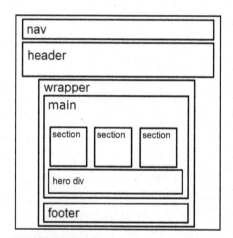

FIGURE 7.50 *Wireframe for Pacific Trails content pages.*

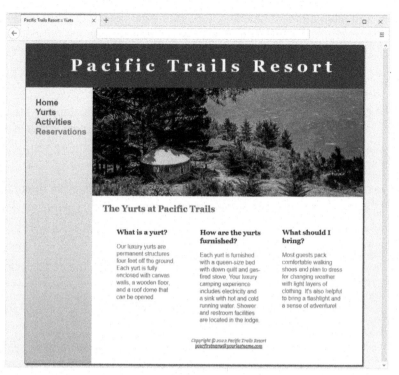

FIGURE 7.51 *Pacific Trails Yurts page.*

Path of Light Yoga Studio Case Study

In this chapter's case study, you will use the existing Path of Light Yoga Studio (Chapter 6) website as a starting point to create a new version of the website with a fixed top navigation bar. Other changes include configuring navigation links within an unordered list, configuring the header area text to be a hyperlink to the Home page, and reworking the text on the content pages. Figure 7.52 displays a wireframe with the new layout.

FIGURE 7.52 *Path of Light Yoga Studio content page layout.*

You have five tasks in this case study:

1. Create a new folder for the Path of Light Yoga Studio website.
2. Edit the yoga.css external style sheet.
3. Edit the Home page (index.html).
4. Edit the Classes page (classes.html).
5. Edit the Schedule page (schedule.html).

Task 1: Create a folder called ch7yoga to contain your Path of Light Yoga Studio website files. Copy the files from the Chapter 6 Case Study ch6yoga folder.

Task 2: Configure the CSS. Launch a text editor and open the yoga.css external style sheet file. Edit the CSS as follows:

- **The universal selector.** Set the box-sizing property to border-box.
  ```
  * { box-sizing: border-box; }
  ```

- **The header element selector.** Add a declaration to configure 50 pixels of top margin.
- **The nav element selector.** This is the area that will be fixed at the top of the page. Configure fixed position, set top to 0, set left to 0, and set z-index to 9999. Also configure 100% width and 50px height.
- **Configure the navigation area.** Use descendant selectors to configure the unordered list and anchor elements *within the nav element*.
  - **Style the unordered list.** Configure the `nav ul` selector with no list markers, zero margin, 2em right padding, and 1.2em font size.
  - **Style the unordered list items.** Configure the `nav li` selector with inline display and 2em left padding.
  - **Style unvisited navigation hyperlinks.** Configure the `:link` pseudo-class with #3F2860 text color.
  - **Style visited navigation hyperlinks.** Configure the `:visited` pseudo-class with #497777 text color.
  - **Style interactive hyperlinks.** Set #A26100 text color for the `:hover` pseudo-class.
- **Configure hyperlinks in the header area.** Use descendant selectors to configure hyperlinks within the header element with no underline. Configure styles to set white (#FFFFFF) text color for the `:link` and `:visited` pseudo-classes. Configure styles to set #EDF5F5 text color for the `:hover` pseudo-class.
- **The footer element selector.** Add a style declaration to clear right float.
- **The onethird class selector.** Configure a style rule to set a left float, 33% width, 2em left padding, and 2em right padding.
- **The onehalf class selector.** Configure a style rule to set a left float, 50% width, 2em left padding, and 2em right padding.
- **The home class selector.** Change the value of the height property from 40vh to 50vh (50% of the viewport height).
- **The content class selector.** Change the value of the height property from 200px to 250px.
- **The mathero id selector.** Add a declaration to clear floats.
- **The loungehero id selector.** Add a declaration to clear floats.

Save your yoga.css file. Use the CSS validator (http://jigsaw.w3.org/css-validator) to check your syntax. Correct and retest if necessary.

Task 3: Edit the Home Page. Open the index.html file in a text editor. Configure the "Path of Light Yoga Studio" text in the header area to be a hyperlink to the Home page (index.html). Configure the navigation hyperlinks using an unordered list. Remove the ` ` special characters. Save your file and test it in a browser. It should be similar to the page shown in Figure 7.53. When you

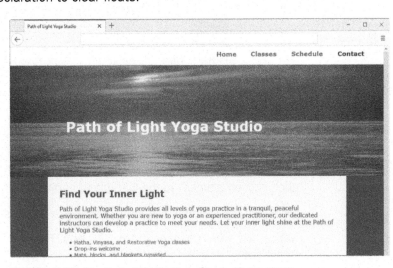

FIGURE 7.53 *The Path of Light Yoga Studio Home page.*

scroll the page vertically in the browser window, you should notice the fixed navigation bar.

Task 4: Edit the Classes Page. Launch a text editor and open the classes.html file. Modify the page in a similar manner as the Home page.

Examine the wireframe in Figure 7.52 and notice that there are three sections within the main element. Remove the tags that configure the description list from the page. Notice the text content is a series of yoga class titles and yoga class descriptions. Configure each yoga class title within an h3 element. Configure each yoga class description within a paragraph element. Code a section element to contain each yoga class title and yoga class description pair. Assign each section to the CSS class named `onethird`.

Save your file and test it in a browser. It should be similar to the page shown in Figure 7.54.

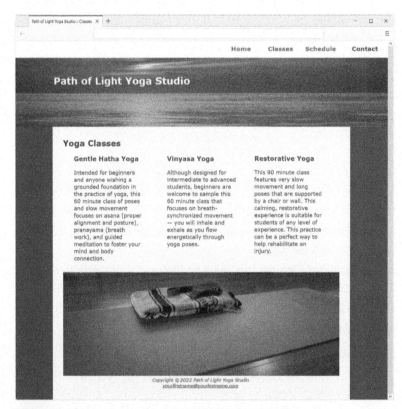

FIGURE 7.54 *The new Path of Light Yoga Studio Classes page.*

Task 5: Edit the Schedule Page. Launch a text editor and open the schedule.html file. Modify the page in a similar manner as the Home page.

View the wireframe in Figure 7.52 and notice that there are sections within the main element. This web page is a little different, it will only have two sections. Use Figure 7.55 as a guide. Code a section element to contain each pair of h3 and ul elements. Assign each section element to the CSS class named `onehalf`.

Save your file and test it in a browser. It should be similar to Figure 7.55.

Notice how small changes in the CSS and HTML added interest to the Path of Light Yoga Studio website. Interactive hyperlinks and the use of multiple columns for the text content create a more engaging visual experience.

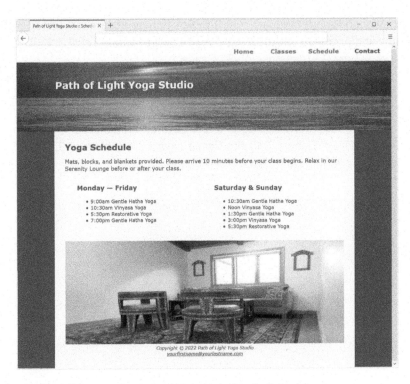

FIGURE 7.55 *The new Path of Light Yoga Studio Schedule page.*

Endnotes

1. "Cascading Style Sheets Level 2 Revision 2 (CSS 2.2) Specification: 9.5 Floats." Edited by Bert Bos, *W3C*, W3C, 12 Apr. 2016, www.w3.org/TR/CSS22/visuren.html#floats.

2. "Cascading Style Sheets Level 2 Revision 2 (CSS 2.2) Specification: 9.5.2 Controlling flow nest to floats: the 'clear' property." Edited by Bert Bos, *W3C*, W3C, 12 Apr. 2016, www.w3.org/TR/CSS22/visuren.html#flow-control.

3. "CSS Overflow Module Level 3." Edited by L. David Baron et al., *W3C CSS Working Group Editor Drafts*, W3C, 16 Jan. 2020, drafts.csswg.org/css-overflow-3/.

4. "CSS Intrinsic & Extrinsic Sizing Module Level 3." Edited by Tab Atkins et al., W3C, 22 May 2019, www.w3.org/TR/css-sizing-3/.

5. Meyer, Eric A. More Eric Meyer on CSS. New Riders, 2004.

6. Zeldman, Jeffrey. Designing with Web Standards (1st ed.). New Riders, 2003.

7. "CSS Display Module Level 3." Edited by Tab Atkins Jr. and Ellika J. Etemad, W3C, 11 July 2019, www.w3.org/TR/css-display-3/.

8. "Selectors Level 3." Edited by Tantek Çelik et al., W3C, 30 Jan. 2018, www.w3.org/TR/selectors-3/.

9. "Cascading Style Sheets Level 2 Revision 1 (CSS 2.1) Specification 13 Paged Media." Edited by Bert Bos et al., W3C, 7 June 2911, www.w3.org/TR/CSS2/page.html#pagebreaks.

10. Shea, Dave. "CSS Sprites: Image Slicing's Kiss of Death." *A List Apart*, 5 Mar. 2004, alistapart.com/article/sprites/.

11. "CSS Positioned Layout Module Level 3 6. Positioning Schemes." Edited by Rossen Atanassov and Arron Eicholz, *CSS Positioned Layout Module Level 3*, W3C, 17 May 2016, www.w3.org/TR/css-position-3/#pos-sch.

12. "CSS Positioned Layout Module Level 6 6. Box Offsets." Edited by Rossen Atanassov and Arron Eicholz, *CSS Positioned Layout Module Level 3*, W3C, 17 May 2016, www.w3.org/TR/css-position-3/#box-offsets-trbl.

13. "CSS Positioned Layout Module Level 3 11. Layered Presentation." Edited by Rossen Atanassov and Arron Eicholz, *CSS Positioned Layout Module Level 3*, W3C, 17 May 2016, www.w3.org/TR/css-position-3/#propdef-z-index.

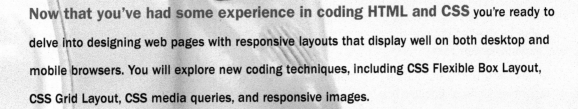

CHAPTER 8

Responsive Layout Basics

Now that you've had some experience in coding HTML and CSS you're ready to delve into designing web pages with responsive layouts that display well on both desktop and mobile browsers. You will explore new coding techniques, including CSS Flexible Box Layout, CSS Grid Layout, CSS media queries, and responsive images.

You'll learn how to...

▎ Describe the purpose of CSS Flexible Box Layout

▎ Configure a Flexbox Container and Flexbox Items

▎ Create a web page that applies CSS Flexible Box Layout

▎ Describe the purpose of CSS Grid Layout

▎ Configure a Grid Container

▎ Configure grid rows, grid columns, grid gaps, grid areas, and grid templates

▎ Create responsive page layouts with CSS Grid Layout

▎ Configure web pages for mobile display using the viewport meta tag

▎ Apply responsive web design techniques with CSS media queries

▎ Apply responsive image techniques with CSS

▎ Apply responsive image techniques with HTML using the img, picture, and source elements

CSS Flexible Box Layout

Since the early days of the Web, designers have striven to configure multicolumn web pages. Back in the 1990s, it was common to use HTML tables to configure a two- or three-column page layout. As browsers offered increased support for CSS, web developers discovered CSS float property techniques like the one you used in Chapter 7 to create the look of multicolumn pages. You will find many pages on the Web configured using CSS float techniques.

However, the quest for more robust and responsive multicolumn layout methods has continued. There are two new CSS layout systems that have recently gained widespread browser support: CSS Flexible Box Layout and CSS Grid Layout. This section introduces CSS Flexible Box Layout.

The purpose of **CSS Flexible Box Layout** (called **flexbox**) is to provide a flexible layout—elements contained within a flex container can be configured in *one* dimension (either horizontally or vertically) in a flexible manner with flexible sizing. In addition to changing the horizontal or vertical organization of elements, flexbox can also be used to change the order of display of the elements. Due to its flexibility, flexbox is well suited for responsive web design.

CSS Flexible Box Layout Module has reached W3C Candidate Recommendation status and is well supported by recent versions of popular browsers.[1]

Configure a Flexible Container

Flexbox is typically used to configure a specific area of a web page rather than the entire page layout. To configure an area on a web page that uses flexbox layout, you need to indicate the **flex container**, which is the element that will contain the flexible area.

The `display` Property

Use the CSS **`display` property** to configure a flex container. The value `flex` indicates a flexible block container. The value `inline-flex` indicates a flexible inline-display container.

For example, to configure an id named `gallery` as a flex container, code the following CSS:

```
#gallery { display: flex; }
```

Each child element of the flex container is a **flex item**. In the following HTML, each img tag is considered a flex item within the div element assigned to the `gallery` id.

```
<div id="gallery">
  <img src="bird1.jpg" width="200" height="150" alt="Red Crested Cardinal">
  <img src="bird2.jpg" width="200" height="150" alt="Rose-Breasted Grosbeak">
  <img src="bird3.jpg" width="200" height="150" alt="Gyrfalcon">
  <img src="bird4.jpg" width="200" height="150" alt="Rock Wren">
  <img src="bird5.jpg" width="200" height="150" alt="Coopers Hawk">
  <img src="bird6.jpg" width="200" height="150" alt="Immature Bald Eagle">
</div>
```

Figure 8.1 shows a page that uses flexbox to display an image gallery. By default, the flex area will have a horizontal flow direction and be configured as one horizontal row. If the content does not fit in the browser area, the browser may either try to reduce the size of some of the objects or display a scroll bar as shown in Figure 8.1. To try this out, launch a browser to display the example in the student files (chapter8/flex1.html).

Even though there are six images within the flexible area, the items in the flex area will not automatically wrap to another line if the browser window is not large enough to display them all. Next, let's explore a property that will correct this issue.

FIGURE 8.1 *A flex area with default properties.*

The `flex-wrap` Property

The **flex-wrap property** configures whether flex items are displayed on multiple lines. Values for this property include `nowrap`, `wrap`, and `wrap-reverse`. The default value is `nowrap`, which configures single-line display for horizontal flow flex containers and a single-column display for vertical flow flex containers. The value `wrap` will allow flex items to display on multiple lines for horizontal flow flex containers and to display on multiple columns for vertical flow flex containers. The `wrap-reverse` value provides for wrapping and displays the flex items in reverse order.

The flex items in Figure 8.2 (student files chapter8/flex2.html) wrap to the next line. The following CSS configures the flex container:

FIGURE 8.2 *Flex Items wrap to the next line.*

```
#gallery { display: flex;
           flex-wrap: wrap; }
```

The `flex-direction` Property

Configure the flow direction of the flex items with the **flex-direction property**. The value `row` is the default and configures a horizontal flow direction, `column` configures a vertical flow direction, `row-reverse` configures a horizontal flow with the flex items in reverse order, and `column-reverse` configures a vertical flow with the flex items in reverse order.

More About Flex Containers

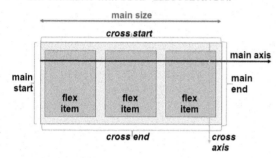

FIGURE 8.3 *Horizontal flow direction.*

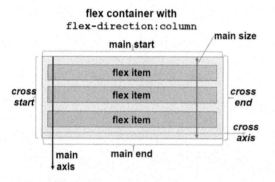

FIGURE 8.4 *Vertical flow direction.*

Flow Direction

Flex containers can be configured with either horizontal flow or vertical flow. Figure 8.3 shows a diagram of a flex container configured with horizontal flow direction. The **main size** is the width of the flex container content area. The **main axis** is the direction of the flow (in this case horizontal). The **main start** indicates the beginning of the flex area. The **main end** indicates the end of the flex area. The **cross axis** is the direction of the wrap (if any exists).

Figure 8.4 shows a diagram of a flex container configured with vertical flow direction. The **main size** is the height of the flex container content. The **main axis** is the direction of the flow (in this case vertical). The **main start** indicates the beginning of the flex area. The **main end** indicates the end of the flex area. The **cross axis** is the direction of the wrap (if any exists).

The `justify-content` Property

Use the **`justify-content` property** to configure how the browser should display extra space along the main axis in the flex container. Values for this property are shown in Table 8.1.

TABLE 8.1 *Values for the* `justify-content` *Property of a Flex Area*

Value	Purpose
`flex-start`	Default. Flex items begin at main start
`flex-end`	Flex items begin at main end
`center`	Flex items display centered in the flex container with equal empty space before the first flex item and after the last flex item
`space-between`	Flex items are evenly distributed in the flex container. The first flex item begins at main start. The last flex item is placed at main end
`space-around`	Flex items are evenly distributed in the flex container with space before the first flex item and after the last flex item

As you examine Figure 8.5, which displays a series of flex containers with horizontal flow, observe how each value of the `justify-content` property configures both the placement of the flex items and the space between the flex items. The student files (chapter8/flexj.html) contain the sample code.

Setting the `justify-content` property to `space-between` or `space-around` causes the browser to automatically calculate and display empty space between the flex items.

The `align-items` Property

The **align-items property** configures the way the browser displays extra space along the *cross-axis* of the container. Values include `flex-start`, `flex-end`, `center`, `baseline`, and `stretch`. The `align-items` property can be used along with the `justify-content` property to vertically and horizontally center content. For example, to configure a 400px high header element with a vertically and horizontally centered flex item (student files chapter8/flexc.html), code the following CSS:

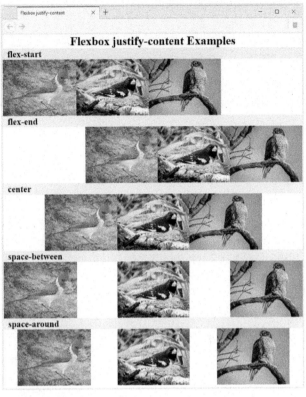

FIGURE 8.5 *The* `justify-content` *property.*

```
header { height: 400px; display: flex;
         justify-content: center;
         align-items: center; }
```

The `flex-flow` Property

The **flex-flow property** is a shorthand property that configures both the flex-direction and the flex-wrap. To configure an id named `demo` as a flexible container with a horizontal flow that wraps, code the following CSS:

```
#demo { display: flex; flex-flow: row wrap; }
```

Flexbox and the `gap` Property

The **gap**, **row-gap**, and **column-gap** properties configure space between rows (`row-gap`) and space between adjacent items (`column-gap`) using a numeric length or percentage value.[2] The shorthand gap property configures both the row-gap and column-gap. At the time this was written, only the Firefox and Chrome browsers supported the gap properties for flexbox, but wider support was expected soon. To configure an id named `demo` as a flexible container with a horizontal flow that wraps and has a 1em gap between rows and adjacent items, code the following CSS:

```
#demo { display: flex; flex-flow: row wrap; gap: 1em; }
```

Flexbox Image Gallery

  Hands-On Practice 8.1

You'll configure an image gallery with flexbox properties in this Hands-On Practice. Create a new folder called ch8flex1. Copy the starter1.html file from the chapter8 folder in the student files into your ch8flex1 folder. Copy the following files from the chapter8/starters folder into your ch8flex1 folder: bird1.jpg, bird2.jpg, bird3.jpg, bird4.jpg, bird5.jpg, and bird6.jpg.

1. Launch a text editor and open the starter1.html file. Add the following HTML below the opening main tag to create a div assigned to the `gallery` id that contains six images.

```
<div id="gallery">
  <img src="bird1.jpg" width="200" height="150" alt="Red Crested Cardinal">
  <img src="bird2.jpg" width="200" height="150" alt="Rose-Breasted Grosbeak">
  <img src="bird3.jpg" width="200" height="150" alt="Gyrfalcon">
  <img src="bird4.jpg" width="200" height="150" alt="Rock Wren">
  <img src="bird5.jpg" width="200" height="150" alt="Coopers Hawk">
  <img src="bird6.jpg" width="200" height="150" alt="Immature Bald Eagle">
</div>
```

The div is the flex container. Each img element is a flex item in the flex container. Save the file with the name index.html.

2. Edit the index.html file and configure CSS between the style tags in the head section. Configure an id named `gallery`. Set the `display` property to `flex`, `flex-direction` property to `row`, `flex-wrap` to `wrap`, and `justify-content` to `space-around`. The code follows:

```
#gallery { display: flex;
           flex-direction: row;
           flex-wrap: wrap;
           justify-content:
             space-around; }
```

Save the file and test it in a browser. Your page should look similar to Figure 8.6. Observe that while the browser configured empty space between the flex items on each row (the main axis), there is no empty space in the vertical (cross axis) area between each row element.

3. Next, you'll configure the flex items to have a margin, which will force some empty space

FIGURE 8.6 *The first version of the gallery.*

between the rows. Recall that a flex item is a child element of the flex container. In our page, each img element is a flex item. Edit the index.html file and code CSS above the closing style tag for the img selector that sets a 1em margin and a box-shadow.

```
img { margin: 1em;
      box-shadow: 10px 10px 10px #777; }
```

Save the file and test in a browser. As you resize your browser smaller and larger, your page should be similar to Figures 8.7, 8.8, and 8.9. A sample solution is in the student files chapter8/8.1 folder.

FIGURE 8.7 *Two rows of flex items.*

FIGURE 8.8 *Each row now has two items.*

FIGURE 8.9 *As the browser is resized, more items fit on the first row.*

You might be wondering why this hands-on practice didn't implement the gap property instead of the margin property to add empty space between flex items. At the time this was written, using the gap property with flexbox was not yet supported in all modern browsers. The student files (chapter8/8.1/gap.html) has a sample page that applies the gap property to add space between flex items.

As you display your image gallery page and resize the browser viewport, you may notice that the flexbox gallery is flexible and responsive to browser size, although the flex items are not necessarily displayed in a grid—that's what CSS Grid Layout can do and you'll explore that later in the chapter. In the next section, you'll delve more into configuring flexible sizes for flex items.

 FAQ How were image galleries configured before there was flexbox?

We relied on the float property or the inline-block property to configure image galleries before flexbox was developed. Wondering how this was done? Sample pages are in the student files chapter8/faq folder.

Configure Flex Items

By default, all items contained within a flex container are flexible in size and are allocated the same amount of display area in the flex container. Use the **flex property** to customize the size of each flex item and indicate whether it can grow (flex grow factor) or shrink (flex shrink factor) depending on the size of the browser viewport. The flex property can be set to the keyword `none`, the keyword `initial`, or a list of up to three values that configure the `flex-grow`, `flex-shrink`, and `flex-basis` properties. Table 8.2 describes these properties.

TABLE 8.2 *The* `flex` *Properties*

Flex Properties	Description
`flex-grow`	A positive number that determines the growth of the flex item relative to the other items in the flex container. Default value is 0.
`flex-shrink`	A positive number that determines how much the flex item will shrink relative to the other items in the flex container. Default value is 1.
`flex-basis`	Configures the initial dimension along the main axis of the flex item.

Value	Purpose
`content`	Indicates the width of the item's content.
`auto`	Default value, a specified width or if there is no specified width, the width of the item's content.
Positive numeric value	A value indicating the width of the item in units or percentage.

It's not always necessary to list all three values when configuring the flex property. Table 8.3 describes some common situations encountered when configuring a flex item.[3]

TABLE 8.3 *Flex Item Examples*

Flex Item Situation	Shorthand Notation	Equivalent
Fully Flexible Item *free space evenly distributed*	`flex: auto;`	`flex: 1 1 auto;`
Fully Inflexible Item	`flex: none;`	`flex: 0 0 auto;`
Partially Inflexible Item *shrinks to minimum size if needed*	`flex: initial;`	`flex: 0 1 auto;`
Proportional Flexible Item *The item takes up the specified proportion of free space in the container.*	`flex: positive number;` For example: `flex: 3;`	`flex: 3 1 0;`

Proportional Flexible Item

Let's focus on the last row of Table 8.3. One of the most powerful ways to use the flex property is to configure proportional flexible items. Setting one numeric value for the flex property sets the flex grow factor. If you configure an element with `flex: 2;` it will take up twice as much space within the container element as the others. Since the values work in proportion to the whole, you may find it helpful to use flex values that add up to 10. Examine the three-column page layout in Figure 8.10 and notice how the nav, main, and aside elements are organized in a row within another element that will serve as a flex container. The CSS to configure the proportion of the flexible area allocated to each column could be as follows:

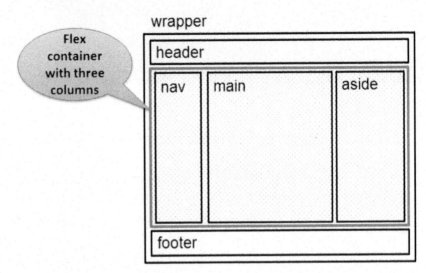

FIGURE 8.10 *Three-column page layout with the flex container indicated.*

```
nav    { flex: 1; }
main   { flex: 7; }
aside  { flex: 2; }
```

The `order` Property

Use the **order property** to display the flex items in a different order than they are coded. The order property accepts numeric values. The default value is 0.

Be aware that the W3C cautions that web designers should use the order property only for visual reordering. The change in order should not change the meaning or intent of the content because accessibility software such as screen readers will render the content in the order it was coded.

 Browsers applying the flexbox layout system ignore the float property when it is applied to a flex item. However, any floats that are applied to content within a flex item are still rendered by the browser.

In the next section, you'll get to practice configuring a flex container and flex items.

Practice with Flexbox

 Hands-On Practice 8.2 ————————————————

In this Hands-On Practice, you'll begin with a web page using the float layout technique from Chapter 7 and apply flexbox properties to configure a three-column layout similar to Figure 8.10.

Create a new folder called ch8flex2. Copy the starter2.html file from the chapter8 folder in the student files into your ch8flex2 folder. Copy the lighthouse.jpg and light.gif files from the chapter8/starters folder into your ch8flex2 folder.

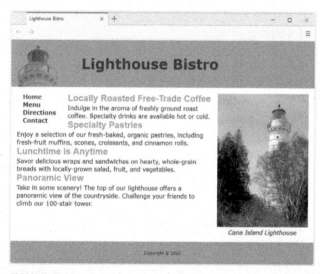

1. Open the starter2.html file in a browser. It should look similar to Figure 8.11. Launch a text editor and open the starter2.html file. Observe the HTML and notice that there is a div assigned to the id named `content` that contains the nav, aside, and main elements in that order.

FIGURE 8.11 *The web page before flexbox is configured.*

2. Your goal is to configure the layout of the `content` div with flexbox. Configure CSS with a flex container assigned to the id named `content`. The nav, main, and aside elements are children of the div and are the flex items. You'll configure them with different color backgrounds to emphasize the three columns. To prevent the nav element from growing in size, set the nav element's flex value to `none`. Set the main element's flex value to 6 and the aside element's flex value to 4. Add the following CSS below the opening style tag to configure the flex container and flex items:

```
#content { display: flex; }
nav      { flex: none;
           background-color: #B3C7E6; }
main     { flex: 6;
           min-width: 20em;
           background-color: #FFFFFF; }
aside    { flex: 4;
           background-color: #EAEAEA; }
```

Save the file with the name index.html and test in a browser. Your page should be similar to Figure 8.12. Notice that there are three columns but that the aside (area with lighthouse image) displays to the left of the main content text area because that is the order of the HTML. If this is how the owners of Lighthouse Bistro want, that's great. However, if they would prefer the main content text to display between the nav area and the aside area, you'll need to use the order property to change the order of the display of the flexbox items.

FIGURE 8.12 *Flexbox properties have been applied.*

3. Launch a text editor and open the index.html file. You will add CSS to configure the order of the left to right display of the flex items nav, main, and aside. You'll use the order property and assign values for each flex item. The CSS follows:

```
nav    { order: 1; }
main   { order: 2; }
aside  { order: 3; }
```

Save the file and test in a browser. Your page should be similar to Figure 8.13. A sample solution is in the student files chapter8/8.2 folder.

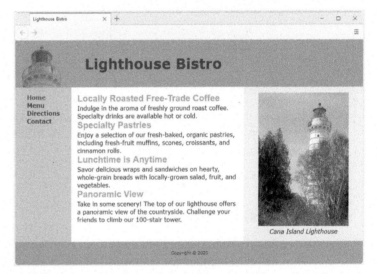

FIGURE 8.13 *The flexbox* order *property has been applied.*

You've just been introduced to flexbox, but there is more to explore. Check out these resources:

▶ http://css-tricks.com/snippets/css/a-guide-to-flexbox/

▶ https://developer.mozilla.org/en-US/docs/Web/CSS/CSS_Flexible_Box_Layout

▶ https://scotch.io/tutorials/a-visual-guide-to-css3-flexbox-properties

CSS Grid Layout

You have used both the CSS float property and CSS Flexible Box Layout (flexbox) to create multicolumn web pages. There is another new emerging layout system: CSS Grid Layout. The purpose of **CSS Grid Layout** is to configure a two-dimensional grid-based layout. The grid can be created as either fixed-size or flexible and contains one or more grid items that can be individually defined as fixed-size or flexible. Unlike flexbox which is intended for one-dimensional page layout, CSS Grid Layout is optimized for two-dimensional page layout.

CSS Grid Layout has reached W3C Candidate Recommendation status and is well-supported by recent versions of popular browsers.[4] Browsers that do not support grid layout ignore the style rules associated with grid properties.

Configure a Grid Container

To configure an area on a web page that uses CSS Grid Layout, you need to define the **grid container**, which is the element that will contain the grid area.

The `display` Property

Use the CSS **`display` property** to configure a grid container. The value `grid` indicates a block container. The value `inline-grid` indicates an inline-display container. For example, to configure an id named `gallery` as a grid container, code the following CSS:

```
#gallery { display: grid; }
```

Designing a Grid

A **grid** is comprised of horizontal and vertical **grid lines** that delineate **grid rows** and **grid columns** (generically referred to as **grid tracks**). A **grid cell** (which is a **grid item**) is the intersection of a grid row and a grid column. A **grid area** is a rectangle that can contain one or more grid items. A **grid gap** is optional and indicates an empty area or gutter between items in the grid container.

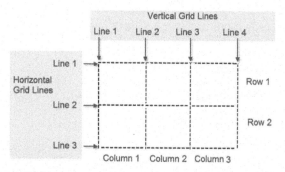

A first step is to visualize the grid, usually by sketching out the layout. Figure 8.14 shows a wireframe for a grid that shows grid lines, three columns, and two rows. A grid of this type could be used to display an image gallery.

FIGURE 8.14 *A grid with three columns and two rows.*

Each child element of the grid container is a **grid item**. In the following HTML, each img element within the gallery div is considered to be a grid item.

```
<div id="gallery">
  <img src="bird1.jpg" width="200" height="150" alt="Red Crested Cardinal">
  <img src="bird2.jpg" width="200" height="150" alt="Rose-Breasted Grosbeak">
  <img src="bird3.jpg" width="200" height="150" alt="Gyrfalcon">
  <img src="bird4.jpg" width="200" height="150" alt="Rock Wren">
  <img src="bird5.jpg" width="200" height="150" alt="Coopers Hawk">
  <img src="bird6.jpg" width="200" height="150" alt="Immature Bald Eagle">
</div>
```

Configure Grid Columns and Grid Rows

A basic method to configure grid rows and columns is to use the **grid-template-columns property** and the **grid-template-rows property** to tell the browser how to reserve space for the columns and rows in the grid. These properties accept a variety of values which will be introduced in the next section. In this example, we'll use pixel units.

In our image gallery example, we'll configure the grid-template-columns property to display three columns with a fixed width in two rows with a fixed height. The CSS to configure the grid follows:

```
#gallery { display: grid;
           grid-template-columns: 220px 220px 220px;
           grid-template-rows: 170px 170px; }
```

The code above explicitly creates a grid with three columns and two rows. Figure 8.15 shows a display of this grid in the browser (see the student files chapter8/grid1.html).

Observe that this very basic grid is fixed—it does not change when you resize the browser window. Grid layout becomes powerful when it is flexible, and the grid can change dimensions based on the browser viewport.

The empty space between the grid items in Figure 8.15 is configured by setting the size of the rows and columns to be larger than the size of the images in the image gallery.

FIGURE 8.15 *A basic grid.*

For this basic image gallery grid, other methods could have been used to configure empty space between items. One alternate method would be to set padding and/or margin for the img element selector. Another alternate method is to create a grid gap or gutter between items, which will be introduced in the next section along with techniques to configure a flexible grid.

Grid Columns, Rows, and Gap

You've observed that the `grid-template-columns` and `grid-template-rows` properties with pixel unit values inform the browser to reserve space for each row and column in a grid. Table 8.4 lists other commonly used values for these properties.[5]

TABLE 8.4 *Commonly Used Values to Configure Columns and Rows*

Value	Description
numeric length unit	Configures a fixed size with a length unit such as px or em Example: `220px`
numeric percentage	Configures a percentage size; Example: 20%
numeric `fr` unit	Configures a **flex factor** unit (denoted by `fr`) that directs the browser to allocate a fractional part of the remaining space
`auto`	Configures a size to hold the maximum content
`minmax` (min, max)	Configures a size range greater or equal to min value and less than or equal to max value. The max can be set to a flex factor
`repeat` (repetition amount, format value)	Repeats the column or row the number of times specified by the repetition amount numeric value or keyword and uses the format value to configure the column or row. The `auto-fill` keyword indicates to repeat but stop before an overflow Example: `repeat(autofill, 250px)`

Grid Layout and the `gap` Property

The **gap**, **column-gap**, and **row-gap** properties inform the browser to provide empty space or gutters between grid tracks. Table 8.5 describes the properties and their purpose.

TABLE 8.5 *Gap Properties*

Property	Description
`column-gap`	Value: Numeric length or percentage Defines a gap between the columns in a grid or flexbox container
`row-gap`	Value: Numeric length or percentage Defines a gap between the rows in a grid or flexbox container
`gap`	Value: `row-gap` value `column-gap` value Shorthand property. Providing just one value sets both the `row-gap` and the `column-gap`

The `order` Property

Use the **`order` property** to display the grid items in a different order than they are coded. The order property accepts a numeric value. The default value is 0. Be aware that the W3C cautions that web designers should use the order property only for visual reordering. The change in order should not change the meaning or intent of the content because accessibility software such as screen readers will render the content in the order it was coded.

Hands-On Practice 8.3

You'll explore two more ways to configure the image gallery grid displayed in Figure 8.15 in this Hands-On Practice. Create a new folder called ch8grid1. Copy the starter3.html file from the chapter8 folder in the student files into your ch8grid1 folder. Copy the following files from the chapter8/starters folder into your ch8grid1 folder: bird1.jpg, bird2.jpg, bird3.jpg, bird4.jpg, bird5.jpg, and bird6.jpg.

1. Launch a text editor and open the starter3.html file. Review the HTML and note that it contains a div assigned to the `gallery` id with six img elements for your image gallery. The div is the grid container. Each img element is a grid item since it is a child element of the div. Save the file with the name index.html.

2. Edit the index.html file and configure CSS between the style tags in the head section. Configure an id named `gallery`. Set the `display` property to `grid`. To divide the available browser space into three columns of 200 pixels each, set the `grid-template-columns` property to `repeat(3, 200px)`. To cause the browser to automatically generate rows as needed, set the `grid-template-rows` property to `auto`. Configure the base size of the gutters between row and column tracks by setting the `grid-gap` (and `gap`) properties to 2em. The CSS code follows:

```
#gallery { display: grid;
           grid-template-columns: repeat(3, 200px);
           grid-template-rows: auto;
           grid-gap: 2em; gap: 2em; }
```

Save the file and test it in a browser. Your page should look similar to Figure 8.15 (student files chapter8/8.3/a.html).

FIGURE 8.16 *The grid stretches as you widen the browser.*

3. Configure the image gallery grid to be responsive and automatically change the number of columns and rows displayed as the browser viewport is resized. Use the `auto-fill` keyword in the `repeat()` function to direct the browser to fill the viewport with as many columns as it can without overflowing. Edit the index.html file and change `repeat(3, 200px)` to `repeat(auto-fill, 200px)`.

Save the file and test it in a browser. Your page will look like Figure 8.15 when the browser viewport is just large enough to display three images in a row. As you widen the browser viewport, more columns will display in a single row, similar to Figure 8.16. As you narrow the browser viewport, the number of columns will decrease as shown in Figure 8.17. A sample solution is in the student files (chapter8/8.3/index.html).

FIGURE 8.17
Responsive grid.

Two-Column Grid Page Layout

Figure 8.18 shows a sample grid wireframe for a two-column page layout with grid lines, rows, and columns indicated. Recall that the first step in configuring CSS Grid Layout is to create a wireframe.

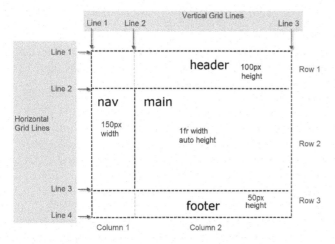

FIGURE 8.18 *Two-column CSS Grid Layout.*

Configure Grid Columns and Rows

Using the wireframe as a guide, configure the `grid-template-columns` and `grid-template-rows` properties. Recall that values include pixel units, percentages, keywords such as `auto`, and flex factor units. A new unit, `fr`, denotes a flex factor and directs the browser to allocate portions of the remaining space. Use `1fr` to allocate all remaining space.

The grid wireframe shown in Figure 8.18 contains two columns and three rows.

▶ The header is 100px in height, takes up the first row, and spans two columns.

▶ The nav is 150px wide and is located in the first column of the second row.

▶ The main content is located in the second column of the second row and needs to be large enough to hold whatever content is provided. The width of the main content is `1fr` and occupies all available space left after the nav element is rendered with a 150px width. The height of the main content is the `auto` value and will expand to contain whatever content is present.

▶ The footer takes up the third row, is 50px in height, and spans three columns.

Now let's configure the grid. Code styles for an id named `mygrid`. Set the `display` property to `grid`. Use the `grid-template-columns` property to set the first column to 150px and the second column to `1fr`. Use the `grid-template-rows` property to set the first row to 160px, the second row to `auto`, and the third row to `auto`. The CSS follows:

```
#mygrid { display: grid;
          grid-template-columns: 150px 1fr;
          grid-template-rows: 100px auto auto; }
```

Configure Grid Items

The next step is to indicate what elements should be placed in each grid item and grid area. There are a variety of techniques than can be used to configure grid items. We'll focus on using the grid-row and grid-column properties. The **grid-row property** configures the area in rows that is reserved for the item in the grid. The **grid-column property** configures the area in columns that is reserved for the item in the grid. A variety of values are accepted by these properties, such as grid line numbers and grid line names.[6]

Grid Line Numbers

In this example we will assign the `grid-row` and `grid-column` properties for each grid item with a starting grid line number and an ending grid line number separated by a / character. View Figure 8.18 and observe that the header area begins at vertical grid line 1 and ends at vertical grid line 3 in the first row (the grid track between horizontal grid line 1 and horizontal grid line 2). The CSS to configure the header is:

```
header { grid-row: 1 / 2;
         grid-column: 1 / 3; }
```

Each grid item shown in Figure 8.18 is configured in a similar manner. Configure the navigation, main, and footer elements with the following CSS:

FIGURE 8.19 *Web page with CSS Grid Layout.*

```
nav    { grid-row: 2 / 3; grid-column: 1 / 2; }
main   { grid-row: 2 / 3; grid-column: 2 / 3; }
footer { grid-row: 3 / 4; grid-column: 1 / 3; }
```

Figure 8.19 shows a web page that uses this grid layout, which you will create in the next Hands-On Practice. Let's first review below how to configure a grid page layout with grid line numbers.

Configure a Grid Page Layout with Grid Line Numbers

To create a grid page layout using grid line numbers, configure the following:

1. Use the display property to declare the grid container. Each child element of the grid container will be an item in the grid. If needed, use the grid-template-columns and grid-template-rows properties to specify the row and column dimensions. Recall that the browser will use the content to automatically determine the column width and row height unless you otherwise configure it.

2. Using your wireframe sketch as a guide, carefully determine the vertical grid lines and horizontal grid lines. Use the grid-rows and grid-column properties to indicate the vertical and horizontal grid lines for each grid item in the grid.

In this Hands-On Practice you'll configure a web page with a two-column grid layout. Use the grid wireframe shown in Figure 8.20 to create the layout for the web page shown in Figure 8.21.

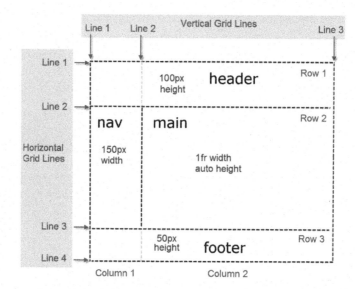

FIGURE 8.20 *Grid two-column layout.*

FIGURE 8.21 *Web page using grid layout.*

Create a new folder called ch8grid2. Copy the starter4.html file from the chapter8 folder in the student files into your ch8grid2 folder. Copy light2.jpg file from the chapter8/starters folder into your ch8grid2 folder.

1. Launch a browser and display your starter4.html file. It should look similar to Figure 8.22—which shows the web page configured with float methods from Chapter 7 and is not yet using grid layout.

2. Open your starter4.html file in a text editor. Save the file as index.html.

   Our grid layout will follow the two-column grid wireframe in Figure 8.20. Use the horizontal grid lines to indicate the rows. Use the vertical grid lines to indicate the columns.

   You will edit the CSS in the head section. Add style rules in the head section to configure an id named `wrapper` as a grid container with a two-column, three-row grid layout. The CSS follows:

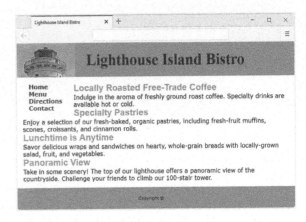

FIGURE 8.22 *The web page without grid layout.*

```
#wrapper { display: grid;
          grid-template-columns: 150px 1fr;
          grid-template-rows: 100px auto 50px; }
```

3. Next, use the wireframe in Figure 8.20 as a guide to configure the HTML element selectors and their starting and ending grid lines. Use the horizontal grid line numbers to configure the `grid-row` property. Use the vertical grid line numbers to configure the grid-column property. The CSS follows:

```
header { grid-row: 1 / 2; grid-column: 1 / 3; }
nav    { grid-row: 2 / 3; grid-column: 1 / 2; }
main   { grid-row: 2 / 3; grid-column: 2 / 3; }
footer { grid-row: 3 / 4; grid-column: 1 / 3; }
```

Save your file. Launch a browser that supports grid layout and test your page. Your page should look similar to Figure 8.21. A sample solution is in the student files chapter8/8.4 folder.

As you completed this Hands-On Practice you may have noticed that it is a bit tedious to determine and code all those vertical and horizontal grid line numbers. Perhaps you wondered if there is another method to configure a grid with CSS. The next section introduces you to a method to configure a grid template without all those line numbers!

Browsers applying the grid layout system ignore the float property when it is applied to a grid item. However, any floats that are applied to content within a grid item are still rendered by the browser.

Layout with Grid Areas

Recall that a grid area is a rectangle that can contain one or more grid items. The rectangle is bounded by grid lines. In our previous examples, you have configured grid areas by indicating grid line numbers. CSS Grid Layout provides a way to name the grid areas which eliminates keeping track of all those grid line numbers! Let's take a closer look.

The grid-area Property

FIGURE 8.23 A grid wireframe.

The **grid-area property** associates a grid item with a named area of the grid.[7] The wireframe in Figure 8.23 specifies the layout for the web page shown in Figure 8.24. Note the four grid areas in Figure 8.23: header, nav, main, and footer.

Use the grid-area property to associate a CSS selector (either an HTML element selector, class selector or id selector) with each of these named grid areas.

We will use HTML element selectors in this example. To associate the HTML elements header, nav, main, and footer with uniquely named grid areas, code the following CSS:

```
header { grid-area: header; }
nav    { grid-area: nav; }
main   { grid-area: main; }
footer { grid-area: footer; }
```

The grid-template-areas Property

FIGURE 8.24 Web page configured with grid template areas.

Use the **grid-template-areas property** to visually indicate the placement of the named grid areas on the grid.[8] The syntax is different from what you have used before:

▶ The value of this property is a series of strings that contain named grid areas.

▶ Each string is a row on the grid.

▶ The number of named grid areas in the string indicates the number of columns in the row.

▶ If a column in a row should be skipped, use the period "." symbol.

▶ Each row must indicate the same number of columns.

To configure the grid in Figure 8.23, start out by declaring the grid with the `display` property and assigning the `grid-template-columns` and `grid-template-rows` properties.

Then, use the `grid-template-areas` property to write the named grid area values row by row within quotation marks. The first row is the header named area which takes up two columns. The second row contains the nav area in the first column and the main area in the second column. The third row contains the footer area across two columns. The CSS follows:

```
#wrapper { display: grid;
          grid-template-columns: 150px 1fr;
          grid-template-rows:    100px auto 50px;
          grid-template-areas:

                    "header header"
                    "nav    main"
                    "footer footer"; }
```

The web page shown in Figure 8.24 uses this layout (also chapter8/grid/area.html in the student files).

Configuring a Grid with an Empty Area

It is possible to have empty areas within a grid layout, as shown in the wireframe in Figure 8.25. The third row indicated the footer area in the second column and nothing in the first column. This a situation where the "." symbol comes in handy. Figure 8.26 shows the web page created with this layout (also in the student files chapter8/grid/area2.html).

The following CSS will configure the `grid-template-areas` property for the wireframe:

```
#wrapper {
    display: grid;
    grid-template-columns: 150px 1fr;
    grid-template-rows:    100px auto 50px;
    grid-template-areas:   "header header"
                           "nav main"
                           ". footer" ;

    }
```

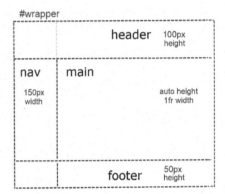

FIGURE 8.25 *This grid layout reserves empty space in the third row.*

FIGURE 8.26 *Web page with empty space in the third row.*

#wrapper

	header	100px height
nav	main	aside
150px width	auto height 1fr width	30% width
	footer	50px height

FIGURE 8.27 *Three-column grid layout.*

FIGURE 8.28 *The web page configured with the* grid-template-areas *property.*

You will work with grid template areas in this Hands-On Practice. Create a new folder called ch8grid3. Copy the starter5.html file from the chapter8 folder in the student files into your ch8grid3 folder. Copy header.jpg and scenery.jpg files from the chapter8/starters folder into your ch8grid3 folder. You will configure two grid layout for the wireframe in Figure 8.27. The web page is displayed in Figure 8.28.

Launch a text editor and open the starter5.html file. Save the file with the name index.html. You will edit the CSS in the head section: use the grid-area property to associate HTML elements with named grid areas, declare the grid, use the grid-template-columns and grid-template-rows properties to indicate the row and column dimensions, and use the grid-template-areas property to describe the layout of the grid.

1. With the Figure 8.27 wireframe in mind, add style rules in the head section to associate the HTML elements with the named grid areas (header, nav, main, aside, and footer) using the grid-area property. The CSS follows:

```
header { grid-area: header; }
nav    { grid-area: nav; }
main   { grid-area: main; }
aside  { grid-area: aside; }
footer { grid-area: footer; }
```

2. Configure a grid layout for the Figure 8.27 wireframe. Add style rules in the head section to configure an id named wrapper as a grid container with a three-column, three-row grid layout. Specify three columns (width set to 150px, 1fr, and 30%) with the grid-template-columns property.

Specify three rows (height set to 100px, auto, and 50px) with the grid-template-rows property. Also configure the grid-template-areas property with the placement of the header, nav, main, aside, and footer grid areas indicated. The CSS follows:

```
#wrapper { display: grid;
        grid-template-columns: 150px 1fr 30%;
        grid-template-rows: 100px auto 50px;
        grid-template-areas:
                    "header header header"
                    "nav    main    aside"
                    "footer footer footer" ; }
```

Save the file. Launch a browser that supports grid layout and test your page. Your page should look similar to Figure 8.28. A sample solution is in the student files (chapter8/8.5/index.html).

The `grid-template` Property

The **`grid-template` property** is a shorthand property that combines the `grid-template-areas`, `grid-template-rows`, and `grid-template-columns` properties.[9] The value of this property begins with a series of strings for each row followed by the height of each row. The string for each row is the named grid areas for that row enclosed in quotation marks. The width of the columns is indicated on a final row that begins with a "`/`" symbol.

To create the grid for the wireframe in Figure 8.27 (corresponding web page in Figure 8.28), the following steps are involved:

1. Configure each named grid area with the `grid-area` property.
2. Identify the `#wrapper` selector as the grid container.
3. Configure the `grid-template` property based on the Figure 8.27 wireframe. The first row is the header area which takes up three columns and is 100px in height. The second row contains the nav area in the first column, main area in the second column, and the aside area in the third column. Configure the height of this row as `auto` to use up all available space. The third row contains the footer area across three columns and is 50 pixels high. Note that the named areas that comprise each row are contained within quotation marks. The last row begins with a "`/`" symbol followed by the actual values for the column widths (in this case 150px, 1fr, and 30%).

The CSS follows:

```
header { grid-area: header; }
nav    { grid-area: nav; }
main   { grid-area: main; }
aside  { grid-area: aside; }
footer { grid-area: footer; }

#wrapper { display: grid;
           grid-template:
                  "header header header" 100px
                  "nav main aside " auto
                  "footer footer footer" 50px
                  / 150px 1fr   30%;
```

A sample page with this coding technique is in the student files (chapter8/8.5/grid3.html). This shorthand technique provides a convenient way to indicate not only the locations of the named grid areas but also the dimensions of the columns and rows in one property. There are many different methods available for configuring a CSS grid layout system. The rest of this chapter configures grid page layout with the `grid-template` property and named grid areas.

Progressive Enhancement with Grid

A design strategy for using grid layout is to first configure the web page layout so it displays well in nonsupporting browsers, next use a new technique called a CSS feature query to check for grid support, and then configure the grid layout to be used by supporting browsers.

CSS Feature Query

A **feature query** is a conditional that can be used to test for support of a CSS property, and if support is found, apply the specified style rules.[10] A feature query is coded using the **@supports() rule**. You code the property and value you are checking for within the parentheses. For example, to check for grid layout support, code the following CSS:

```
@supports ( display: grid) {
}
```

All the style rules needed for grid layout are coded between the "{" opening brace and the "}" ending brace. You'll get some experience with this in the next Hands-On Practice.

 Hands-On Practice 8.6 ———————————————————————

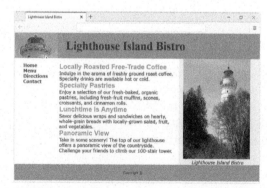

FIGURE 8.29 *The web page displayed in a browser*

You'll use a feature query to progressively enhance an existing web page with grid layout in this Hands- On Practice. Create a new folder called ch8feat Copy the starter6.html file from the chapter8 folder in the student files into your ch8feat folder. Copy the lighthouse.jpg and light2.jpg files from the chapter8/starters folder into your ch8feat folder.

1. Open your starter6.html file in a browser and the display should be similar to Figure 8.29, which is the display before any code for grid layout is added. This is a three-column layout with the nav, main, and aside elements floating next to each other.

2. Next, launch a text editor and open the starter6.html file. Our grid layout will follow the wireframe in Figure 8.30. Notice that there are three rows and three columns. The header takes up the entire first row. The nav, main, and aside elements are in the second row. The footer takes up

the entire third row. Observe Figure 8.30 and note the heights and widths indicated. You will add an `@supports` rule to the CSS before the ending style tag to check for grid support. You will place code to configure a three-column grid layout. The CSS follows:

```
@supports (display: grid) {
        grid-template:
            "header header header" 100px
            "nav main aside" auto
            "footer footer footer" 50px
            / 150px 1fr 300px ; }
    header { grid-area: header; }
    nav    { grid-area: nav; }
    main   { grid-area: main; }
    aside  { grid-area: aside; }
    footer { grid-area: footer; }

}
```

FIGURE 8.30 *Grid layout.*

Save your file and test it in a browser that supports grid layout. Your page should look similar to Figure 8.31. Notice that the page looks a bit odd with the main content area text taking up only part of the middle area.

3. Open your file in a text editor and notice that the main element selector has 50% width set— this is causing the awkward display in Figure 8.31. You need to rework this when grid layout is implemented. It's easy to do this by adding a style rule to the `@supports` feature query that resets the width. Add the following style rule to the styles within the `@supports` feature query:

FIGURE 8.31 *The layout needs improvement.*

```
main { width: 100%; }
```

Save your file. Launch a browser that supports grid layout and test your page. It should look similar to Figure 8.29. Be mindful that sometimes when you work with this technique you might also need to reset margins or padding—it depends on the specific web page and CSS that was originally coded. Note that we didn't need to write any new styles resetting the float property because CSS grid layout ignores the float property.

To recap, we applied the principle of progressive enhancement. We began with a web page that was configured with old-fashioned three-column layout using the `float` property. Next, we configured grid layout within a feature query (which nonsupporting browsers will ignore). Then we looked for any styles that were causing a display issue (the `width` property for the main element in this case) and coded new styles within the feature query to correct the display. The result is a web page that looks good on both supporting and nonsupporting browsers. A sample solution is in the student files chapter8/8.6 folder.

You've just been introduced to grid layout, but there is more to explore. Check out these resources: https://css-tricks.com/snippets/css/complete-guide-grid/ and https://developer.mozilla.org/en-US/docs/Web/CSS/CSS_Grid_Layout.

Centering with Flexbox and Grid

FIGURE 8.32 *Centered text.*

In Chapter 6, you were introduced to the technique of horizontally centering a block display element with CSS by setting its margin property to the value `auto`. However, until flexbox and grid layout, it has been difficult to vertically center an element within a browser viewport. Figures 8.32 shows a web page with text that is centered both vertically and horizontally. Locate the page in the student files (chapter8/center.html) and display it in a browser. As you resize the browser viewport, the text remains both vertically and horizontally centered.

To accomplish this layout, configure the container element with the following CSS:

- ▶ `display: flex;`
- ▶ `min-height: 100vh;` (This indicates 100% of the viewport height.)
- ▶ `justify-content: center;`
- ▶ `align-items: center;`
- ▶ *Optional:* `flex-wrap: wrap;` (This will allow multiple centered flex items.)

 Hands-On Practice 8.7 ───────────────────────────────

You'll explore creating web pages with both horizontally and vertically centered content in this Hands-On Practice. Create a new folder called ch8center. Copy the template.html file from the chapter8 folder in the student files into your ch8center folder. Copy the lake.jpg file from the chapter8/starters folder into your ch8center folder.

1. Launch a text editor and open your template.html file. Change the title of the web page to "Centered Heading". Edit the HTML and configure a header element, h1 element, and main element between the opening and closing body tags. The code follows:

   ```
   <header>
     <h1>Centered Heading</h1>
   </header>
   <main>
     Additional page content and navigation go here
   </main>
   ```

2. Continue editing the file and configure the CSS. Code opening and closing style tags in the head section. Next, code style rules within the style tags. Configure the body element selector with zero margin. Configure the header element selector as a flex

container with `justify-content` set to `center`, `align-items` set to `center`, minimum height `100vh`, and `#227093` background color for the header element selector. Configure the h1 element selector with white, Arial font. The code follows:

```
<style>
    body { margin: 0; }
    header { display: flex; min-height: 100vh;
             justify-content: center; align-items: center;
             background-color: #227093; }
    h1 { color: #FFFFFF; font-family: Arial, sans-serif; }
</style>
```

Save your file as index.html and display it in a browser. Your display should be similar to Figure 8.32. Resize the browser window and see how the h1 text remains centered in the viewport. Scroll down the page to see the text within the main element. You can compare your work to chapter8/center.html in the student files.

3. Next, add a background image that covers the entire browser viewport. Open index.html in a text editor and add styles for the header element selector that will configure the lake.jpg background image with 100% size that does not repeat. Remove the style rule for the background color. The new styles are shown in blue:

FIGURE 8.33 *Centered text with background image.*

```
header { display: flex; min-height: 100vh;
         justify-content: center; align-items: center;
         background-image: url(lake.jpg);
         background-size: 100% 100%;
         background-repeat: no-repeat; }
```

Save your file and display it in a browser. Your display should be similar to Figure 8.33. As you resize the browser window, notice how the h1 text remains centered and the background image changes in size. Scroll down the page to see the text within the main element. You can compare your work to chapter8/8.7/index.html in the student files.

4. There is another way to accomplish this layout: when grid or flex layout is being used, setting the margin property to `auto` causes the browser to both vertically and horizontally center an item. You will demonstrate this handy behavior as you rework the CSS.

a. Open index.html in a text editor. Remove the `justify-content` and `align-items` style rules. Add a style rule for the h1 element selector that sets margin to `auto`. Save your file and display it in a browser. Your display should still be similar to Figure 8.33. You can compare your work to the sample in the student files (chapter8/8.7/flex.html).

b. Open index.html in a text editor. Change the value of the display property from `flex` to `grid`. Save your file and display it in a browser. Your display should still be similar to Figure 8.33. You can compare your work to chapter8/8.7/grid.html in the student files.

You explored several layout techniques in this Hands-On Practice. The new flexbox and grid layout systems offer a wide variety of page layout options for web developers.

Viewport Meta Tag

There are multiple uses for meta tags. You've used the meta tag since Chapter 1 to configure the character encoding on a web page. In this section, we'll explore the **viewport meta tag**, which was created as an Apple extension that helps with displays on mobile devices such as iPhones and Android smartphones by setting the width and scale of the viewport. Figure 8.34 shows the display of a web page in a desktop browser.

Figure 8.35 displays a screen shot of the same web page displayed on an Android device. Examine Figure 8.35 and notice that the mobile device zoomed out to display the entire web page on the tiny screen. The text on the web page is difficult to read.

Figure 8.36 shows the same web page after the viewport meta tag was added to the head section of the document with the `width` and `initial scale` directives.[11] Setting the `initial-scale` directive to the value 1 caused the mobile browser to avoid zooming out on the web page and to display it in a more usable manner. The code is shown below:

```
<meta name="viewport"
content="width=device-width,
initial-scale=1.0">
```

FIGURE 8.34 *A web page displayed in a desktop browser.*

FIGURE 8.35 *Mobile display of a web page without the viewport meta tag.*

FIGURE 8.36 *The viewport meta tag helps with mobile display.*

Code the viewport meta tag with the HTML `name="viewport"` and `content` attributes. The value of the HTML `content` attribute can be one or more **directives** (also referred to as properties by Apple), such as the `device-width` directive and directives that control zooming and scale. Table 8.6 lists viewport meta tag directives and their values.[12]

TABLE 8.6 *Viewport Meta Tag Directives*

Directive	Values	Purpose
width	Numeric value or `device-width` which indicates actual width of the device screen	The width of the viewport in pixels
height	Numeric value or `device-height` which indicates actual height of the device screen	The height of the viewport in pixels
initial-scale	Numeric multiplier; set to 1 for 100% initial scale	Initial scale of the viewport
minimum-scale	Numeric multiplier	Minimum scale of the viewport
maximum-scale	Numeric multiplier	Maximum scale of the viewport
user-scalable	`yes` allows scaling, `no` disables scaling	Determines whether a user can zoom in or zoom out

Now that you've scaled the page to be readable, what about styling it for optimal mobile use? That's where CSS comes into play. You'll explore CSS Media Queries in the next section.

If a web page displays a phone number, wouldn't it be handy for a person using a smartphone to be able to tap on the phone number and place a call or send an SMS (short message service) text message? It's very easy to configure a telephone hyperlink or SMS hyperlink for use by smartphones.

According to RFC 3966,[13] you can configure a telephone hyperlink by using a telephone scheme: Begin the `href` value with `tel:` followed by the phone number. For example, to configure a telephone hyperlink on a web page for use by mobile browsers, code as follows:

```
<a href="tel:888-555-5555">Call 888-555-5555</a>
```

RFC 5724[14] indicates that an SMS scheme hyperlink intended to send a text message can be configured by beginning the `href` value with `sms:` followed by the phone number, as shown in the following code:

```
<a href="sms:888-555-5555">Text 888-555-5555</a>
```

Not all mobile browsers and devices support telephone and text hyperlinks, but expect increased use of this technology in the future.

CSS Media Queries

Recall from Chapter 3 that the term **responsive web design** refers to progressively enhancing a web page for different viewing contexts (including smartphones and tablets) through the use of coding techniques including fluid layouts, flexible images, and media queries.

For examples of the power of responsive web design techniques, review Figures 3.45, 3.46, and 3.47, which are actually the same .html web page file that was configured with CSS to display differently, depending on the viewport size detected by media queries. The Media Queries website displays a gallery of websites that demonstrate responsive web design.[15] The screen captures in the gallery show web pages displayed with the following browser viewport widths: 320px (smartphone display), 768px (tablet portrait display), 1024px (netbook display and tablet landscape display), and 1600px (large desktop display).

What's a Media Query?

A **media query** is made up of a media type (such as screen) and a logical expression that determines the capability of the device that the browser is running on, such as screen resolution and orientation (portrait or landscape).[16] When the media query evaluates as true, the media query directs browsers to CSS you have coded and configured specifically for those capabilities.

FIGURE 8.37 *CSS media queries help to configure the page for mobile display.*

Media Query Example Using a Link Element

Figure 8.37 shows the same web page as Figure 8.35, but it looks quite different because of a link element that includes a media query and is associated with a style sheet configured for optimal mobile display on a popular smartphone. The HTML is shown below:

```
<link href="lighthousemobile.css" rel="stylesheet"
      media="(max-width: 480px) ">
```

The code sample above will direct browsers to an external stylesheet that has been configured for optimal display on the most popular smartphones. The media type value `only` is a keyword that will hide the media query from outdated browsers. The media type value `screen` targets devices with screens. Commonly used media types and keywords are listed in Table 8.7.

The `max-width` media feature is set to 480px. While there are many different screen sizes for smartphones these days, a maximum width of 480px will target the display size of many popular models. A media query may test for both minimum and maximum values. For example,

```
<link href="lighthousetablet.css" rel="stylesheet"
      media="(min-width: 768px) and (max-width: 1024px) ">
```

TABLE 8.7 *Commonly Used Media Types*

Media Type	Value Purpose
all	All devices (default)
screen	Screen display of web page
speech	devices that "read out" a web page such as screenreaders
print	Printout of web page

Media Query Example Using an @media Rule

A second method of using media queries is to code them directly in your CSS using an **@media rule**. Begin by coding @media followed by the media type and logical expression. Then enclose the desired CSS selector(s) and declaration(s) within a pair of braces. Table 8.8 lists commonly used media query features.[17] The sample code below configures a different background image specifically for a narrow width display such as a smartphone.

```
@media (max-width: 480px) {
  header { background-image: url(mobile.gif); }
}
```

TABLE 8.8 *Commonly Used Media Query Features*

Features	Values	Criteria
max-device-height	Numeric value	The height of the screen size of the output device in pixels is smaller or equal to the value
max-device-width	Numeric value	The width of the screen size of the output device in pixels is smaller or equal to the value
min-device-height	Numeric value	The height of the screen size of the output device in pixels is greater than or equal to the value
min-device-width	Numeric value	The width of the screen size of the output device in pixels is greater than or equal to the value
max-height	Numeric value	The height of the viewport in pixels is smaller than or equal to the value; (reevaluated when screen is resized)
min-height	Numeric value	The height of the viewport in pixels is greater than or equal to the value; (reevaluated when screen is resized)
max-width	Numeric value	The width of the viewport in pixels is smaller than or equal to the value; (reevaluated when screen is resized)
min-width	Numeric value	The width of the viewport in pixels is greater than or equal to the value; (reevaluated when screen is resized)
orientation	portrait or landscape	The orientation of the device

Mobile First

Many web developers follow a responsive design layout strategy called Mobile First, a term coined by Luke Wroblewski almost a decade ago.[18] The Mobile First process begins by first configuring a page layout that works well in smartphones (you can test with a small browser window). This provides the quickest display for mobile devices. Next, resize the browser viewport to be larger until the design "breaks" and needs to be reworked for a pleasing display—this is the point where you need to code a media query. If appropriate, continue resizing the browser viewport to be larger until the design breaks and code additional media queries.

Responsive Layout with Media Queries

Hands-On Practice 8.8

Small Display	Medium Display	Large Display
wrapper	wrapper	wrapper

FIGURE 8.38 *Three wireframe layouts.*

You'll practice a Mobile First strategy for responsive design in this Hands-On Practice. First, you will configure a page layout that works well in smartphones (test with a small browser window). Then you'll resize the browser viewport to be larger until the design "breaks" and code media queries and additional CSS as appropriate using traditional float layout techniques (as introduced in Chapter 7). Figure 8.38 shows wireframes for three different layouts.

Create a new folder called ch8resp. Copy the starter8.html file from the chapter8 folder in the student files into your ch8resp folder. Copy the lighthouse.jpg and light.gif files from the chapter8/starters folder into your ch8resp folder.

1. Launch a text editor and open your starter8.html file. View the HTML and notice that a div assigned to the `wrapper` id has child elements of header, nav, main, aside, and footer as shown.

```
<div id="wrapper">
  <header> … </header>
  <nav> … </nav>
  <main> … </main>
  <aside> … </aside>
  <footer> … </footer>
</div>
```

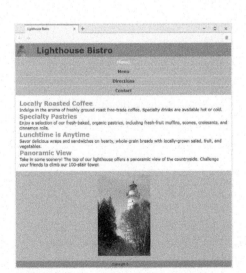

FIGURE 8.39 *Normal flow full-width block elements.*

Observe the CSS and note the `wrapper` id's child elements (header, nav, main, aside, and footer) do not have the float property associated with them. Browsers render this page using normal flow with each element displayed under the preceding element, similar to the Small Display wireframe in Figure 8.38. Notice also that there are no minimum widths assigned. This layout will work well on a small display such as a smartphone. Save the file with the name index.html.

2. Display your index.html file in a desktop browser. If your browser viewport is a typical size, it will look a bit awkward and similar to Figure 8.39. Don't worry though, we intend this layout to be

displayed on narrow mobile screens—so resize your browser to be narrower until your display is similar to Figure 8.40, which simulates the mobile display.

3. There is one more item needed for a more pleasing and usable display on an actual mobile device: the viewport meta tag. Launch a text editor and open index.html. Add a viewport meta tag in the head section of the document below the meta tag. The HTML follows:

```
<meta name="viewport" content="width=device-width,
initial-scale=1.0">
```

Save your file. If you display it in a desktop browser, it will look the same. An example is in the student files (chapter8/8.8/step3.html). Figure 8.41 shows a smartphone display of the page.

4. In the past, web developers would try to target specific devices (such as smartphone and tablet). However, the modern process is to determine the condition for the media query by widening the browser until the display begins to "break" or look awkward. Display your index.html file in a browser—first narrow it and then gradually widen it. The point where it starts to seem awkward is around 600px wide, so that's what we'll code for our media query.

FIGURE 8.40
*Smartphone
display simulation.*

You'll configure the layout to follow the Medium Display wireframe in Figure 8.38. Observe the layout: horizontal header, horizontal navigation, adjacent main and aside elements, and horizontal footer.

Launch a text editor and open index.html. Code a CSS media query after the other style rules to change the display when the `min-width` of the viewport is 600px. Add style rules within the media query that will create a horizontal navigation area with `inline-block` display, width, padding, centered text, and no border; configure the main element selector with a left float and a width of 55%, set a 55% left margin for the aside element selector, and configure the footer element selector to clear floats. The CSS follows:

```
@media (min-width: 600px) {
    nav li { display: inline-block;
             width: 7em;
             padding: 0.5em;
             border: none; }
    nav ul   { text-align: center; }
    main     { float: left;
               width: 55%; }
    aside    { margin-left: 55%; }
    footer   { clear: both; }
}
```

FIGURE 8.41
*Smartphone
display.*

FIGURE 8.42 *Implementing the Medium Display wireframe.*

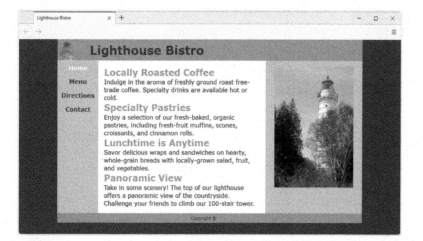

FIGURE 8.43 *Implementing the Large Display wireframe on a desktop browser.*

Save the file and launch it in a browser. You should be able to resize your browser viewport and obtain a display similar to Figure 8.42. An example is in the student files (chapter8/8.8/step4.html).

5. Repeat the process to determine the condition for the next breakpoint. When displayed in a browser, the web page seems to be a bit awkward around 1024px, so that's what you'll code for in the next media query. Configure the layout to follow the Large Display wireframe in Figure 8.38. Observe the layout: horizontal header; adjacent nav, main and aside elements; and horizontal footer.

   Launch a text editor and open index.html. Code a CSS media query after the other style rules to change the display when the `min-width` of the viewport is 1024px. Add style rules within the media query that will set left float for the nav element, a centered `wrapper` id with 80% width and 1200px maximum width, and a #000066 background color for the body element selector. The CSS follows:

```
@media (min-width: 1024px){
  nav li { display: block; }
  nav ul { text-align: left; }
  nav { float: left; }
  #wrapper { width: 80%; margin: auto; max-width: 1200px; }
  body { background-color: #000066; }
}
```

Save the file and test it in a browser. You should be able to resize your browser viewport and obtain a display similar to Figure 8.43. A sample solution is in the student files (chapter8/8.8/index.html). In this Hands-On Practice, you applied media queries to a web page and configured float layout. Since float layout will be in use for quite some time, it's good to be familiar with it. In the next Hands-On Practice, you will follow a more modern approach which is applying media queries and configuring grid layout.

There is no single correct way to configure a media query. When web developers first began writing media queries, there were very few mobile devices and they could be targeted with pixel-perfect precision. While this is no longer the case, web developers often use the `max-width` and/or `min-width` features to determine the size of the viewport being used. Here is a typical media query to target a smartphone display which checks for a maximum width value of 480 pixels:

```
@media (max-width: 480px) {
}
```

Websites exist that list mobile device dimensions and suggested breakpoints for media queries.[19,20] However, today, there are a huge number of different mobile devices with various screen resolutions, so a modern approach is to focus on the responsive display of your content and then configure media queries as needed for your content to reflow on a variety of screen sizes. You will need to test your responsive web pages to find the best choices for your specific content. Check for long line lengths or too much empty space on the page—that's probably a signal that a new media query is needed.

Most of the examples in this chapter use pixel values for media query conditions, but some web developers prefer to use em unit values. The first media query in Hands-On Practice 8.8 could have been written to check for `min-width` of 40em:

```
@media (min-width: 40em) {
}
```

An example file with em unit media queries is in the student files (chapter8/8.8/emunit.html).

Visit the following resources for more information about media queries:

▶ https://developers.google.com/web/fundamentals/design-and-ux/responsive/

▶ https://www.smashingmagazine.com/2018/02/media-queries-responsive-design-2018/

▶ https://developer.mozilla.org/en-US/docs/Web/CSS/Media_Queries/Using_media_queries

Google Chrome Dev Tools can be helpful when testing responsive web pages. Visit these resources to get started:

▶ https://developers.google.com/web/tools/chrome-devtools/device-mode/#responsive

▶ https://developers.google.com/web/tools/chrome-devtools/device-mode/#device

Responsive Grid Layout with Media Queries

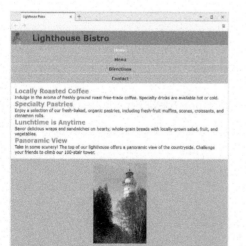

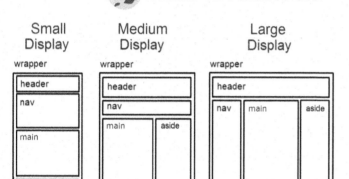

Small Display

wrapper
- header
- nav
- main
- aside
- footer

Medium Display

wrapper
- header
- nav
- main | aside
- footer

Large Display

wrapper
- header
- nav | main | aside
- footer

FIGURE 8.44 *Three wireframe layouts.*

Hands-On Practice 8.9

You'll practice a Mobile First strategy for responsive design using grid layout and media queries in this Hands-On Practice. First, you will configure a page layout that works well in smartphones (test with a small browser window). Then you'll resize the browser viewport to be larger until the design "breaks" and code media queries and additional CSS as appropriate using grid layout for the page and flexbox layout for the navigation area. Figure 8.44 shows wireframes for three different layouts.

Create a new folder called ch8resp2. Copy the starter8.html file from the chapter8 folder in the student files into your ch8resp2 folder. Copy the lighthouse.jpg and light.gif files from the chapter8/starters folder into your ch8resp2 folder.

1. Launch a text editor and open your starter8.html file. View the HTML and notice that a div assigned to the `wrapper` id has child elements of header, nav, main, aside, and footer as shown. The `wrapper` id will be the grid container. The header, nav, main, aside, and footer elements are the grid items.

```
<div id="wrapper">
  <header> … </header>
  <nav> … </nav>
  <main> … </main>
  <aside> … </aside>
  <footer> … </footer>
</div>
```

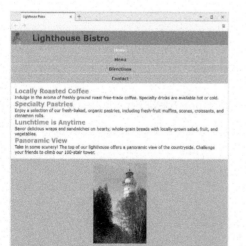

FIGURE 8.45 *The web page initial display.*

Observe the CSS and note that while there are styles that set the visual look of the elements, the CSS does not include any styles for layout. Browsers render this page using normal flow with each element displayed under the preceding element,

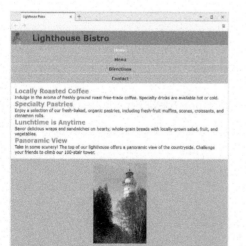

300 Chapter 8 ▶ Responsive Layout Basics

similar to the Small Display wireframe in Figure 8.44. Notice also that there are no minimum widths assigned. This layout will work well on a small display such as a smartphone. Save the file with the name index.html.

2. Display your index.html file in a desktop browser. If your browser viewport is a typical size, it will look a bit awkward and similar to Figure 8.45. Don't worry though, we intend this layout to be displayed on narrow mobile screens—so resize your browser to be narrower until your display is similar to Figure 8.46, which simulates the mobile display.

3. There is one more item needed for a more pleasing and usable display on an actual mobile device: the viewport meta tag. Launch a text editor and open index.html. Add a viewport meta tag in the head section of the document below the meta tag. The HTML follows:

```
<meta name="viewport" content="width=device-width,
initial-scale=1.0">
```

Save your file. If you display it in a desktop browser, it will look the same. Figure 8.41 shows a display of how a smartphone would render this page. An example is in the student files (chapter8/8.7/step3.html).

4. Since the initial display of the web page is rendered well in normal flow, you will only need to configure grid layout when the media queries are triggered. You will determine the condition for the media query by widening the browser until the display begins to "break" or look awkward. Display your index.html file in a browser—first narrow it and then gradually widen it. The point where it starts to seem awkward is around 600px wide, so that's what we'll code for our first media query and grid layout.

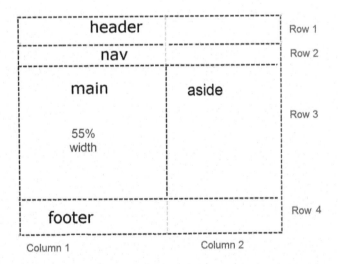

FIGURE 8.46
Smartphone display simulation.

You'll configure the layout to follow the Medium Display wireframe in Figure 8.44. Observe the layout: horizontal header, horizontal navigation, adjacent main and aside elements, and horizontal footer.

Launch a text editor and open index. html. Since the browser display seems a bit awkward around 600px wide, code a media query below the existing CSS that checks for 600px min-width. You will add style rules within the media query. Figure 8.47 shows a grid layout wireframe that corresponds to the Medium Display shown in Figure 8.44.

Figure 8.48 depicts the browser rendering of the page. Notice that the navigation area is now horizontal instead of vertical. You will use flexbox layout to configure this.

header		Row 1
nav		Row 2
main 55% width	aside	Row 3
footer		Row 4
Column 1	Column 2	

FIGURE 8.47 *Grid layout for Medium Display.*

FIGURE 8.48 *Implementing the Medium Display wireframe in grid layout.*

Configure the nav ul element selector to be a flexbox container and set the `flex-direction` to `row`, `flex-wrap` to `nowrap`, and `justify-content` to `space-around`. Also code CSS to eliminate the bottom border on the li elements in the navigation area.

Next, assign grid areas for header, nav, aside, main, and footer to HTML element selectors. Configure a grid assigned to the `wrapper` id. Use the `grid-template` property to describe a grid that contains grid areas header, nav, main, aside, and footer. Use the grid layout in Figure 8.47 as a guide. Set the first column in the grid to 55% width. The CSS follows:

```
@media (min-width: 600px) {
    nav ul { display: flex; flex-flow: row nowrap;
         justify-content: space-around; }
    nav ul li { border-bottom: none; }
header   { grid-area: header; }
nav      { grid-area: nav; }
main     { grid-area: main; }
aside    { grid-area: aside; }
footer   { grid-area: footer; }
#wrapper { display: grid;
         grid-template:
         "header header"
         "nav       nav"
         "main     aside"
         "footer footer"
         / 55%; }
}
```

Save your file and test it in a browser. It should look similar to Figure 8.48. An example is in the student files (chapter 8/8.9/step4.html).

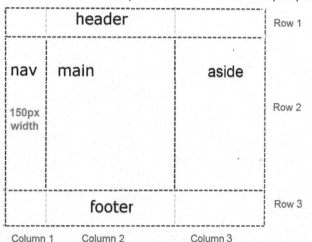

5. Repeat the process to determine the condition for the next breakpoint. When displayed in a browser, the web page seems to be a bit awkward around 1024px, so that's what you'll code for in the next media query. Configure the layout to follow the Large Display wireframe in Figure 8.44. Observe the layout—horizontal header; adjacent nav, main and aside elements; and horizontal footer. Figure 8.49 shows a grid layout wireframe for the Large Display.

FIGURE 8.49 *Grid layout for Large Display.*

Figure 8.50 depicts the browser rendering of the page. Notice the changes in the page: a dark blue background showing on either side of the centered web page and a vertical navigation area.

Launch a text editor and open index.html. Code a CSS media query after the other style rules to change the display when the `max-width` of the viewport is at least 1024px. Add style rules within the media query that will set a dark blue background color for the body element selector,

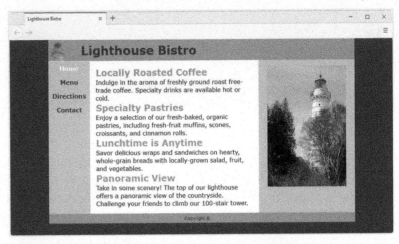

FIGURE 8.50 *Implementing the Large Display wireframe with grid layout on a desktop browser.*

center the `wrapper` id, and configure the nav ul element selector as a flexbox container with `flex-direction` set to `column` and `flex-wrap` set to `nowrap`.

Next, assign grid areas for header, nav, aside, main, and footer to HTML element selectors. Configure a grid assigned to the `wrapper` id. Use the `grid-template` property to describe a grid that contains grid areas header, nav, main, aside, and footer. Use the grid layout in Figure 8.49 as a guide. Set the first column in the grid to 150px width. The CSS follows:

```
@media (min-width: 1024px) {
        body { background-color: #000066; }
        nav ul { display: flex;
                flex-direction: column; flex-wrap: nowrap; }
header     { grid-area: header; }
nav        { grid-area: nav; }
main       { grid-area: main; }
aside      { grid-area: aside; }
footer     { grid-area: footer; }
#wrapper { width: 80%; margin: auto; max-width: 1200px;
           display: grid;
           grid-template:
           "header header header"
           "nav      main    aside"
           "footer footer    footer"
           / 150px; }
        }
```

Save the file and test it in a browser. You should be able to resize your browser viewport and obtain a display similar to Figure 8.50. A sample solution is in the student files (chapter8/8.9/index.html).

Flexible Images with CSS

A **flexible image** is a fluid image that will not break the page layout as the browser viewport is resized.[21] Flexible images (often referred to as **responsive images**), along with fluid layouts and media queries, are the components of responsive web design. You will be introduced to several different coding techniques to configure responsive images in this chapter.

The most widely supported technique to configure an image as flexible requires a change to the HTML and additional CSS to style the flexible image.

1. Edit the img elements in the HTML. Remove the `height` and `width` attributes.
2. Configure the `max-width: 100%;` style declaration in the CSS. If the width of the image is less than the width of the container element, the image will display with its actual dimensions. If the width of the image is greater than the width of the container element, the image will be resized by the browser to fit in the container (instead of hanging out over the margin).
3. To keep the dimensions of the image in proportion and maintain the aspect ratio of the image, Bruce Lawson suggests to also set the `height: auto;` style declaration in the CSS.[22]

Background images can also be configured for a more fluid display at various viewport sizes. Although it's common to code a `height` property when configuring a background image with CSS, the result is a somewhat nonresponsive background image. Explore configuring other CSS properties for the container such as `font-size`, `line-height`, and `padding` in percentage values. The `background-size: cover;` property can also be useful. You'll typically see a more pleasing display of the background image in various-sized viewports. Another option is to configure different image files to use for backgrounds and use media queries to determine which background image is displayed. A disadvantage to this option is that multiple files are downloaded although only one file is displayed. You'll apply flexible image techniques in the next Hands-On Practice.

 Hands-On Practice 8.10 ───────────────────────────────

In this Hands-On Practice, you'll work with a web page that demonstrates responsive web design. Figure 8.51 depicts the default single column page display for small viewports, the two-column page display that triggers when the viewport has a minimum width of 38em units, and the three-column page display that triggers when the viewport has a minimum width of 65em units. You will edit the CSS to configure flexible images.

Create a folder named flexible8. Copy the starter10.html file from the chapter8 folder in the student files into the flexible8 folder and rename it index.html. Copy the following images from the student files chapter8/starters folder into the flexible8 folder: header.jpg and pools.jpg. Launch a browser and view index.html as shown in Figure 8.52. View the code in a text editor and notice that the `height` and `width` attributes have already been removed from the HTML. View the CSS and notice that the web page uses a grid layout with a flexbox navigation area. Edit the embedded CSS.

Desktop Browser   Tablet Display Width   Smartphone Display Width

FIGURE 8.51 *The web page demonstrates responsive web design techniques.*

1. Locate the h1 element selector. Add declarations to set the font size to 300%, and bottom padding to 1em. The CSS follows:

```
h1 { text-align: center;
     font-size: 300%;
     padding-bottom: 1em;
     text-shadow: 3px 3px 3px #E9FBFC; }
```

2. Locate the header element selector. Add the `background-size: cover;` declaration to cause the browser to scale the background image to fill the container. The CSS follows:

```
header { background-image: url(header.jpg);
         background-repeat: no-repeat;
         background-size: cover; }
```

3. Add a style rule for the img element selector that sets maximum width to 100% and height to the value auto. The CSS follows:

```
img { max-width: 100%;
      height: auto; }
```

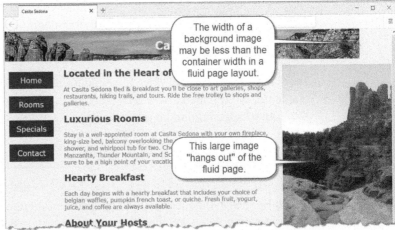

4. Save the index.html file. Test your index.html file in a desktop browser. As you resize the browser window, you'll see your page respond and look similar to the screen captures in Figure 8.51. The web page demonstrates responsive web design with the following techniques: fluid layout, media queries, and flexible images. A suggested solution is in the student files chapter8/8.10 folder.

FIGURE 8.52 *The web page before the images are configured to be flexible.*

Picture Element

The purpose of the **picture element** is to provide a method for a browser to display different images depending on specific criteria indicated by the web developer.[23] The picture element begins with the `<picture>` tag and ends with the `</picture>` tag. The picture element is a container element that is coded along with source elements and a fallback img element to provide multiple image files that can be chosen for display by the browser.

Source Element

The **source element** is a self-contained, or void, tag that is used together with a container element. The picture element is one of several elements (see the video and audio elements in Chapter 11) that can contain one or more source elements. When used with a picture element, multiple source elements are typically configured to specify different images. The browser downloads only the image it chooses to display based on the criteria provided. Code the source elements between the opening and closing picture tags. Table 8.9 lists attributes of the source element when coded within a picture element container.

TABLE 8.9 *Attributes of the Source Element*

Attribute	Value
srcset	Required. Provides image choices for the browser in a comma-separated list. Each item can contain the image URL (required), optional maximum viewport dimension, and optional pixel density for high resolution devices.
media	Optional. Media query to specify conditions for browser display.
sizes	Optional. Numeric or percentage value to specify the dimensions of the image display. May be further configured with a media query.
type	Optional. Image MIME type.

There are many potential ways to configure responsive images with the picture and source elements. We will focus on a basic technique that uses the media and type attributes attribute to specify conditions for display.

 Hands-On Practice 8.11 ————————————————————————

In this Hands-On Practice, you will configure responsive images with the picture, source, and img elements as you create the page shown in Figure 8.53.

Create a new folder named ch8picture. Copy the files from the chapter8/image folder into your ch8picture folder. Launch a text editor and open the file located at chapter8/template.html in the student files. Save the file as index.html in your ch8picture folder. Modify the file to configure a web page as indicated:

FIGURE 8.53 *Responsive image with the picture element.*

1. Configure the text, Picture Element, within an h1 element and within the title element.

2. Code the following in the body of the web page:

```
<picture>
    <source type="image/webp" media="(max-width: 480px)"
        srcset="small.webp">
    <source type="image/webp" media="(max-width: 800px)"
        srcset="medium.webp">
    <source type="image/webp"  srcset="large.webp">
    <source media="(max-width: 480px)" srcset="small.jpg">
    <source media="(max-width: 800px)" srcset="medium.jpg">
    <source srcset="large.jpg">
    <img src="fallback.jpg" alt="waterwheel">
</picture>
```

Save your file and test your page. Notice how a different image is displayed depending on browser support of WebP and the width of the browser viewport. The browser requests the first image it encounters that it can support and that meets the criteria in the media query. If the browser supports the picture element, source element, and WebP images: the small.webp file is displayed if the viewport width 480px or less, the medium.webp file is displayed if the viewport width is 800px or less, and the large.webp file is displayed if the viewport width is greater than 800px. If the browser supports the picture element and source element but does not support WebP images: the small.jpg file is displayed if the viewport width is 480px or less, the medium.jpg file is displayed if the viewport width is 800px or less, and the large.jpg file is displayed if the viewport width is greater than 800px. If the browser does not support the picture and source elements, the fallback.jpg file is displayed.

As you test, try resizing and refreshing the browser display. You may need to resize the browser, close it, and launch it again to test for display of the different images. A suggested solution is in the student files chapter8/8.11 folder.

This Hands-On Practice provided a very basic example of responsive images with the picture element. The picture and source element responsive image technique is intended to eliminate multiple image downloads that can occur with CSS flexible image techniques. The browser downloads only the image it chose to display based on the criteria provided.

Responsive Img Element Attributes

This section introduces three attributes used with the img element to support responsive images: the sizes, scrset, and loading attributes.

The sizes Attribute

The purpose of the img element's **sizes attribute** is to inform the browser as it processes the srcset attribute about how much of the viewport should be used to display the image. The default value of the sizes attribute is 100vw, which indicates 100% of the viewport width is available to display the image. The value of the sizes attribute can be a percentage of the viewport width or a specific pixel width (such as 400px). The sizes attribute can also contain one or more media queries along with the width for each condition.[24]

The srcset Attribute

The purpose of the img element's **srcset attribute** is to provide a method for a browser to display different images depending on specific criteria indicated by the web developer. The value of the srcset attribute provides image choices for the browser in a comma-separated list. Each list item can contain the image URL (required), optional maximum viewport dimension, and optional pixel density for high resolution devices.[25]

There are many potential ways to configure responsive images with the img element, sizes, attribute, and srcset attribute. We will focus on a basic technique that uses the browser viewport dimension to specify conditions for display.

 Hands-On Practice 8.12 ———————————————————————

In this Hands-On Practice, you will configure responsive images with the img element's sizes and srcset attributes as you create the page shown in Figure 8.54.

Create a new folder named ch8image. Copy the large.jpg, medium.jpg, small.jpg, and fallback.jpg files from the chapter8/image folder into your ch8image folder. Launch a text editor and open the file located at chapter8/template.html in the student files. Save the file as index.html in your ch8image folder. Modify the file to configure a web page as indicated:

1. Configure the text "Img Element" within an h1 element and within the title element.

FIGURE 8.54 *Responsive image with the image element's srcset attribute.*

2. Code the following in the body of the web page:

```
<img src="fallback.jpg" alt="waterwheel"  sizes="100vw"
    srcset="large.jpg 1200w, medium.jpg 800w, small.jpg 320w">
```

Save your file and test your page. Notice how a different image is displayed depending on the width of the browser viewport. If the viewport's minimum width is 1200px or greater, the large.jpg image is shown. If the viewport's minimum width is 800px or greater but less than 1200px, the medium.jpg image is displayed. If the viewport's minimum width is 320px greater but less than 800px, the small.jpg image is shown. If none of these criteria are met, the fallback.jpg image should be displayed.

As you test, try resizing and refreshing the browser display. You may need to resize the browser, close it, and launch it again to test for display of the different images. Browsers that do not support the image element's new sizes and srcset attributes will ignore these attributes and display the fallback.jpg image. A suggested solution is in the student files chapter8/8.12 folder.

This Hands-On Practice provided a very basic example of responsive images with the img element and new sizes and srcset attributes which (like the picture element responsive image technique) is intended to eliminate multiple image downloads that can occur with CSS flexible image techniques. The browser downloads only the image it chose to display based on the criteria provided.

The loading Attribute

The img element's **loading attribute** informs the browser how quickly to request an image. The value `eager` is the default and indicates the image should be downloaded immediately. The value `lazy` indicates the browser should defer the request until a condition is met, such as whether the image is currently in or near the browser viewport.[26] If a page displays a gallery of images, it could be useful to configure `loading="lazy"` on images further down the page so that only images currently in or near the browser viewport are immediately requested for download. A long web page with multiple images could request lazy loading on images that will not initially display in the browser viewport. Since only the images currently needed in the browser viewport will be initially requested for download, a benefit of lazy loading for web pages with multiple images is a decrease in perceived load time. An example page that implements lazy loading is in the student files chapter8/lazy folder.

Testing Mobile Display

The best way to test the mobile display of a web page is to publish it to the Web and access it from a mobile device, as shown in Figure 8.55. (See Chapter 12 for an introduction to publishing a website with FTP.)

Testing with a Desktop Browser

If you don't have a smartphone and/or are unable to publish your files to the Web—no worries—as you've seen in this chapter (also see Figure 8.56), you can approximate the mobile display of your web page using a desktop browser and verify the placement of your media queries.

▶ If you have coded media queries within your CSS, display your page in a desktop browser and then reduce the width and height of the viewport until it approximates a mobile screen size.

▶ If you have coded media queries within a link tag, edit the web page and temporarily modify the link tag to point to your mobile CSS style sheet. Then, display your page in a desktop browser and reduce the width and height of the viewport until it approximates a mobile screen size.

FIGURE 8.55 *Testing the page with a smartphone.*

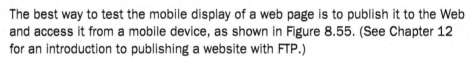

Responsive Testing Tools

It can be helpful to know the size of your browser viewport as you are testing a responsive web page. The following free tools can help you view your web pages in specific viewport dimensions.

▶ Chris Pederick's Web Developer Extension
Available for Firefox and Chrome
https://chrispederick.com/work/web-developer
Select Resize > Display Window Size

▶ Viewport Dimensions Extension
Available for Chrome at
https://github.com/CSWilson/Viewport-Dimensions

▶ Responsive Design Checker
https://www.responsivedesignchecker.com/

FIGURE 8.56 *Approximating the mobile display with a browser.*

Built-in Browser Tools

Most popular browsers offer built-in tools for developers. For example, the Chrome and Firefox browsers each offer developer tools that not only display various viewport sizes, but also will display your web page in the dimensions of selected mobile devices and offer simulations including network throttling (download speed, upload speed, minimum latency), and touch events.

▶ Google Chrome Dev Tools: Device Mode
  https://developers.google.com/web/tools/chrome-devtools/device-mode/

▶ Firefox Responsive Design Mode
  https://developer.mozilla.org/en-US/docs/Tools/Responsive_Design_Mode

▶ Microsoft Edge (Chromium) Developer Tools
  https://docs.microsoft.com/en-us/microsoft-edge/devtools-guide-chromium

More Mobile Testing Options

Companies such as Browserstack.com and CrossBrowserTesting.com provide robust manual and automated testing across a wide variety of devices, operating systems, and browsers. For example, Browserstack offers testing with over 2,000 actual devices, browsers, and operating systems.[27] The testing platforms typically offer a free trial period followed by the purchase of a monthly or annual service plan.

If you are a software developer or information systems major, you may want to explore the SDKs (Software Developer Kits) for the iOS and Android platforms. Each SDK includes a mobile device emulator.

Figure 8.57 shows the variety of mobile screen resolutions reported in a recent year by StatCounter Global Stats.[28]

At the time this data was collected, the most popular mobile screen resolution was 360x640, at 26%.

Visit https://gs.statcounter.com for the most up-to-date information.

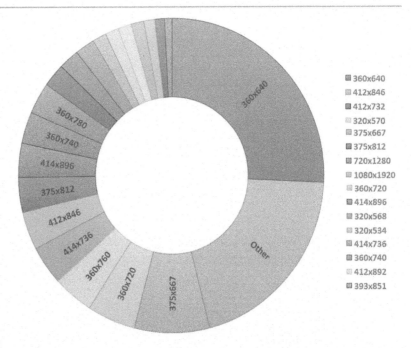

FIGURE 8.57 *A wide variety of mobile screen resolutions.*

Review and Apply

Review Questions

Multiple Choice. Choose the best answer for each item.

1. Which meta tag is used to configure display for mobile devices?
   a. viewport
   b. handheld
   c. mobile
   d. screen

2. Which of the following values would you assign to the display property to configure a flexbox container?
   a. `grid`
   b. `flexbox`
   c. `flex`
   d. `inline`

3. Which of the following properties configures proportional flexible items?
   a. `align-items`
   b. `flex`
   c. `flex-wrap`
   d. `justify-content`

4. Which of the following is a container element that is coded along with source elements and a fall-back img element to provide multiple image files that can be chosen for display by the browser?
   a. photo
   b. picture
   c. figure
   d. sourceset

5. Which of the following properties configure how the browser should display extra space along the main axis in the flex container?
   a. `align-items`
   b. `flex-flow`
   c. `flex-direction`
   d. `justify-content`

6. Which of the following is optimized for responsive two-dimensional page layout?
   a. CSS Absolute Positioning
   b. CSS Display Layout
   c. CSS Flexible Box Layout
   d. CSS Grid Layout

7. Which of the following would occur if a browser does not support grid or flexbox layout?
   a. The browser will display a warning message.
   b. The browser will ignore properties associated with grid and flexbox.
   c. The device will shut down.
   d. The browser will display a blank page.

8. Which of the following is a conditional that can be used to test for support of a CSS property?
   a. viewport query
   b. media query
   c. property query
   d. feature query

9. Which of the following properties configures empty space between grid tracks?
   a. `gutter`
   b. `align`
   c. `gap`
   d. `flex-direction`

10. Which of the following attributes of the img element will provide a method for the browser to display different files based on specific criteria?
    a. href
    b. srcset
    c. sizes
    d. alt

Review Answers

1. a   2. c   3. b   4. b   5. d   6. d   7. b   8. d   9. c   10. b

Hands-On Exercises

1. Write the CSS to configure the nav element selector as a flex container with rows that wrap.

2. Write the CSS for a feature query that checks for support of CSS grid layout.

3. Write the CSS to configure a grid for an id named container that has two columns and two rows. The first row is 100 pixels high. The first grid item in the second row takes up 75% of the width. It may be helpful to draw a wireframe of the grid layout before you write the code.

4. Write the CSS to configure a media query that triggers when the screen size is 1024 pixels or less.

5. Create a web page that displays eight of your favorite photos (or eight photos supplied by your instructor). The page layout should implement flexbox to configure a responsive display. Include your name in an e-mail address in the page footer area.

6. Create a web page that displays eight of your favorite photos (or eight photos supplied by your instructor). Also include a header and footer area on the page. The page layout should implement grid layout to configure a responsive display. Include your name in an e-mail address in the page footer area.

7. Draw a wireframe for the home page of your school's website. Write the CSS to configure a grid layout for the wireframe.

Focus on Web Design

Now that you've had some practice creating responsive web pages, it's a good idea to explore resources on the web about responsive web design best practices. Use the following URLs as a starting point as you research this topic. Write a one-page, double-spaced summary that describes four recommended practices of responsive web design.

- ▶ https://designmodo.com/responsive-design-examples/
- ▶ https://www.uxpin.com/studio/blog/best-practices-examples-of-excellent-responsive-design/
- ▶ https://www.impactbnd.com/blog/responsive-design-best-practices
- ▶ https://www.toptal.com/designers/responsive/responsive-design-best-practices
- ▶ https://fireart.studio/blog/how-to-design-responsive-website-best-practices/

Case Study

You will continue the case studies from Chapter 7 as you configure the layout of the websites to use a mobile first, responsive design.

Pacific Trails Resort Case Study

In this chapter's case study, you will use the existing Pacific Trails Resort (Chapter 7) website as a starting point to create a new version of the website. The new version will utilize a responsive layout with media queries that displays well on desktop browsers and mobile devices. The updated design features a Mobile First approach, a flexbox navigation layout, and a grid layout for large device display. In addition, you'll modify the content page columns to use flexbox layout instead of float. This is a lot to accomplish but we'll take a step-by-step approach.

First, you will examine the HTML structure of the pages and configure a page layout that works well in smartphones (test with a small browser window). Then you'll resize the browser viewport to be larger until the design "breaks" and code media queries, a feature query, and additional CSS as appropriate using grid layout for the page and flexbox layout for the navigation area. Figure 8.58 shows wireframes for three different layouts. The Home page web page displays will be similar to Figure 8.59.

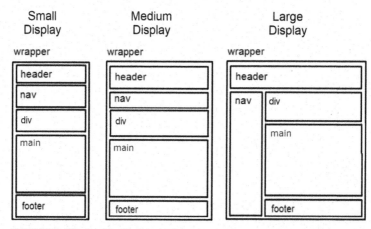

FIGURE 8.58 *Pacific Trails wireframes.*

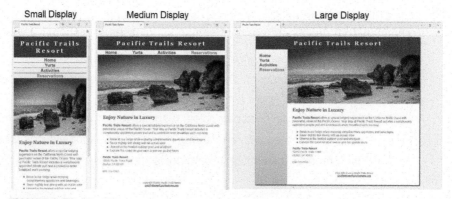

FIGURE 8.59 *The Home page.*

You have five tasks in this case study:

1. Create a new folder for the Pacific Trails Resort website.

2. Review the HTML structure and edit the pacific.css external style sheet to configure a single-column (smartphone) display.

3. Configure the CSS needed for pleasing display of the web pages on medium sized mobile devices.

4. Configure the CSS and HTML needed for a pleasing display of the web pages on large mobile devices and desktops.

5. Add a viewport meta tag to each web page.

Task 1: Create a folder called ch8pacific to contain your Pacific Trails Resort website files. Copy the files from the Chapter 7 Case Study ch7pacific folder into the ch8pacific folder.

Task 2: Configure a Small Single-Column Layout. Open the index.html file in a text editor. View the HTML and notice that a div assigned to the `wrapper` id has child elements of header, nav, div, main, and footer as shown.

```
<div id="wrapper">
  <header> … </header>
  <nav> … </nav>
  <div> … </div>
  <main> … </main>
  <footer> … </footer>
</div>
```

a. Configure the CSS. Launch a text editor and open the pacific.css style sheet. Edit the styles to achieve a layout that displays well on small devices using normal flow (no floats) with full-width block elements.

1. Remove any style declarations that configure float and left margin from all selectors.

2. Edit the styles for the nav element selector. Remove the `width` declaration. Set padding to 0. Configure centered text.

3. Edit the styles for the `#wrapper` id selector. Remove the `width`, `min-width`, `max-width`, `margin`, `border`, and `box-shadow` declarations.

4. Code a style rule for the `nav li` selector. Set a 1 pixel dark blue solid bottom border.

5. Edit the styles for the header element selector. Remove the style declarations for padding and height.

6. Edit the styles for the h1 element selector. Remove the declaration for `font-size`. Set top and bottom padding to 0.5em.

7. Remove the section element selector and all style declarations.

8. Save your pacific.css file. Use the CSS validator (https://jigsaw.w3.org/css-validator) to check your syntax. Correct and retest if necessary.

b. Test the web pages. Display your index.html file in a browser. If your browser viewport is a typical size, the page will look a bit awkward and similar to Figure 8.60. This layout is intended for narrow mobile screens.

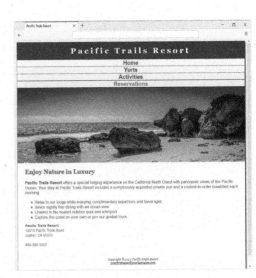

FIGURE 8.60 *Normal flow full-width block elements.*

Resize your browser to be narrower until your display is similar to the Small Display shown in Figure 8.59, which simulates mobile display. Test the yurts.html and activities.html files in a similar manner—they should be similar to the Small Display in Figures 8.61 and 8.62.

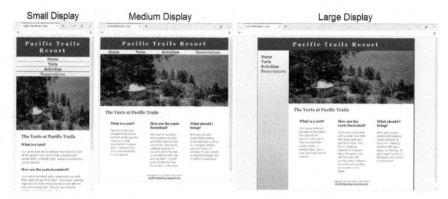

FIGURE 8.61 *The Yurts page.*

FIGURE 8.62 *The Activities page.*

Task 3: Configure a Medium Layout. Display your index.html file in a browser—first narrow it and then gradually widen it. The point where it starts to seem awkward in the navigation area is around 600px wide, so that's what you'll code for our media query. When the media query is triggered, you'll configure the layout to follow the Medium Display wireframe in Figure 8.58, which has a horizontal navigation bar.

a. Configure the CSS. Launch a text editor and open the pacific.css style sheet. Place your cursor below the existing styles. Configure a media query that is triggered when the minimum width is 600px or greater. Code the following styles within the media query.

1. Configure styles for the `nav ul` selector. Configure a flex container for a row that does not wrap. Also configure styles to cause the browser to display empty space before, between, and after the flex items.

2. Configure styles for the `nav li` selector. Eliminate the bottom border (hint: use `border-bottom: none;`).

3. Code a style rule for the section element selector. Set 2em left and right margin.

4. Save your pacific.css file. Check your syntax with the CSS validator (https://jigsaw.w3.org/cssvalidator). Correct and retest if necessary.

b. Test the web pages. Display your index.html file in a browser. You should be able to resize your browser viewport and obtain a display similar to the Medium Display in Figure 8.59. When you test the Yurts and Activities pages, you will notice that their Medium Display is still different from Figures 8.61 and 8.62.

c. You need to make a few changes to configure the Yurts and Activities pages so that their content within the section elements uses three columns similar to the layout in Figures 8.61 and 8.62. A quick way to accomplish this would be to revert back to the Chapter 7 Case Study CSS—configure the section element selector with right float and 33% width. Another way to create the three-column look is to configure a flex container for the three section elements, which is what you will do in this case study.

1. Modify the HTML. The `flow` class will be used to configure a flex container. Edit the yurts.html file. Code a div assigned to a class named `flow` that contains all section elements. Save the file. Edit the activities.html file. Code a div assigned to a class named `flow` that contains all section elements. Save the file.

2. Modify the CSS. Edit the pacific.css file. Code a new style rule within the media query for the `flow` class selector. Configure a flex container. The flex direction is `row`. Add a declaration to the styles for the section element selector and set the `flex` property to 1. Save the pacific.css file. Check your syntax with the CSS validator (https://jigsaw.w3.org/cssvalidator). Correct and retest if necessary.

3. Display your index.html, yurts.html, and activities.html files in a browser. The display of the index.html file should remain unchanged. However, the Yurts page should be similar to Figure 8.61, and the Activities page should be similar to Figure 8.62.

Task 4: Configure a Large Grid Layout. Display your index.html file in a browser and gradually widen—you may notice that the page seems to get a bit awkward around 1024px, so that's what you'll code for the next media query. When the media query is triggered, you'll configure the grid layout for supporting browsers in Figure 8.63, which corresponds to the Large Display wireframe in Figure 8.58. Nonsupporting browsers (such as Internet Explorer) will render a web page layout with horizontal navigation.

a. Configure the CSS. Launch a text editor and open the pacific.css style sheet. Place your cursor below the existing styles. Configure a media query that is triggered when the minimum width is 1024px or greater. Code the following styles within the media query.

1. Configure styles for the `#wrapper` selector. Set auto margin, 80% width, a 1px solid dark blue border, and a 3px dark blue drop shadow.

2. Configure styles for the nav element selector. Configure left text alignment and 1em left padding.

3. Next, get ready for some progressive enhancement with grid layout. Configure a feature query (use `@supports`) to check for grid display). Configure the following styles within the feature query.

   i. Configure styles for the `nav ul` selector. Set the `flex-direction` property to the value `column`. Set the top padding to 1em.

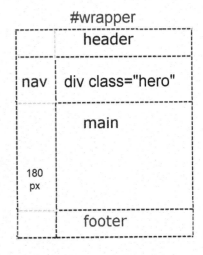

FIGURE 8.63 *Grid layout for Large Display.*

ii. Notice the grid areas in Figure 8.63. Code style rules to configure each child element of the #wrapper with the `grid-area` property to correspond to Figure 8.63. Use Hands-On Practice 8.5 as a guide to configuring the header, nav, main, and footer element selectors. Configure the `hero` class selector as a grid-area as shown: `.hero { grid-area: hero; }`

iii. Code styles for the #wrapper selector. Set the display property to grid and use Figure 8.63 as a guide to configure the `grid-template` property.

4. Save your pacific.css file.

b. **Modify the HTML.** Launch a text editor and edit the index.html, yurts.html, and activities.html files. The class named `hero` is part of the grid layout. Locate the div that contains the hero image on each page. Add `class="hero"` to the div tag. Save your files.

c. Display your index.html file in a browser. You should be able to resize your browser viewport and obtain a layout similar to the Large Display in Figure 8.58. The display of the Home page should be similar to Figure 8.59. Test the yurts.html and activities.html files in a similar manner. The Yurts page should be similar to Figure 8.61. The Activities page should be similar to Figure 8.62.

Task 5: Add a Viewport Meta Tag. Launch a text editor and edit the index.html, yurts.html, and activities.html files. Configure a viewport meta tag in the head section of each page that configures the width to the device-width and sets the initial-scale to 1.0. Save your files. When you test them in a browser, the display will be unchanged, but the viewport meta tag will improve the display on a mobile device.

You have accomplished a great deal as you completed this case study. The design is now responsive and displays well on devices with various size viewports. You have also configured a modern grid layout which includes a flexbox layout navigation area and a flexbox layout with three columns on the content pages. Pacific Trails Resort is responsive and mobile!

Path of Light Yoga Studio Case Study

In this chapter's case study, you will use the existing Path of Light Yoga Studio (Chapter 7) website as a starting point to create a new version of the website. The new version will utilize a responsive layout with media queries that displays well on desktop browsers and mobile devices. You'll practice a Mobile First strategy for responsive design. First, you will examine the HTML structure of the pages and configure a page layout that works well in smartphones (test with a small browser window). Then you'll resize the browser viewport to be larger until the design "breaks" and code media queries and additional CSS as appropriate using flexbox layout for the navigation area. Figure 8.64 shows wireframes for three different layouts. The Home page displays will be similar to Figure 8.65.

You have five tasks in this case study:

1. Create a new folder for the Path of Light Yoga Studio website.

2. Edit the yoga.css external style sheet to configure a single-column (smartphone) display.

3. Configure the HTML and CSS needed for pleasing display of the web pages on medium sized mobile devices.

4. Configure the CSS needed for a pleasing display of the web pages on large mobile devices and desktops.

5. Add a viewport meta tag to each web page.

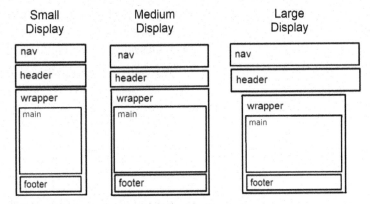

FIGURE 8.64 *Path of Light Yoga Studio wireframes.*

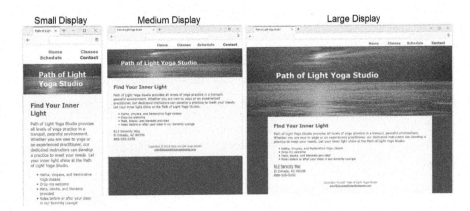

FIGURE 8.65 *The Home page.*

Task 1: Create a folder called ch8yoga to contain your Path of Light Yoga Studio website files. Copy the files from the Chapter 7 Case Study ch7yoga folder into the ch8yoga folder.

Task 2: Configure a Small Single-Column Layout. Open the index.html file in a text editor. View the HTML and notice that the nav and header elements precede a div assigned to the `wrapper` id with child elements of main and footer as shown.

```
<nav> … </nav>
<header> … </header>
<div id="wrapper">
   <main> … </main>
   <footer> … </footer>
</div>
```

a. Configure the CSS. Launch a text editor and open the yoga.css style sheet. Edit the styles to achieve a layout that displays well on small devices using normal flow (no floats) with full-width block elements.

1. Edit the styles for the body element selector. Remove the `max-width` and `min-width` declarations.

2. Edit the styles for the nav element selector. Set the height to `auto`. Set right padding to 0.

3. Edit the styles for the `#wrapper` id selector. Remove the margin and width declarations.

4. Edit the styles for the header element selector. Set font size to 90% and minimum height to 200px.

5. Remove the `.onethird` class selector and style declarations.

6. Remove the `.onehalf` class selector and style declarations.

7. Edit the styles for the `.home` class selector. Remove the `font-size` and `min-height` declarations. Set the height to 20vh (20% of the viewport height), top padding to 2em, and left padding to 10%.

8. Edit the styles for the `.content` class selector. Remove the `padding-bottom` declaration. Set left padding to 10% and height to 20vh.

9. Edit the styles for the `nav li` selector. Set left padding to 4em.

10. Code styles for the section element selector. Set left and right padding to .5em.

11. The hero image will not display on small devices. Edit styles for the `#mathero` and `#loungehero` selectors. Set display to `none`. Also remove the `clear` declarations.

12. Code styles for a `#flow` id selector. Set display to `block`. Note: this selector will be used to configure a flex container in Task 3.

13. Save your yoga.css file. Use the CSS validator (https://jigsaw.w3.org/css-validator) to check your syntax. Correct and retest if necessary.

b. Test the web pages. Display your index.html file in a browser. This layout is intended for narrow mobile screens. Resize your browser to be narrower until your display is similar to the Small Display shown in Figure 8.65, which simulates mobile display. Test the classes.html and schedule.html files in a similar manner.

Task 3: Configure a Medium Layout. First, you will edit the HTML to configure a more pleasing display on a wider viewport. Next, you will code 600px as the breakpoint for the first CSS media query. When the media query is triggered, you'll configure the layout to follow the Medium Display wireframe in Figure 8.64 and your pages should look similar to Figures 8.65, 8.66, and 8.67.

a. Edit the HTML. You need to rework the content area on the Classes and Schedule pages.

1. Launch a text editor and open classes.html. Locate the section elements and remove the `class="onethird"` code from each. Code a div assigned to an id named `flow` that contains all section elements. Save the file.

Small Display   Medium Display   Large Display

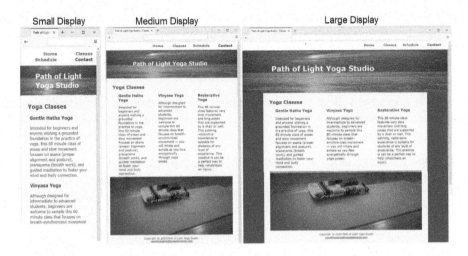

FIGURE 8.66 *The Classes page.*

Small Display   Medium Display   Large Display

FIGURE 8.67 *The Schedule page.*

2. Launch a text editor and open schedule.html. Locate the section elements and remove the `class="onehalf"` code from each. Code a div assigned to an id named `flow` that contains both section elements. Save the file.

b. Configure the CSS. Launch a text editor and open the yoga.css style sheet. Place your cursor below the existing styles. Configure a media query that is triggered when the minimum width is 600px or greater. Code the following styles within the media query.

1. Code styles for the `nav ul` selector. Configure a flex container for a row that does not wrap. Also set `justify-content` to `flex-end`.

2. Code styles for the section element selector. Set 2em left and right padding. Set the flex property to the value 1.

3. The hero images will display on medium and large displays. Code styles for the `#mathero` and `#loungehero` selectors. Set display to `block` with 1em bottom margin.

4. Code styles for the `#flow` id selector. Configure a flex container. The flex direction is `row`. Set the `flex-wrap` property to `wrap`.

5. Save your yoga.cs file. Check your syntax with the CSS validator (https://jigsaw.w3.org/cssvalidator). Correct and retest if necessary.

c. Test the web pages. Display your index.html file in a browser. You should be able to resize your browser viewport and obtain a display similar to the Medium Display in Figure 8.65. Test the classes.html and schedule.html files in a similar manner.

Task 4: Configure a Large Grid Layout. Set 1024px as the breakpoint for the next media query. When the media query is triggered, configure the layout to follow the Large Display wireframe in Figure 8.64. When the task is complete, your pages should look similar to Figures 8.65, 8.66, and 8.67.

a. Configure the CSS. Launch a text editor and open the yoga.css style sheet. Place your cursor below the existing styles. Configure a media query that is triggered when the minimum width is 1024px or greater. Code the following styles within the media query.

1. Code styles for the header element selector. Set font size to 120%.

2. Configure styles for the `.home` class selector. Set the height to 50% of the viewport height (50vh), 5em top padding, and 8em left padding.

3. Code styles for the `.content` class selector. Set the height to 30% of the viewport height (30vh), 2em top padding, and left padding to 8em.

4. Code styles for the `#wrapper` id selector. Set the area to be horizontally centered (hint: `margin: auto;`) with 80% width.

5. Save your yoga.css file. Check your syntax with the CSS validator (https://jigsaw.w3.org/cssvalidator). Correct and retest if necessary.

b. Test the web pages. Display your index.html file in a browser. You should be able to resize your browser viewport and obtain a display similar to the Large Display in Figure 8.65. Test the classes.html and schedule.html files in a similar manner.

Task 5: Add a Viewport Meta Tag. Launch a text editor and edit the index.html, classes.html, and schedule.html files. Configure a viewport meta tag in the head section of each page that configures the width to the device-width and sets the initial-scale to 1.0. Save your files. When you test them in a browser, the display will be unchanged, but the viewport meta tag will improve the display on a mobile device.

You have accomplished a great deal as you completed this case study. The design is now responsive and displays well on devices with various size viewports. You have also configured a modern layout which includes a flex layout navigation area and another flex container on the content pages. Path of Light Yoga Studio is responsive and mobile!

Endnotes

1. "CSS Flexible Box Layout Module Level 1." Edited by Tab Atkins et al., *W3C*, 19 Nov. 2018, www.w3.org/TR/css-flexbox-1/.

2. "CSS Box Alignment Module Level 3: 8. Gaps Between Boxes." Edited by Elika J. Etemad and Tab Atkins Jr., *W3C*, 21 Apr. 2020, www.w3.org/TR/2020/WD-css-align-3-20200421/#gaps.

3. "CSS Flexible Box Layout Module Level 1: 7. Flexibility." Edited by Tab Atkins et al., *W3C*, 19 Nov. 2018, www.w3.org/TR/css-flexbox-1/#flexibility.

4. "CSS Grid Layout Module Level 1." Edited by Tab Atkins et al., *W3C*, 14 Dec. 2017, www.w3.org/TR/css-grid-1/.

5. "CSS Grid Layout Module Level 1: 7.2. Explicit Track Sizing: the Grid-Template-Rows and Grid-Template-Columns Properties: grid-template-columns." Edited by Tab Atkins et al., *W3C*, 14 Dec. 2017, www.w3.org/TR/css-grid-1/#track-sizing.

6. "CSS Grid Layout Module Level 1: 8.3. Line-Based Placement: the Grid-Row-Start, Grid-Column-Start, Grid-Row-End, and Grid-Column-End Properties." Edited by Tab Atkins et al., *W3C*, 14 Dec. 2017, www.w3.org/TR/css-grid-1/#typedef-grid-row-start-grid-line.

7. "CSS Grid Layout Module Level 1: 8.4. Placement Shorthands: the Grid-Column, Grid-Row, and Grid-Area Properties." Edited by Tab Atkins et al., *W3C*, 14 Dec. 2017, www.w3.org/TR/css-grid-1/#propdef-grid-area.

8. "CSS Grid Layout Module Level 1: 7.3. Named Areas: the Grid-Template-Areas Property." Edited by Tab Atkins et al., *W3C*, 14 Dec. 2017, www.w3.org/TR/css-grid- 1/#grid-template-areas-property.

9. "CSS Grid Layout Module Level 1: 7.4. Explicit Grid Shorthand: the Grid-Template Property." Edited by Tab Atkins et al., *W3C*, 14 Dec. 2017, www.w3.org/TR/css-grid- 1/#propdef-grid-template.

10. Baron, L. David. "CSS Conditional Rules Module Level 3." *W3C*, 4 Apr. 2013, www.w3.org/TR/css3-conditional/.

11. "Configuring the Viewport." *Safari Web Content Guide*, Apple, 12 Dec. 2016, developer.apple.com/library/archive/documentation/AppleApplications/Reference/SafariWebContent/UsingtheViewport/UsingtheViewport.html.

12. "Supported Meta Tags: Table 1 Viewport Properties." *Safari HTML Reference*, Apple, 15 July 2014, developer.apple.com/library/archive/documentation/AppleApplications/Reference/Safari HTMLRef/Articles/MetaTags.html#//apple_ref/doc/uid/TP40008193.

13. Schulzrinn, H. "The Tel URI for Telephone Numbers." *IETF Tools*, The Internet Society, Dec. 2004, tools.ietf.org/html/rfc3966.

14. Wilde, E., and A. Vaha-Sipila. "URI Scheme for Global System for Mobile Communications (GSM) Short Message Service (SMS)." *IETF Tools*, IETF Trust, Jan. 2010, tools.ietf.org/html/rfc5724.

15. "Media Queries." Edited by Eivind Uggedal, *Media Queries*, mediaqueri.es/.

16. "Media Queries Level 4." Edited by Florian Rivoal and Tab Atkins Jr., W3C, 5 Sept. 2017, www.w3.org/TR/mediaqueries-4.

17. "Media Queries: 4. Media Features." Edited by Håkon Lie et al., *Media Queries*, *W3C*, 19 June 2012, www.w3.org/TR/css3-mediaqueries/#media1.

18. Wroblewski, Luke, and Jeffrey Zeldman. *Mobile First*. A Book Apart, 2011.

19. Coyier, Chris. "Media Queries for Standard Devices." *CSS-Tricks*, 20 Feb. 2019, css- tricks.com/snippets/css/media-queries-for-standard-devices/.

20. "Mobile Viewports for Responsive Experiences." *Mobile Viewports for Responsive Experiences*, Adobe, 2020, docs.adobe.com/content/help/en/target/using/experiences/vec/mobile-viewports.html.

21. Marcotte, Ethan. *Responsive Web Design*. A Book Apart, 2014.

22. Lawson, Bruce. "Responsive Web Design: Preserving Images' Aspect Ratio." *Bruce Lawson*, 18 June 2012, www.brucelawson.co.uk/2012/responsive-web-design- preserving-images-aspect-ratio/.

23. "HTML 5.1 2nd Edition: 4.7.2 The picture element." *W3C*, 2 Oct. 2017, www.w3.org/TR/html51/semantics-embedded-content.html#the-picture-element.

24. "HTML 5.1 2nd Edition: 4.7.5 The img element." *W3C*, 2 Oct. 2017, www.w3.org/TR/html51/semantics-embedded-content.html#the-img-element.

25. "HTML Living Standard: 4.8.4.2.1 Srcset attributes." *WHATWG*, WHATWG, 10 Sept. 2020, html.spec.whatwg.org/#adaptive-images.

26. "HTML Living Standard: 2.6.7 Lazy Loading Attributes." *WHATWG*, WHATWG, 28 Apr. 2020, html.spec.whatwg.org/multipage/urls-and-fetching.html#lazy-loading-attribute.

27. "2000+ Desktop & Mobile Browsers for Cross-Browser Testing." *BrowserStack*, www.browserstack.com/list-of-browsers-and-platforms/live.

28. "Mobile Screen Resolution Stats Worldwide." *StatCounter Global Stats*, StatCounter, Mar. 2020, gs.statcounter.com/screen-resolution-stats/mobile/worldwide.

Table Basics

In this chapter, you'll become familiar with coding HTML tables to organize information on a web page.

You'll learn how to...

▌ Describe the recommended use of a table on a web page

▌ Configure a basic table with the table, table row, table header, and table cell elements

▌ Configure table sections with the thead, tbody, and tfoot elements

▌ Increase the accessibility of a table

▌ Style an HTML table with CSS

▌ Describe the purpose of CSS structural pseudo-classes

Table Overview

The purpose of a table is to organize information. In the past, before Cascading Style Sheets (CSS) was well supported by browsers, tables were also used to format web page layouts. An HTML table is composed of rows and columns, like a spreadsheet. Each individual table cell is at the intersection of a specific row and column.

- Each table begins with a `<table>` tag and ends with a `</table>` tag.
- Each table row begins with a `<tr>` tag and ends with a `</tr>` tag.
- Each cell (table data) begins with a `<td>` tag and ends with a `</td>` tag.
- Table cells can contain text, graphics, and other HTML elements.

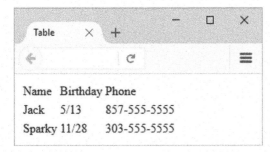

FIGURE 9.1 *Table with three rows and three columns.*

Figure 9.1 shows a sample table with three rows and three columns. The sample HTML for the table shown in Figure 9.1 follows:

```
<table>
  <tr>
    <td>Name</td>
    <td>Birthday</td>
    <td>Phone</td>
  </tr>
  <tr>
    <td>Jack</td>
    <td>5/13</td>
    <td>857-555-5555</td>
  </tr>
  <tr>
    <td>Sparky</td>
    <td>11/28</td>
    <td>303-555-5555</td>
  </tr>
</table>
```

Notice how the table is coded row by row. Also, each row is coded cell by cell. This attention to detail is crucial to the successful use of tables. An example can be found in the student files (chapter9/table1.html).

The Table Element

A **table element** is a block display element that contains tabular information. The table begins with a `<table>` tag and ends with a `</table>` tag.

The `border` Attribute

In earlier versions of HTML (such as HTML4 and XHTML), the purpose of the `border` attribute was to configure a visible table `border` by coding a table element with the `border` attribute set to the value 1 (border="1"). Although you'll see many existing web pages using this attribute, the `border` attribute is obsolete in HTML5. The modern approach is to configure CSS to style the border of a table. The following CSS configures a border around a table and around each table cell:

```
table, td, th { border: 1px solid #000; }
```

Table Captions

The **caption element** is often used with a data table to describe its contents. The table shown in Figure 9.2 (student files chapter9/table2.html) uses `<caption>` tags to set "Bird Sightings" as the caption. The caption element is coded on the line immediately after the opening `<table>` tag. The HTML for the table follows:

```
<table>
  <caption>Bird Sightings</caption>
  <tr>
    <td>Name</td>
    <td>Date</td>
  </tr>
  <tr>
    <td>Bobolink</td>
    <td>5/25/10</td>
  </tr>
  <tr>
    <td>Upland Sandpiper</td>
    <td>6/03/10</td>
  </tr>
</table>
```

Bird Sightings

Name	Date
Bobolink	5/25/20
Upland Sandpiper	6/03/20

FIGURE 9.2 *The caption for this table is Bird Sightings.*

? FAQ What about other attributes that I've seen coded on table tags, like `cellpadding`, `cellspacing`, and `summary` attributes?

Earlier versions of HTML (such as HTML4 and XHTML) provided a variety of attributes for configuring the table element, including `cellpadding`, `cellspacing`, `border`, `bgcolor`, `align`, `width`, and `summary`.[1] These attributes are considered invalid and obsolete in HTML5. It is preferred to configure presentational display characteristics (such as alignment, border, width, cell padding, cell spacing, and background color) within CSS instead of with HTML attributes. Although the `summary` attribute supported accessibility and served to describe the table, the W3C suggests using one of the following techniques to replace the `summary` attribute and provide context for a table: configure descriptive text in the caption element, provide an explanatory paragraph directly on the web page, or simplify the table.[2] You'll get practice configuring tables with CSS later in this chapter.

Table Rows, Cells, and Headers

VideoNote
Configure a Table

The **table row element** configures a row within a table on a web page. The table row begins with a `<tr>` tag and ends with a `</tr>` tag.

The **table data element** configures a cell within a row in a table on a web page. The table cell begins with a `<td>` tag and ends with a `</td>` tag. See Table 9.1 for common attributes of the table data cell element.

TABLE 9.1 *Commonly Used Attributes of the Table Data and Table Header Cell Elements*

Attribute	Value	Purpose
`colspan`	Numeric	The number of columns spanned by a cell
`headers`	The id value(s) of a column or row heading cell	Associates the table data cells with table header cells; may be accessed by screen readers
`rowspan`	Numeric	The number of rows spanned by a cell
`scope`	`row` or `column`	The scope of the table header cell contents (row or column); may be accessed by screen readers

The **table header element** is similar to a table data element and configures a cell within a row in a table on a web page. Its special purpose is to configure column and row headings. Text displayed within a table header element is centered and bold. The table header element begins with a `<th>` tag and ends with a `</th>` tag. See Table 9.1 for common attributes of the table header element.

Name	Birthday	Phone
Jack	5/13	857-555-5555
Sparky	11/28	303-555-5555

FIGURE 9.3 *Using `<th>` tags to indicate column headings.*

Figure 9.3 shows a table with column headings configured by `<th>` tags (student files chapter9/table3.html). As you view the following HTML for the table, notice that the first row uses `<th>` instead of `<td>` tags:

```
<table>
  <tr>
    <th>Name</th>
    <th>Birthday</th>
    <th>Phone</th>
  </tr>
  <tr>
    <td>Jack</td>
    <td>5/13</td>
    <td>857-555-5555</td>
  </tr>
  <tr>
    <td>Sparky</td>
    <td>11/28</td>
    <td>303-555-5555</td>
  </tr>
</table>
```

 Hands-On Practice 9.1

Create a web page similar to Figure 9.4 that describes two schools you have attended. Use "School History" as the caption. The table has three rows and three columns. The first row will have table header elements with the headings School Attended, Years, and Degree Awarded. You will complete the second and third rows with your own information within table data elements.

School History		
School Attended	**Years**	**Degree Awarded**
Schaumburg High School	2016—2020	High School Diploma
Harper College	2020—2021	Web Developer Certificate

FIGURE 9.4 *School History table.*

To get started, launch a text editor and open the student files chapter9/template.html. Modify the title element. Use table, table row, table header, table data, and caption elements to configure a table similar to Figure 9.4.

Hints: The table has three rows and three columns. Use the table header element for the cells in the first row. Configure the table and cell borders with embedded CSS in the head section of the web page:

```
table, td, th { border: 1px solid #000; }
```

Save your file and display it in a browser. It should look similar to Figure 9.4. A sample solution is found in the student files (chapter9/9.1).

Span Rows and Columns

This spans two columns

| Column 1 | Column 2 |

FIGURE 9.5 *Table with a row that spans two columns.*

You can alter the gridlike look of a table by applying the `colspan` and `rowspan` attributes to table data or table header elements. As you get into more complex table configurations like these, be sure to sketch the table on paper before you start typing the HTML.

The **colspan attribute** specifies the number of columns that a cell will occupy. Figure 9.5 shows a table cell that spans two columns.

The HTML for the table follows:

```
<table>
  <tr>
    <td   colspan="2">This spans two columns</td>
  </tr>
  <tr>
    <td>Column 1</td>
    <td>Column 2</td>
  </tr>
</table>
```

| This spans two rows | Row 1 Column 2 |
| | Row 2 Column 2 |

FIGURE 9.6 *Table with a column that spans two rows.*

The **rowspan attribute** specifies the number of rows that a cell will occupy. An example of a table cell that spans two rows is shown in Figure 9.6.

The HTML for the table follows:

```
<table>
  <tr>
    <td   rowspan="2">This spans two rows</td>
    <td>Row 1 Column 2</td>
  </tr>
  <tr>
    <td>Row 2 Column 2</td>
  </tr>
</table>
```

An example of the tables in Figures 9.5 and 9.6 can be found in the student files (chapter9/table4.html).

Hands-On Practice 9.2

You will create a web page with the table shown in Figure 9.7 in this Hands-On Practice.

Launch a text editor and open chapter9/ template.html in the student files. Modify the title element. Use table, table row, table head, and table data elements to configure the table.

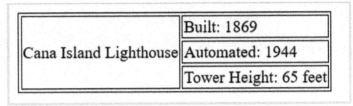

FIGURE 9.7 *Practice with the rowspan attribute.*

1. Code the opening `<table>` tag.

2. Begin the first row with a `<tr>` tag.

3. The table data cell with "Cana Island Lighthouse" spans three rows. Code a table data element. Use the `rowspan="3"` attribute.

4. Code a table data element that contains the text "Built: 1869".

5. End the first row with a `</tr>` tag.

6. Begin the second row with a `<tr>` tag. This row will only have one table data element because the cell in the first column is already reserved for "Cana Island Lighthouse".

7. Code a table data element that contains the text "Automated: 1944".

8. End the second row with a `</tr>` tag.

9. Begin the third row with a `<tr>` tag. This row will only have one table data element because the cell in the first column is already reserved for "Cana Island Lighthouse".

10. Code a table data element that contains the text "Tower Height: 65 feet".

11. End the third row with a `</tr>` tag.

12. Code the closing `</table>` tag.

13. Configure the table and cell borders with embedded CSS in the head section of the web page:

```
table, td, th { border: 1px solid #000; }
```

Save the file and view it in a browser. A sample solution is found in the student files (chapter9/9.2). Notice how the "Cana Island Lighthouse" text is vertically aligned in the middle of the cell—this is the default vertical alignment. You can modify the vertical alignment using CSS—see the section "Style a Table with CSS" later in this chapter.

Configure an Accessible Table

 Tables can be useful to organize information on a web page, but what if you couldn't see the table and were relying on assistive technology like a screen reader to read the table to you? You'd hear the contents of the table just the way it is coded—row by row, cell by cell. This might be difficult to understand. This section discusses coding techniques to improve the accessibility of tables.

Bird Sightings

Name	Date
Bobolink	5/25/20
Upland Sandpiper	6/03/20

FIGURE 9.8 *This simple data table uses <th> tags and the caption element to provide for accessibility.*

Table Header and Caption Elements

For a simple informational data table like the one shown in Figure 9.8, the W3C recommends the following:

▶ Use table header elements (<th> tags) to indicate column or row headings.[3]

▶ Use the caption element to provide a text title or caption for the table.[4]

An example web page is in the student files (chapter9/table5.html). The HTML follows:

```
<table>
<caption>Bird Sightings</caption>
  <tr>
    <th>Name</th>
    <th>Date</th>
  </tr>
  <tr>
    <td>Bobolink</td>
    <td>5/25/10</td>
  </tr>
  <tr>
    <td>Upland Sandpiper</td>
    <td>6/03/10</td>
  </tr>
</table>
```

The headers and id Attributes

However, for more complex tables, the W3C recommends specifically associating the table data cell values with their corresponding headers.[5] The technique that is recommended uses the id attribute (usually in a <th> tag) to identify a specific header cell and the **headers attribute** in a <td> tag. The code to configure the table in Figure 9.8 using headers and id attributes is as follows (also found in the student files chapter9/table6.html):

```
<table>
<caption>Bird Sightings</caption>
  <tr>
    <th id="name">Name</th>
    <th id="date">Date</th>
  </tr>
  <tr>
    <td headers="name">Bobolink</td>
    <td headers="date">5/25/10</td>
  </tr>
  <tr>
    <td headers="name">Upland Sandpiper</td>
    <td headers="date">6/03/10</td>
  </tr>
</table>
```

The scope Attribute

The **scope attribute** specifies the association of table cells and table row or column, headers. It is used to indicate whether a table cell is a header for a column (scope="col"), row (scope="row"), a group of columns (scope="colgroup") , or a group of rows (scope="rowgroup"). An example of the code for the table in Figure 9.8 that uses this attribute is as follows (student files chapter9/table7.html):

```
<table>
<caption>Bird Sightings</caption>
  <tr>
    <th scope="col">Name</th>
    <th scope="col">Date</th>
  </tr>
  <tr>
    <td>Bobolink</td>
    <td>5/25/10</td>
  </tr>
  <tr>
    <td>Upland Sandpiper</td>
    <td>6/03/10</td>
  </tr>
</table>
```

The scope attribute can be useful for a complex table with multiple levels of row and column headings. However, the W3C recommends using multiple smaller, basic tables instead of a large, complex table.[6] The student files (chapter9/table8.html) contains an example of a complex table utilizing the scope attribute and two simple tables that convey similar information.

Style a Table with CSS

Before CSS was well supported by browsers, it was common practice to configure the visual esthetic of a table with HTML attributes. The modern approach is to use CSS to style a table. Table 9.2 lists corresponding CSS properties with HTML attributes used to style tables.

TABLE 9.2 *Configuring Tables with HTML Attributes and CSS Properties*

HTML Attribute	CSS Property
align	To align a table, configure the `width` and `margin` properties for the table element selector. To center a table, use `table { width: 75%; margin: auto; }` To align content within table cells, use `text-align`
width	`width`
height	`height`
cellpadding	`padding`
cellspacing	`border-spacing` configures space between cell borders with a numeric value (px or em) or percentage. If you set a value to 0, omit the unit. One value configures both horizontal and vertical spacing. When two values are used, the first value configures the horizontal spacing and the second value configures the vertical spacing. `border-collapse` configures the border area. The values are `separate` (default) and `collapse` (removes extra space between table and cell borders).
bgcolor	`background-color`
valign	`vertical-align`
border	`border`, `border-style`, `border-spacing`
none	`background-image`
none	`caption-side` specifies caption placement. Values are `top` (default) and `bottom`

 Hands-On Practice 9.3

Lighthouse Island Bistro Specialty Coffee Menu

Specialty Coffee	Description	Price
Lite Latte	Indulge in a shot of espresso with steamed, skim milk	$3.50
Mocha Latte	Choose dark or mile chocolate with steamed milk	$4.00
MCP Latte	A lucious mocha latte with caramel and pecan syrup	$4.50

FIGURE 9.9 *The table before CSS.*

In this Hands-On Practice, you will code CSS style rules to configure an informational table on a web page. Create a folder named ch9table. Copy the starter.html file from the chapter9 folder in the student files to your ch9table folder. Display the file in a browser. The page should look similar to Figure 9.9.

Launch a text editor and open the starter.html file from your ch9table folder. Locate the style tags in the head section. You will code embedded CSS in this Hands-On Practice. Position your cursor on the blank line between the style tags.

1. Configure the table element selector to be centered, have a dark blue, 5 pixel border, and have a width of 600px:

   ```
   table { margin: auto; border: 5px solid #000066; width: 600px; }
   ```

   Save the file as menu.html and display your page in a browser. Notice that there is a dark blue border surrounding the entire table.

2. Configure the td and th element selectors with a border, padding, and Arial or the default sans-serif font typeface:

   ```
   td, th { border: 1px solid #000066; padding: 5px;
            font-family: Arial, sans-serif; }
   ```

Save the file as menu.html and display your page in a browser. Each table cell should now be outlined with a dark blue border and should display text in a sans-serif font. Continue editing the CSS.

3. Eliminate the empty space between the borders of the table cells with the **border-spacing property**. Add a `border-spacing: 0;` declaration to the table element selector. Save the file and display your page in a browser.

4. Configure the caption to be displayed with Verdana or the default sans-serif font type-face, bold font weight, font size 1.2em, and 5 pixels of bottom padding:

   ```
   caption { font-family: Verdana, sans-serif; font-weight: bold;
             font-size: 1.2em; padding-bottom: 5px; }
   ```

5. Let's experiment and configure background colors for the rows instead of cell borders. Modify the style rule for the td and th element selectors, remove the border declaration, and set `border-style` to none:

   ```
   td, th { padding: 5px; font-family: Arial, sans-serif;
            border-style: none; }
   ```

6. Create a new class called `altrow` that sets a background color:

   ```
   .altrow { background-color: #EAEAEA; }
   ```

7. Modify the <tr> tags in the HTML: assign the second and fourth <tr> tags to the `altrow` class. Save the file. Display it in a browser. The table area should look similar to Figure 9.10.

Notice how the background color of the alternate rows adds subtle interest to the web page. Compare your work with the sample located in the student files (chapter9/9.3).

Lighthouse Island Bistro Specialty Coffee Menu

Specialty Coffee	Description	Price
Lite Latte	Indulge in a shot of espresso with steamed, skim milk	$3.50
Mocha Latte	Choose dark or mile chocolate with steamed milk	$4.00
MCP Latte	A lucious mocha latte with caramel and pecan syrup	$4.50

FIGURE 9.10 *Rows are configured with alternating background colors.*

CSS Structural Pseudo-classes

In the previous section, you configured CSS and applied a class to every other table row to configure alternating background colors, often referred to as "zebra striping." You may have found this to be a bit inconvenient and wondered if there was a more efficient method. Well, there is! CSS **structural pseudo-classes** allow you to select and apply classes to elements based on their position in the structure of the document, such as every other row. CSS pseudo-classes are supported by current versions of popular browsers. Table 9.3 lists common CSS structural pseudo-classes and their purpose.[7]

To apply a pseudo-class, write it after the selector. The following code sample will configure the first item in an unordered list to display with red text.

```
li:first-of-type { color: #FF0000; }
```

TABLE 9.3   *Common CSS Structural Pseudo-classes*

Pseudo-class	Purpose
:first-of-type	Applies to the first element of the specified type
:first-child	Applies to the first child of an element
:last-of-type	Applies to the last element of the specified type
:last-child	Applies to the last child of an element
:nth-of-type(n)	Applies to the "nth" element of the specified type
	Value: an integer, odd, or even

 Hands-On Practice 9.4

In this Hands-On Practice, you will rework the table you configured in Hands-On Practice 9.3 to use CSS structural pseudo-classes to configure color.

1. Launch a text editor and open the menu.html file in your ch9table folder (also found in the student files chapter9/9.3). Save the file as menu2.html.

2. View the source code and notice that the second and fourth tr elements are assigned to the `altrow` class. You won't need this class assignment when using CSS structural pseudo-classes. Delete `class="altrow"` from the tr elements.

3. Examine the embedded CSS and locate the `altrow` class. Change the selector to use a structural pseudo-class that will apply the style to the even-numbered table rows. Replace `.altrow` with `tr:nth-of-type (even)` as shown in the following CSS declaration:

```
tr:nth-of-type(even) { background-color: #EAEAEA; }
```

4. Save the file. Display your page in a browser. The table area should look similar to the one shown in Figure 9.11.

5. Let's configure the first row to have a dark blue background (#006) and light gray text (#EAEAEA) with the `:first-of-type` structural pseudo-class. Add the following to the embedded CSS:

Lighthouse Island Bistro Specialty Coffee Menu

Specialty Coffee	Description	Price
Lite Latte	Indulge in a shot of espresso with steamed, skim milk	$3.50
Mocha Latte	Choose dark or mile chocolate with steamed milk	$4.00
MCP Latte	A lucious mocha latte with caramel and pecan syrup	$4.50

FIGURE 9.11 *The table at Step 4.*

```
tr:first-of-type { background-color: #006; color: #EAEAEA }
```

6. Save the file. Display your page in a browser. The table area should look similar to the one shown in Figure 9.12. A sample solution is available in the student files (chapter9/9.4).

Lighthouse Island Bistro Specialty Coffee Menu

Specialty Coffee	Description	Price
Lite Latte	Indulge in a shot of espresso with steamed, skim milk	$3.50
Mocha Latte	Choose dark or mile chocolate with steamed milk	$4.00
MCP Latte	A lucious mocha latte with caramel and pecan syrup	$4.50

FIGURE 9.12 *CSS pseudo-class selectors style the table rows.*

Configuring the First Letter

The CSS `::first-letter` pseudo-element[8] configures the first letter of an element, as shown in Figure 9.13. The code follows:

```
p::first-letter { font-size: 3em;
                  font-weight: bold; color: #F00; }
```

See chapter9/letter.html in the student files for an example.

FIGURE 9.13 *Configure the first letter with CSS.*

Explore the topic of pseudo-elements further. Find out about the `:before`, `:after`, and `:first-line` pseudo-elements at the following resources:

▶ https://css-tricks.com/pseudo-element-roundup
▶ https://www.hongkiat.com/blog/pseudo-element-before-after

Configure Table Sections

There are many configuration options when coding tables. Table rows can be put together into three types of table row groups: table head with `<thead>`, table body with `<tbody>`, and table footer with `<tfoot>`.

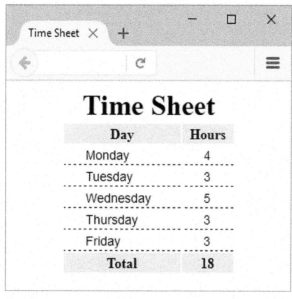

FIGURE 9.14 *CSS configures the thead, tbody, and tfoot element selectors.*

These groups can be useful when you need to configure the areas in the table in different ways, using either attributes or CSS. The `<tbody>` tag is required if you configure a `<thead>` or `<tfoot>` area, although you can omit either the table head or table footer if you like.

The following code sample (see chapter9/tfoot.html in the student files) configures the table shown in Figure 9.14 and demonstrates the use of CSS to configure a table head, table body, and table footer with different styles.

The CSS styles a centered 200-pixel-wide table with a caption that is rendered in a large, bold font; a table head section with a light gray (#EAEAEA) background color; and a table body section styled with slightly smaller text (.90em) using a sans-serif font; table body td element selectors set to display with some left padding and a dashed bottom border; and a table footer section that has centered, bolded text and a light gray background color (#EAEAEA). The CSS follows:

```
table { width: 200px;
        margin: auto; }
table, th, td { border-style: none; }
caption { font-size: 2em;
          font-weight: bold; }
thead { background-color: #EAEAEA; }
tbody { font-family: Arial, sans-serif;
        font-size: .90em; }
tbody td { border-bottom: 1px #000033 dashed;
           padding-left: 25px; }
tfoot { background-color: #EAEAEA;
        font-weight: bold;
        text-align: center; }
```

The HTML for the table follows:

```
<table>
<caption>Time Sheet</caption>
<thead>
  <tr>
    <th id="day">Day</th>
    <th id="hours">Hours</th>
  </tr>
</thead>
<tbody>
  <tr>
    <td headers="day">Monday</td>
    <td headers="hours">4</td>
  </tr>
  <tr>
    <td headers="day">Tuesday</td>
    <td headers="hours">3</td>
  </tr>
  <tr>
    <td headers="day">Wednesday</td>
    <td headers="hours">5</td>
  </tr>
  <tr>
    <td headers="day">Thursday</td>
    <td headers="hours">3</td>
  </tr>
  <tr>
    <td headers="day">Friday</td>
    <td headers="hours">3</td>
  </tr>
</tbody>
<tfoot>
  <tr>
    <td headers="day">Total</td>
    <td headers="hours">18</td>
  </tr>
</tfoot>
</table>
```

This example demonstrates the power of CSS in styling documents. The `<td>` tags within each table row group element selector (`thead`, `tbody`, and `tfoot`) inherited the font styles configured for their parent group element selector. Notice how a descendant selector configures padding and border only for `<td>` tags that are contained within the `<tbody>` element. Sample code is located in the student files (chapter9/tfoot.html). Take a few moments to explore the web page code and display the page in a browser.

Review and Apply

Review Questions

Multiple Choice. Choose the best answer for each item.

1. Which HTML tag pair is used to begin and end a table row?

   a. `<td> </td>`   **b.** `<tr> </tr>`

   c. `<table> </table>`   **d.** `<th> </th>`

2. Which CSS declaration removes extra space between table and cell borders?

   a. `display: none;`

   b. `border-style: none;`

   c. `border-spacing: 0;`

   d. `border-collapse: 0;`

3. Which HTML tag pair is used to group rows in the footer of a table?

   a. `<footer> </footer>`

   b. `<tr> </tr>`

   c. `<tfoot> </tfoot>`

   d. `<trfoot> </trfoot>`

4. Which of the following configures a table cell with bold, centered text?

   a. table row element

   b. table data element

   c. table header element

   d. table element

5. Which HTML tag pair is used to specify table headings?

   a. `<td> </td>`   **b.** `<th> </th>`

   c. `<head> </head>`   **d.** `<tr> </tr>`

6. Which CSS property replaces the use of the HTML cellpadding attribute?

   a. `cell-padding`

   b. `border-spacing`

   c. `padding`

   d. `border`

7. Which HTML element provides a text description of the contents of a table?

   a. `<table>`   **b.** `<caption>`

   c. `<summary>`   **d.** `<thead>`

8. Which of the following is the recommended use of tables on web pages?

   a. configuring the layout of an entire page

   b. organizing information

   c. forming hyperlinks

   d. configuring a resume

9. Which CSS property specifies the background color of a table?

   a. `table-background`

   b. `bgcolor`

   c. `background-color`

   d. `border-spacing`

10. Which HTML attribute associates a table data cell with a table header cell?

    a. `head`   **b.** `headers`

    c. `align`   **d.** `rowspan`

Review Answers

1. b   2. c   3. c   4. c   5. b   6. c   7. b   8. b   9. c   10. b

Hands-On Exercises

1. Write the HTML for a two-column table that contains the names of your friends and their birthdays. The first row of the table should span two columns and contain the following heading: "Birthday List". Include at least two people in your table.

2. Write the HTML for a three-column table to describe the courses you are taking this semester. The columns should contain the course number, course name, and instructor name. The first row of the table should use th tags and contain descriptive headings for the columns. Use the table row grouping tags <thead> and <tbody> in your table.

3. Use CSS to configure a table that has a red border around both the entire table and the table cells. Write the HTML to create a table with four rows and two columns. The cell in the first column of each row will contain one of the following terms: contrast, repetition, alignment, and proximity. The corresponding cell in the second column of each row will contain a description of the term as it applies to web design (see Chapter 3).

4. Create a web page about your favorite sports team with a two-column table that lists the positions and starting players. Use embedded CSS to style the table border, background color, and center the table on the web page. Place an e-mail link to yourself in the footer area. Save the file as sport9.html.

5. Create a web page about your favorite movie that uses a two-column table containing details about the movie. Use embedded CSS to style the table border and background color. Include the following in the table:

   - Title of the movie
   - Director or producer
   - Leading actor
   - Leading actress
   - Rating (R, PG-13, PG, G, and NR)
   - A brief description of the movie
   - An absolute link to a review about the movie

   Place an e-mail link to yourself on the web page. Save the page as movie9.html.

Focus on Web Design

Good artists view and analyze many paintings. Good writers read and evaluate many books. Similarly, good web designers view and scrutinize many web pages. Explore the Web and find two web pages—one that is appealing to you and one that is unappealing to you. Print out each page. Create a web page that answers the following questions for each of your examples:

   a. What is the URL of the website?
   b. Does this page use tables? If so, for what purpose—page layout, organization of information, or another reason?
   c. Does this page use CSS? If so, for what purpose—page layout, text and color configuration, or another reason?
   d. Is this page appealing or unappealing? Describe three aspects of the page that you find appealing or unappealing.
   e. If this page is unappealing, what would you do to improve it?

Case Study

You will continue the case studies from Chapter 8 as you configure a table on one of the web pages.

Pacific Trails Resort Case Study

In this chapter's case study, you will use the Pacific Trails Resort existing website (Chapter 8) as a starting point and add an informational table to the Yurts page on the Pacific Trails website. Your new page will be similar to Figure 9.15 when you have completed this case study.

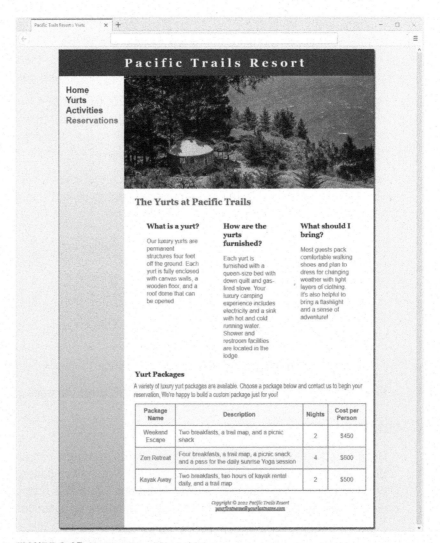

FIGURE 9.15 *Yurts page with a table.*

You have three tasks in this case study:

1. Create a new folder for this Pacific Trails case study.
2. Modify the style sheet (pacific.css) to configure style rules for the new table.
3. Modify the Yurts page to use a table to display information, as shown in Figure 9.15.

Task 1: Create a folder called ch9pacific to contain your Pacific Trails Resort website files. Copy the files from the Chapter 8 Case Study ch8pacific folder into the new ch9pacific folder.

Task 2: Configure the CSS. You will add styles to configure the table on the Yurts page. Launch a text editor and open the pacific.css external style sheet file. You will add new style rules above the media queries.

- **Configure the table.** Code a new style rule for the table element selector that configures a table with a 2 pixel solid blue border (#3399CC) and no cellspacing (use `border-collapse: collapse;`).
- **Configure the table cells.** Code a new style rule for the td and th element selectors that sets padding to 0.5em and configures a 2 pixel solid blue border (#3399CC).
- **Center the td content.** Code a new style rule for the `td` element selector that centers text.
- **Configure the text class.** Notice that the content in the table data cells that contain the text description is not centered. Code a new style rule for a class named `text` that will override the td style rule and left-align the text.
- **Configure alternate-row background color.** The table looks more appealing if the rows have alternate background colors but is still readable without them. Apply the `:nth-of-type` CSS pseudo-class to configure the odd table rows with a light blue background color (#F5FAFC).

Save the pacific.css file.

Task 3: Update the Yurts Page. Open the yurts.html page for the Pacific Trails Resort website in a text editor.

- Add a blank line above the closing main tag.
- Configure the text "Yurt Packages" within an h3 element.
- Below the new h3 element, configure a paragraph with the following text:

  A variety of luxury yurt packages are available. Choose a package below and contact us to begin your reservation. We're happy to build a custom package just for you!

- You are ready to configure the table. Position your cursor on a blank line under the paragraph and code a table with four rows and four columns. Use the table, th, and td elements. Assign the td elements that contain the detailed descriptions to the class named `text`. The content for the table is as follows:

Package Name	Description	Nights	Cost per Person
Weekend Escape	Two breakfasts, a trail map, and a picnic snack	2	$450
Zen Retreat	Four breakfasts, a trail map, a picnic snack, and a pass for the daily sunrise Yoga session	4	$600
Kayak Away	Two breakfasts, two hours of kayak rental daily, and a trail map	2	$500

Save your yurts.html file. Launch a browser and test your new page. It should look similar to Figure 9.15. If the page does not display as you intended, review your work, validate the CSS, validate the HTML, modify as needed, and test again.

Path of Light Yoga Studio Case Study

In this chapter's case study, you will use the Path of Light Yoga Studio existing website (Chapter 8) as a starting point and modify the Schedule page to use two HTML tables to display information. Your new page will be similar to Figure 9.16 when you have completed this case study.

FIGURE 9.16 *Schedule page with HTML tables.*

You have three tasks in this case study:

1. Create a new folder for this Path of Light Yoga Studio case study.
2. Modify the style sheet (yoga.css) to configure style rules for the new tables.
3. Modify the Schedule page to use tables to display information as shown in Figure 9.16.

Task 1: Create a folder called ch9yoga to contain your Path of Light Yoga Studio website files. Copy the files from the Chapter 8 Case Study ch8yoga folder into the new ch9yoga folder.

Task 2: Configure the CSS. You will add styles to configure the tables on the Schedule page. Launch a text editor and open the yoga.css external style sheet file. You will add the new style rules above the media queries.

- **Configure the tables.** Code a new style rule for the table element selector that configures a 1 pixel purple (#40407A) border, a 1em bottom margin, and no cellspacing (use `border-collapse: collapse;`).

- **Configure the table cells.** Code a new style rule for the td and th element selectors that configures 0.5em of padding and 1 pixel purple border (#40407A).

- **Configure alternate-row background color.** The table looks more appealing if the rows have alternate background colors but is still readable without them. Apply the `:nth-of-type` CSS pseudo-class to configure the even table rows with a #D7E8E9 background color.

- **Configure table captions.** Code a new style rule that sets a 1em margin, bold text, and 120% font size.

Save the yoga.css file.

Task 3: Update the Schedule Page. Open the schedule.html page for the Path of Light Yoga Studio website in a text editor. The schedule information currently uses the <h3>, <ul>, and <li> elements. You will rework the page to use two tables to display the schedule information instead of the h3, ul, and li tags. Place each table element within an existing section element. Use a caption element within each table. Note that the table rows have two columns. Configure "Time" and "Class" table headings within each table. Refer to Figure 9.16. Save your page and test it in a browser. If the page does not display as you intended, review your work, validate the CSS, validate the HTML, modify as needed, and test again.

Endnotes

1. "HTML Living Standard: 15 Obsolete Features." *HTML Standard*, WHATWG, 2 May 2020, html.spec.whatwg.org/multipage/obsolete.html.

2. "HTML Living Standard: 4.9.1.1 Techniques for Describing Tables." *HTML Standard*, WHATWG, 2 May 2020, html.spec.whatwg.org/multipage/ tables.html#table-descriptions-techniques.

3. Accessibility Guidelines Working Group. "Using table markup to present tabular information." Edited by Alastair Campbell et al., *Techniques for WCAG 2.1*, W3C, 2018, www. w3.org/WAI/WCAG21/Techniques/html/H51.

4. Accessibility Guidelines Working Group. "Using caption elements to associate data table captions with data tables." Edited by Alastair Campbell et al., *Techniques for WCAG 2.1*, W3C, 2018, www.w3.org/WAI/WCAG21/Techniques/html/H39.

5. Accessibility Guidelines Working Group. "Using id and headers attributes to associate data cells with header cells in data tables." Edited by Alastair Campbell et al., *Techniques for WCAG 2.1*, W3C, 2018, www.w3.org/WAI/WCAG21/Techniques/ html/H43.

6. Accessibility Guidelines Working Group. "Using the scope attribute to associate header cells and data cells in data tables." Edited by Alastair Campbell et al., *Techniques for WCAG 2.1*, W3C, 2018, www.w3.org/WAI/WCAG21/Techniques/html/H63.
7. "Selectors Level 3: 6.6.5. Structural pseudo-classes." Edited by Tantek Çelik et al., *W3C*, W3C, 30 Jan. 2018, www.w3.org/TR/selectors-3/#structural-pseudos.
8. "Selectors Level 3: 7.2. The ::First-Letter Pseudo-Element." Edited by Tantek Çelik et al., *W3C*, W3C, 30 Jan. 2018, www.w3.org/TR/selectors-3/#first-letter.

Form Basics

Forms are used for many purposes all over the Web. They are used by search engines to accept keywords and by online stores to process e-commerce shopping carts. Websites use forms to help with a variety of functions—accepting visitor feedback, encouraging visitors to send a news story to a friend or colleague, collecting e-mail addresses for a newsletter, and accepting order information. This chapter introduces a very powerful tool for web developers—forms that accept information from web page visitors.

You'll learn how to...

- Describe common uses of forms on web pages
- Create forms on web pages using the form, input, textarea, and select elements
- Associate form controls and groups using label, fieldset, and legend elements
- Use the CSS float property to style a form
- Use CSS Grid layout to style a form
- Use CSS Flexbox Layout to style a form
- Describe the features and common uses of server-side processing
- Invoke server-side processing to handle form data
- Configure form controls including the e-mail, URL, datalist, range, spinner, calendar, and color-well controls

Form Overview

Every time you use a search engine, place an order, or join an online mailing list, you use a form. A **form** is an HTML element that contains and organizes objects called **form controls**—such as text boxes, check boxes, and buttons—that can accept information from website visitors.

For example, you may have used Google's search form (https://www.google.com) many times but never thought about how it works. The form is quite simple; it contains just three form controls—the text box that accepts the keywords used in the search and two buttons. The "Google Search" button submits the form and invokes a process to search the Google databases and display a results page. The whimsical "I'm Feeling Lucky" button submits the form and displays the top page for your keywords.

Figure 10.1 shows a form used to enter shipping information. This form contains text boxes to accept information such as name and address. Select lists are used to capture information with a limited number of correct values, such as state and country information. When a visitor clicks the "Continue" button, the form information is submitted and the ordering process continues.

Whether a form is used to search for web pages or to place an order, the form alone cannot do all the processing. The form needs to invoke a program or script on the server in order to search a database or record an order. There are usually two components of a form:

1. The HTML form itself, which is the web page user interface.
2. The server-side processing, which works with the form data and sends e-mail, writes to a text file, updates a database, or performs some other type of processing on the server.

Shipping Address

Name:	
Address Line 1:	
Address Line 2:	
City:	
State:	▾ Zip:
Country:	United States ▾

[Continue]

FIGURE 10.1 *This form accepts order information.*

The Form Element

Now that you have a basic understanding of what forms do, let's focus on the HTML to create a form. The **form element** contains a form on a web page. The `<form>` tag specifies the beginning of a form area. The closing `</form>` tag specifies the end of a form area. There can be multiple forms on a web page, but they cannot be nested inside each other. The form element can be configured with attributes that specify what server-side program or file will process the form, how the form information will be sent to the server, and the name of the form. These attributes are listed in Table 10.1.

TABLE 10.1 *Common Attributes of the Form Element*

Attribute	Value	Purpose
action	URL or file name/path of server-side processing script	Required; indicates where to send the form information when the form is submitted; `mailto:youremailaddress` will launch the visitor's default e-mail application to send the form information
autocomplete	on	Default value; browser will use autocompletion to fill form fields
	off	Browser will not use autocompletion to fill form fields
id	Alphanumeric, no spaces; the value must be unique and not used for other id values on the same web page document	Optional; provides a unique identifier for the form
method	get	Default value; the value of `get` causes the form data to be appended to the URL and sent to the web server
	post	The `post` method is more private and transmits the form data in the body of the HTTP response; this method is preferred by the W3C
name	Alphanumeric, no spaces, begins with a letter; choose a form name value that is descriptive but short; for example, OrderForm is better than Form1 or WidgetsRUsOrderForm	Optional; names the form so that it can be easily accessed by client-side scripting languages, such as JavaScript, to edit and verify the form information before the server-side processing is invoked

For example, the HTML below will configure a form called order that uses the post method and invokes a script called demo.php on your web server.

```
<form name="order" method="post" id="order" action="demo.php">
... form controls go here ...
</form>
```

Form Controls

The purpose of a form is to gather information from a web page visitor; form controls are the objects that accept the information. Types of form controls include text boxes, scrolling text boxes, select lists, radio buttons, check boxes, and buttons. HTML5 offers new form controls including those customized for e-mail addresses, URLs, dates, times, numbers, and color selection. HTML elements that configure form controls will be introduced in the following sections.

Input Element and Text Box

The **input element** can configure a wide variety of different form controls, including text boxes, submit buttons, check boxes, radio buttons, date and time pickers, and sliders. This element is not coded as a pair of opening and closing tags. It is considered to be a stand-alone or void element. The `type` attribute specifies the type of form control that the browser should display. As you work thought the chapter, you will explore the form controls listed in Table 10.2.

TABLE 10.2 *Common Type Attribute Values*

Value	Form Control
`text` (default)	Single line text box
`checkbox`	Check box
`color`	Color picker or color well
`date`	Date picker
`email`	Text box that validates for email format
`file`	Button that invokes the operating system dialog box for choosing a file
`hidden`	Does not appear on screen but can contain a value
`number`	Spinner control for numerical selection
`password`	Text box that obscures the characters entered
`radio`	Radio button
`range`	Slider control for numerical selection
`reset`	Button that causes form to return to initial state
`search`	Text box that accepts a search term
`submit`	Button that submits the form
`tel`	Text box that accepts a telephone number
`url`	Text box that validates URL format

Sample Text Box

E-mail:

FIGURE 10.2 *Text box configured with the input element form element.*

Text Box

The input element with `type="text"` configures a **text box.** The text box form control accepts a single line of text or numeric information such a name, e-mail address, or phone number.

A sample text box is shown in Figure 10.2. The HTML for the text box is shown below.

```
E-mail: <input type="text" name="email" id="email">
```

Table 10.3 describes commonly used input element attributes for text boxes. As you view Table 10.3, notice that there is no value indicated for the `required`, `disabled`, `readonly`, and `autofocus` attributes. These attributes are boolean attributes. A boolean attribute is not coded with a value. If a **boolean attribute** is present, the browser behavior is triggered.

TABLE 10.3 *Common Text Box Input Element Attributes*

Attribute	Value	Usage
type	text (default)	Configures a text box
name	Alphanumeric, no spaces, begins with a letter	Uniquely names the form element so that it can be easily accessed by client-side scripting and server-side processing
id	Alphanumeric, no spaces, begins with a letter	Provides a unique identifier for the form element; used with CSS and client-side scripting; the values assigned to the name and id attributes on a form element are usually the same.
size	Numeric	The width of the text box as displayed by the browser; if size is omitted, the browser displays the text box with its own default size
maxlength	Numeric	The maximum length of data accepted by the text box
value	Text or numeric characters	Assigns an initial value to the text box that is displayed by the browser; accepts information typed in the text box; this value can be accessed by client-side scripting languages and server-side processing
required		A boolean attribute that is coded without any value; if present the browser verifies entry of information.
disabled		A boolean attribute that is coded without any value; if present the form control is disabled.
readonly		A boolean attribute that is coded without any value; if present the form control is for display; cannot be edited
autocomplete	on, off, token value	Supporting browsers will use autocompletion to fill in the form control[1]
autofocus		A boolean attribute that is coded without any value; if present, the form control has cursor focus
list	Datalist element value	Associates the form control with a datalist element id
placeholder	Text or numeric	Browser displays a tip or hint to aid the user
accesskey	Keyboard character	Configures a hot key for the form control
tabindex	Numeric	Configures the tab order of the form control

The **required attribute** is a boolean attribute that will cause supporting browsers to perform form validation. Browsers that support the required attribute will automatically verify that the information has been entered in the text box and display an error message when the condition is not met. Sample HTML follows:

```
E-mail: <input type="text"
        name="email" id="email" required>
```

Figure 10.3 shows an error message automatically generated by Firefox that displayed after the user clicked the form's submit button without entering information in the required text box. Browsers that do not support the required attribute will ignore the attribute.

FIGURE 10.3 *The browser displayed an error message.*

Submit Button and Reset Button

The Submit Button

The **submit button** form control is used to submit the form. When clicked, it triggers the action method on the `<form>` tag and causes the browser to send the form data (the name and value pairs for each form control) to the web server. The web server will invoke the server-side processing program or script listed on the form's `action` attribute.

The input element with `type="submit"` configures a submit button. For example,

```
<input type="submit">
```

The Reset Button

The **reset button** form control is used to reset the form fields to their initial values. A reset button does not submit the form.

The input element with `type="reset"` configures a reset button. For example,

```
<input type="reset">
```

Sample Form

A form with a text box, a submit button, and a reset button is shown in Figure 10.4. Common attributes for submit buttons and reset buttons are listed in Table 10.4.

Sample Form

E-mail: _____

[Submit Query] [Reset]

FIGURE 10.4 *The form contains a text box, a submit button, and a reset button.*

TABLE 10.4 *Common Attributes for Submit Buttons and Reset Buttons*

Attribute	Value	Usage
type	submit	Configures a submit button
	reset	Configures a reset button
name	Alphanumeric, no spaces, begins with a letter	Names the form element so that it can be easily accessed by client-side scripting and server-side processing; the name should be unique
id	Alphanumeric, no spaces, begins with a letter	Provides a unique identifier for the form element
value	Text or numeric characters	Configures the text displayed on the button; a submit button displays text "Submit Query" by default; a reset button displays "Reset" by default
accesskey	Keyboard character	Configures a hot key for the form control
tabindex	Numeric	Configures the tab order of the form control

Hands-On Practice 10.1 ──────────────────────────────

You will code a form in this Hands-On Practice. To get started, launch a text editor and open the template file located at chapter10/template.html in the student files. Save the file with the name join.html. You will create a web page with a form similar to the example in Figure 10.5.

1. Modify the title element to display the text: Form Example.

2. Configure an h1 element with the text: Join Our Newsletter.

3. You are ready to configure the form area. A form begins with the form element. Place your cursor on a blank line under the heading you just added and type in a `<form>` tag as follows:

   ```
   <form method="get">
   ```

   As you read through the chapter, you will find that a number of attributes can be used with the `<form>` element. In your first form, we are using the minimal HTML needed to create the form.

FIGURE 10.5 *Example form.*

4. To create the form control for the visitor's e-mail address to be entered, type the following code
   on a blank line below the form element:

   ```
   E-mail: <input type="text" name="email" id="email"><br><br>
   ```

   This places the text "E-mail:" in front of the text box used to enter the visitor's e-mail address. The input element has a `type` attribute with the value of text that causes the browser to display a text box. The `name` attribute assigns the name `email` to the information entered into the text box (the `value`) and could be used by server-side processing. The `id` attribute uniquely identifies the element on the page. The `<br>` elements configure line breaks.

5. Now you are ready to add the submit button to the form on the next line. Add a `value` attribute set to "Sign Me Up!":

   ```
   <input type="submit" value="Sign Me Up!">
   ```

   This causes the browser to display a button with "Sign Me Up!" instead of the default value of Submit Query.

6. Add a blank space after the submit button and code a reset button:

   ```
   <input type="reset">
   ```

7. Next, code the closing form tag:

   ```
   </form>
   ```

Save your file. Test your page in a browser. It should look similar to Figure 10.5.

You can compare your work with the solution found in the student files (chapter10/10.1) folder. Try to enter some information into your form. Try to click the submit button. Don't worry if the form redisplays but nothing seems to happen when you click the button—you haven't configured this form to work with any server-side processing. Connecting forms to server-side processing is demonstrated later in this chapter. The next sections will introduce you to more form controls.

Check Box and Radio Button

The Check Box

The **check box** form control allows the user to select one or more of a group of predetermined items. The input element with `type="checkbox"` configures a check box. Figure 10.6 shows several check boxes—note that more than one check box can be selected by the user. Table 10.5 lists common attributes. The HTML follows:

Sample Check Box

Choose the browsers you use:

- ☐ Google Chrome
- ☐ Firefox
- ☐ Microsoft Edge

FIGURE 10.6 *Check box.*

```
Choose the browsers you use: <br>
<input type="checkbox" name="Chrome" id="Chrome"
       value="yes"> Google Chrome<br>
<input type="checkbox" name="Firefox" id="Firefox"
       value="yes"> Firefox<br>
<input type="checkbox" name="Edge" id="Edge"
       value="yes"> Microsoft Edge<br>
```

TABLE 10.5 *Common Check Box Attributes*

Attribute	Value	Usage
type	checkbox	Configures the check box
name	Alphanumeric, no spaces, begins with a letter	Names the form element so that it can be easily accessed by client-side scripting and server-side processing; the name of each check box should be unique
id	Alphanumeric, no spaces, begins with a letter	Provides a unique identifier for the form element
checked		Boolean attribute; if present, the browser displays the check box as selected
value	Text or numeric characters	Assigns a value to the check box that is triggered when the check box is checked; this value can be accessed by client-side and server-side processing
disabled		Boolean attribute; if present, form control is disabled and will not accept information
autofocus		Boolean attribute; if present, form control has cursor focus
required		Boolean attribute; if present, the browser verifies entry of information
accesskey	Keyboard character	Configures a hot key for the form control
tabindex	Numeric	Configures the tab order of the form control

The Radio Button

The **radio button** form control allows the user to select exactly one (and only one) choice from a group of predetermined items. Each radio button in a group is given the same `name` attribute and a unique `value` attribute. Because the `name` attribute is the same, the elements are identified as part of a group by the browsers and only one may be selected.

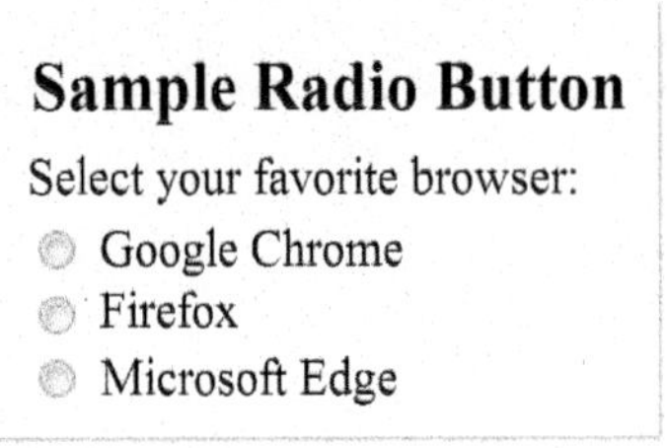

The input element with `type="radio"` configures a radio button. Figure 10.7 shows an example with a radio button group—note that only one radio button can be selected at a time by the user. Common radio button attributes are listed in Table 10.6. The HTML follows:

FIGURE 10.7 *Use radio buttons when only one choice is an appropriate response.*

```
Select your favorite browser:<br>
<input type="radio" name="fav" id="favCH" value="CH"> Google Chrome<br>
<input type="radio" name="fav" id="favFF" value="FF"> Firefox<br>
<input type="radio" name="fav" id="favME" value="ME"> Microsoft Edge<br>
```

Notice that all the `name` attributes have the same value: `fav`. Radio buttons with the same name attribute are treated as a group by the browser. Each radio button in the same group is typically configured with a unique `value` attribute.

TABLE 10.6 *Common Radio Button Attributes*

Attribute	Value	Usage
type	radio	Configures the radio button
name	Alphanumeric, no spaces, begins with a letter	Required; all radio buttons in a group must have the same `name`; names the form element so that it can be easily accessed by client-side scripting and server-side processing
id	Alphanumeric, no spaces, begins with a letter	Provides a unique identifier for the form element
checked		Boolean attribute; if present, the browser displays the radio button as selected
value	Text or numeric characters	Assigns a value to the radio button that is triggered when the radio button is selected; this should be a unique value for each radio button in a group; this value can be accessed by client-side and server-side processing
disabled		Boolean attribute, if present form control is disabled and will not accept information
autofocus		Boolean attribute; if present, form control has cursor focus
required		Boolean attribute; if present, browser verifies entry of information
accesskey	Keyboard character	Configures a hot key for the form control
tabindex	Numeric	Configures the tab order of the form control

Textarea Element

Sample Scrolling Text Box

Comments:

```
Enter comments
```

FIGURE 10.8 *Scrolling text box.*

The **scrolling text box** form control accepts free-form comments, questions, or descriptions. The **textarea element** configures a scrolling text box. The `<textarea>` tag denotes the beginning of the scrolling text box. The closing `</textarea>` tag denotes the end of the scrolling text box. Text between the tags will display in the scrolling text box area. Table 10.7 lists common attributes. The HTML for the scrolling text box in Figure 10.8 follows:

```
Comments:<br>
<textarea name="cm" id="cm" cols="40" rows="2">Enter comments</textarea>
```

TABLE 10.7 *Common Scrolling Text Box Attributes*

Attribute	Value	Usage
name	Alphanumeric, no spaces, begins with a letter	Names the form element so that it can be easily accessed by client-side scripting and server-side processing; the name should be unique
id	Alphanumeric, no spaces, begins with a letter	Provides a unique identifier for the form element
cols	Numeric	Required; configures the width in character columns of the scrolling text box; if `cols` is omitted, the browser displays the scrolling text box with its own default width
rows	Numeric	Required; configures the height in rows of the scrolling text box; if `rows` is omitted, the browser displays the scrolling text box with its own default height
maxlength	Numeric	Maximum number of characters accepted
disabled		Boolean attribute; if present, form control is disabled
readonly		Boolean attribute; if present, form control is for display; cannot be edited
autofocus		Boolean attribute; if present, form control has cursor focus
placeholder	Text or numeric characters	Browser displays a tip or hint to aid the user
required		Boolean attribute; if present, browser verifies entry of information
wrap	hard **or** soft	Browser configures line breaks
accesskey	Keyboard character	Configures a hot key for the form control
tabindex	Numeric	Configures the tab order of the form control

Hands-On Practice 10.2 ─────────────────────

In this Hands-On Practice, you will create a contact form with the following form controls: a First Name text box, a Last Name text box, an E-mail text box, and a Comments scrolling text box. You'll use the form you created in Hands-On Practice 10.1 (see Figure 10.5) as a starting point. Launch a text editor and open the file located at chapter10/10.1/join.html in the student files. Save the file with the name contact.html. The new contact form is shown in Figure 10.9.

FIGURE 10.9 *A typical contact form.*

1. Modify the title element to display the text: Contact Form.

2. Configure the h1 element with the text: Contact Us.

3. A form control for the e-mail address is already coded. Refer to Figure 10.9 and note that you'll need to add text box form controls for the first name and last name above the e-mail form control. Position your cursor after the opening form tag and press the enter key twice to create two blank lines. Add the following code to accept the name of your web page visitor:

```
First Name: <input type="text" name="fname"
               id="fname"><br><br>
Last Name: <input type="text" name="lname"
               id="lname"><br><br>
```

4. Now you are ready to add the scrolling text box form control to the form using a `<textarea>` tag on a new line below the e-mail form control. The code follows:

```
Comments:<br>
<textarea name="comments" id="comments"></textarea><br><br>
```

Save your file and display in a browser to view the default display of a scrolling text box. Note that this default display will differ by browser. Some browsers initially display a vertical scroll bar, while other browsers only render scroll bars after enough text is entered to require them. The developers of browser rendering engines determine the default display of form controls.

5. Let's configure the `rows` and `cols` attributes for the scrolling text box form control. Modify the `<textarea>` tag and set `rows="4"` and `cols="40"` as follows:

```
Comments:<br>
<textarea name="comments" id="comments"
rows="4" cols="40"></textarea><br><br>
```

6. Next, modify the text displayed on the submit button (set the `value` attribute to "Contact"). Save your file. Test your page in a browser. It should look similar to Figure 10.9.

You can compare your work with the solution found in the student files (chapter10/10.2) folder. Try entering some information into your form. Try clicking the submit button. Don't worry if the form redisplays but nothing seems to happen when you click the button—you haven't configured this form to work with any server-side processing. Connecting forms to server-side processing is demonstrated later in this chapter.

Select Element and Option Element

The **select list** form control shown in Figures 10.10 and 10.11 is also known by several other names, including select box, drop-down list, drop-down box, and option box. A select list is configured with one select element and multiple option elements.

The Select Element

The **select element** contains and configures the select list form control. The `<select>` tag denotes the beginning of the select list. The closing `</select>` tag denotes the end of the select list. Attributes configure the number of options to display and whether more than one option item may be selected. Common attributes are listed in Table 10.8.

TABLE 10.8 *Common Select Element Attributes*

Attribute	Value	Usage
name	Alphanumeric, no spaces, begins with a letter	Names the form element so that it can be easily accessed by client-side scripting and server-side processing; the name should be unique
id	Alphanumeric, no spaces, begins with a letter	Provides a unique identifier for the form element
size	Numeric	Configures the number of choices the browser will display; if set to 1, element functions as a drop-down list; scroll bars are automatically added by the browser if the number of options exceeds the space allowed
multiple		Boolean attribute; if present, configures a select list to accept more than one choice; by default, only one choice can be made from a select list
disabled		Boolean attribute; if present, form control is disabled
tabindex	Numeric	Configures the tab order of the form control

The Option Element

The **option element** contains and configures an option item displayed in the select list form control. The `<option>` tag denotes the beginning of the option item. The closing `</option>` tag denotes the end of option item. Attributes configure the value of the option and whether they are preselected. Common attributes are listed in Table 10.9.

TABLE 10.9 *Common Option Element Attributes*

Attribute	Value	Usage
value	Text or numeric characters	Assigns a value to the option; this value can be accessed by client-side scripting and server-side processing
selected		Boolean attribute; if present, configures an option to be initially selected when displayed by a browser
disabled		Boolean attribute; if present, form control is disabled

The HTML for the select list in Figure 10.10 follows:

```
<select size="1" name="favbrowser" id="favbrowser">
  <option>Select your favorite browser</option>
  <option value="Chrome">Chrome</option>
  <option value="Firefox">Firefox</option>
  <option value="Edge">Edge</option>
</select>
```

Select List: One Initial Visible Item

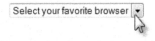

FIGURE 10.10 *A select list with size set to 1 functions as a drop-down box when the arrow is clicked.*

The HTML for the select list in Figure 10.11 follows:

```
<select size="4" name="jumpmenu" id="jumpmenu">
  <option value="index.html">Home</option>
  <option value="products.html">Products</option>
  <option value="services.html">Services</option>
  <option value="about.html">About</option>
  <option value="contact.html">Contact</option>
</select>
```

Select List: Four Items Visible

FIGURE 10.11 *Since there are more than four choices, the browser displays a scroll bar.*

Label Element

 The **label element** is a container tag that associates a text description with a form control. This is helpful to visually challenged individuals using assistive technology such as a screen reader to match up the text descriptions on forms with their corresponding form controls. The label element also benefits individuals without fine motor control. Clicking anywhere on either a form control or its associated text label will set the cursor focus to the form control.

There are two different methods to associate a label with a form control.

1. The first method places the label element as a container around both the text description and the HTML form element. Notice that both the text label and the form control must be adjacent elements. The code follows:

   ```
   <label>E-mail: <input type="text" name="email" id="email"></label>
   ```

2. The second method uses the `for` attribute to associate the label with a particular HTML form element. This is more flexible and it does not require the text label and the form control to be adjacent. The code follows:

   ```
   <label for="email">E-mail: </label>
   <input type="text" name="email" id="email">
   ```

   Notice that the value of the **for attribute** on the label element is the same as the value of the `id` attribute on the input element. This creates the association between the text label and the form control. The input element uses both the `name` and `id` attributes for different purposes. The `name` attribute can be used by client-side scripting and server-side processing. The `id` attribute creates an identifier that can be used by the label element, anchor element, and CSS selectors.

The label element does not display on the web page—it works behind the scenes to provide for accessibility.

Hands-On Practice 10.3 ────────────────────────

In this Hands-On Practice, you will add the label element to the text box and scrolling text area form controls on the form you created in Hands-On Practice 10.2 (shown in Figure 10.9) as a starting point. Launch a text editor and open the file located at chapter10/10.2/contact.html in the student files. Save the file with the name label.html.

1. Locate the text box for the first name. Add a label element to wrap around the input tag as follows:

```
<label>First Name: <input type="text" name="fname"
id="fname">
</label>
```

2. Using the method shown previously, add a label element for the last name and e-mail form controls.

3. Configure the text "Comments:" within a label element. Associate the label with the scrolling text box form control. Sample code follows:

```
<label for="comments">Comments:</label><br>
<textarea name="comments" id="comments"
rows="4" cols="40"></textarea>
```

FIGURE 10.12 *The browser display does not change when the label element is configured.*

4. Save your file. Test your page in a browser. It should look similar to the page shown in Figure 10.12—the label elements do not change the way the page displays, but a web visitor with physical challenges should find the form easier to use.

You can compare your work with the solution found in the student files (chapter10/10.3) folder.

Try entering some information into your form. Try clicking the submit button. Don't worry if the form redisplays but nothing seems to happen when you click the button—you haven't configured this form to work with any server-side processing. Connecting a form to server-side processing is demonstrated later in this chapter.

Fieldset Element and Legend Element

Fieldset and legend elements work together to visually group form controls together and increase the usability of the form.

The Fieldset Element

A technique that can be used to create a more visually pleasing form is to group elements of a similar purpose together using the **fieldset element**, which will cause the browser to render a visual cue, such as an outline or a border, around form elements grouped together within the fieldset. The `<fieldset>` tag denotes the beginning of the grouping. The closing `</fieldset>` tag denotes the end of the grouping.

The Legend Element

Fieldset and Legend

Billing Address
Street:

City: State: Zip:

FIGURE 10.13 *Form controls that are all related to a mailing address.*

The **legend element** provides a description, which can include heading tags, for the fieldset grouping. The `<legend>` tag denotes the beginning of the description. The closing `</legend>` tag denotes the end of the description.

The HTML to create the grouping shown in Figure 10.13 follows:

```
<fieldset>
<legend>Billing Address</legend>
<label>Street: <input type="text" name="street" id="street"
                    size="54"></label><br><br>
<label>City: <input type="text" name="city" id="city"></label>
<label>State: <input type="text" name="state" id="state" maxlength="2"
                    size="5"></label>
<label>Zip: <input type="text" name="zip" id="zip" maxlength="5"
                    size="5"></label>
</fieldset>
```

The grouping and visual effect of the fieldset element creates an organized and appealing web page containing a form. Using the fieldset and legend elements to group form controls enhances accessibility by organizing the controls both visually and semantically. The fieldset and legend elements can be accessed by screen readers and are useful tools to configure groups of radio buttons and check boxes on web pages.

A Look Ahead—Styling a Fieldset Group with CSS

The next section focuses on styling a form with CSS. But how about a quick preview?

Figures 10.13 and 10.14 show the same form elements, but the form in Figure 10.14 is styled with CSS—the same functionality with increased visual appeal. Access the example page at chapter10/fieldset.html in the student files. The style rules follow:

```
fieldset { width: 500px; border: 2px ridge #FF0000;
           font-family: Arial, sans-serif; padding: 10px; }
legend { font-family: Georgia, "Times New Roman", serif;
         font-weight: bold; }
label { padding-left: 10px; }
```

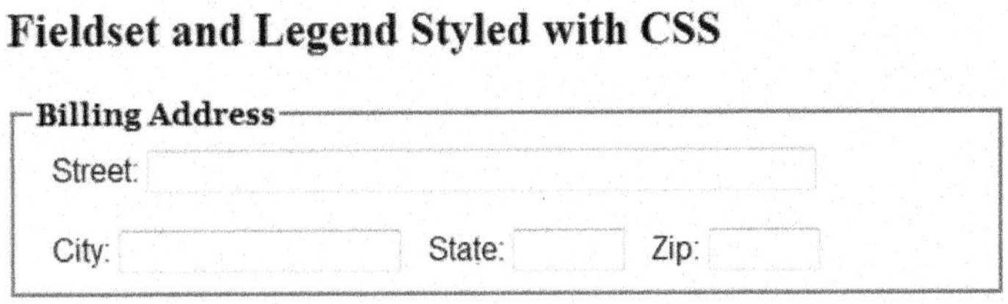

Fieldset and Legend Styled with CSS

FIGURE 10.14 *The fieldset, legend, and label elements are configured with CSS.*

Accessibility and Forms

Using the HTML elements label, fieldset, and legend will increase the accessibility of your web forms. This makes it easier for individuals with vision and mobility challenges to use your form pages. An added benefit is that the use of label, fieldset, and legend elements may increase the readability and usability of the web form for all visitors. Be sure to include contact information (e-mail address and/or phone number) just in case a visitor is unable to submit your form successfully and requires additional assistance.

Some of your website visitors may have difficulty using the mouse and will access your form with a keyboard. The Tab key can be used to move from one form control to another. The default action for the Tab key within a form is to move to the next form control in the order in which the form controls are coded in the web page document. This is usually appropriate. However, if the tab order needs to be changed for a form, use the **tabindex** attribute on each form control.[2]

Another technique that can make your form keyboard-friendly is the use of the **accesskey attribute** on form controls. Assigning accesskey a value of one of the characters (letter or number) on the keyboard will create a shortcut key that your website visitor can press to move the cursor immediately to a form control. Windows users will press the Alt key and the character key. Mac users will press the Ctrl key and the character key. When choosing accesskey values, avoid combinations that are already used by the operating system (such as Alt+F to display the File menu). Testing shortcut keys is crucial. Compliance with WCAG 2.1 requires that there be a way for the use to remap or turn off the shortcut key.[3]

Style a Form with CSS

FIGURE 10.15 *The alignment needs improvement.*

The form in Figure 10.15 looks a little "messy" and you might be wondering how the alignment could be improved. You can configure the layout of a form with CSS. For many years, web developers have typically configured a form with CSS by utilizing the box model and the float property. This section presents that method. In the next section, you'll be introduced to configuring the layout of a form with CSS Grid Layout.

```
form
  ┌──────┐ ┌─────────────────────┐
  │label │ │ text box            │
  └──────┘ └─────────────────────┘
  ┌──────┐ ┌─────────────────────┐
  │label │ │ text box            │
  └──────┘ └─────────────────────┘
  ┌──────┐ ┌─────────────────────────┐
  │label │ │ scrolling text box      │
  └──────┘ └─────────────────────────┘
           ┌─────────────────────┐
           │ submit button       │
           └─────────────────────┘
```

FIGURE 10.16 *Form wireframe.*

When styling a form with CSS, the box model is used to create a series of boxes, as shown in Figure 10.16. The outermost box defines the form area. Other boxes indicate label elements and form controls. CSS is used to configure these components.

 Hands-On Practice 10.4 ─────────

You will style a form with CSS in this Hands-On Practice. To get started, launch a text editor and open the starter.html file from the chapter10 folder in the student files. Save the file with the name contactus.html. When you have completed, your form will be similar to the example in Figure 10.17.

The HTML for the form is shown below for your reference:

```
<form>
  <label for="myName">Name:</label>
  <input type="text" name="myName" id="myName">
  <label for="myEmail">E-mail:</label>
  <input type="text" name="myEmail" id="myEmail">
  <label for="myComments">Comments:</label>
  <textarea name="myComments" id="myComments"
    rows="2" cols="20"></textarea>
  <input type="submit" value="Submit">
</form>
```

FIGURE 10.17 *The form is styled with CSS.*

Configure embedded CSS within the style element as follows:

1. **The form element selector.** Configure with a #EAEAEA background color, Arial or sans serif font, 350px width, and 10 pixels of padding:

```
form { background-color: #EAEAEA;
       font-family: Arial, sans-serif;
       width: 350px; padding: 10px; }
```

2. **The label element selector.** Configure to float to the left, clear left floats, and use block display. Also set width to 100px, 10 pixels of right padding, a 10px top margin, and right-aligned text:

```
label { float: left; clear: left; display: block;
        width: 100px; padding-right: 10px;
        margin-top: 10px; text-align: right; }
```

3. **The input element selector.** Configure with block display and a 10px top margin:

```
input { display: block; margin-top: 10px; }
```

4. **The textarea element selector.** Configure with block display and a 10px top margin:

```
textarea { display: block; margin-top: 10px; }
```

5. **The submit button.** The submit button needs to display under the other form controls, with a 110 pixel margin on the left. You could configure a new id or class and then edit the HTML, but there is a more efficient method. You will configure a new type of selector, an **attribute selector**, which allows you to select using both the element name and attribute value as the criteria. In this case, we need to style input tags that have a type attribute with the value submit in a different manner than the other input tags, so we'll configure an attribute selector for this purpose. The CSS follows:

```
input[type="submit"] { margin-left: 110px; }
```

Save your file and test your page in a browser. It should look similar to Figure 10.17. You can compare your work with the sample in the student files (chapter10/10.4).

The Attribute Selector

Use an **attribute selector** when you need to configure an element that has a specific attribute value, and you would like to avoid creating a new id or class. When you code the selector, type the element name first followed by a set of braces that contain the name and value of the attribute you chose to use for the criteria.[4] For example, the input[type="radio"] selector will configure styles for the radio button form controls but will not configure styles for other input elements.

Form Layout with CSS Grid

CSS Grid Layout offers another method to configure the layout of a form. Figure 10.18 shows a wireframe of a typical form.

When you work with grid layout, it is helpful to create a sketch of the grid you plan to create. Take a moment to examine Figure 10.19, which is a grid layout sketch that corresponds to the wireframe in Figure 10.18. Notice how the horizontal line numbers, vertical line numbers, element names (label, input, and textarea), and attribute values (text and submit) were placed in the grid sketch. It is helpful to create a detailed grid sketch before you begin to code the CSS that configures the placement of the elements in the grid columns and rows.

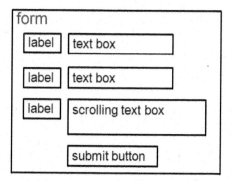

FIGURE 10.18 *Form wireframe.*

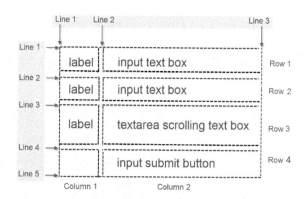

FIGURE 10.19 *The grid for the form.*

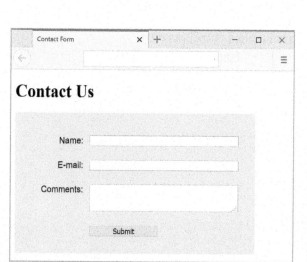

FIGURE 10.20 *The form is styled with CSS grid layout.*

 Hands-On Practice 10.5

In this Hands-On Practice, you will use form you created in Hands-On Practice 10.4 and code a CSS feature query to configure grid layout in supporting browsers. Browsers that do not support grid layout will display the form as originally styled with CSS. Browsers that support grid will follow the grid layout styles. To get started, launch a text editor and open your Hands-on Practice 10.4 file (chapter10/10.4/contactus.html). Save the file with the name contact2.html. When you have completed, your form will be similar to the example in Figure 10.20.

The HTML for the form is shown below for your reference:

```
<form>
  <label for="myName">Name:</label>
  <input type="text" name="myName" id="myName">
  <label for="myEmail">E-mail:</label>
  <input type="text" name="myEmail" id="myEmail">
  <label for="myComments">Comments:</label>
  <textarea name="myComments" id="myComments"
    rows="2" cols="20"></textarea>
  <input type="submit" value="Submit">
</form>
```

1. Locate the closing style tag. You will code CSS above the closing style tag and below the existing CSS in the file. Configure a feature query that will test for grid layout support:

```
@supports ( display: grid) {


}
```

2. Configure the following CSS within the feature query to configure the form using grid layout.

   a. **The form element selector.** Configure declarations to set the display property to grid with auto rows and two columns (6em and 1fr). Also set 1em grid gap, #EAEAEA background color, Arial or sans serif font, 60% width, 20em minimum width, and 2em padding:

```
form { display: grid;
       grid-template-rows: auto;
       grid-template-columns: 6em 1fr;
       gap: 1em;
       background-color: #EAEAEA;
       font-family: Arial, sans-serif;
       width: 60%; min-width: 20em;
       padding: 2em; }
```

   b. **The submit button.** Review the grid sketch in Figure 10.19 and notice how the elements in the grid are placed one right after another to fill the grid except for the submit button, which is in the second column of the grid. Use an attribute selector to target the submit button and explicitly place it in the second column of the grid. Also set the width to 10em. Also set the left margin to 0. The CSS follows:

```
input[type="submit"] { grid-column: 2 / 3;
                        width: 10em; margin-left: 0; }
```

Save your file and test your page in a browser. It should look similar to Figure 10.20. You can compare your work with the sample in the student files (chapter10/10.5).

Server-Side Processing

VideoNote

Connect a Form to Server-side Processing

As you've coded and tested the forms in this chapter, you may have noticed that when you click the submit button, the form just redisplays—the form doesn't "do" anything. This is because the forms haven't been configured to invoke server-side processing.

Your web browser requests web pages and their related files from a web server. The web server locates the files and sends them to your web browser. Then the web browser renders the returned files and displays the requested web pages. Figure 10.21 illustrates the communication between the web browser and the web server.

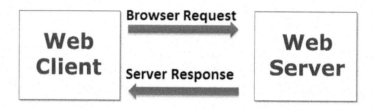

FIGURE 10.21 *The web browser (client) communicates with the web server.*

Sometimes a website needs more functionality than static web pages—possibly a site search, order form, e-mail list, database display, or other type of interactive, dynamic processing. This is when server-side processing is needed. Early web servers used a protocol called **Common Gateway Interface** (CGI) to provide this functionality. CGI is a protocol, or standard method, for a web server to pass a web page user's request (which is typically initiated through the use of a form) to an application program and to accept information to send to the user. The web server typically passes the form information to a small application program that is run by the operating system and processes the data, and it usually sends back a confirmation web page or message. Perl and C are popular programming languages for CGI applications.

Server-side scripting is a technology in which a server-side script is run on a web server to dynamically generate web pages. Examples of server-side scripting technologies include PHP, Ruby on Rails, Sun JavaServer Pages, and ASP.NET. Server-side scripting differs from CGI in that it uses **direct execution**—the script is run either by the web server itself or by an extension module to the web server.

A web page invokes server-side processing by either an attribute on a form or by a hyperlink—the URL of the script is used. Any form data that exists is passed to the script. The script completes its processing and may generate a confirmation or response web page with the requested information. When invoking a server-side script, the web developer and the server-side programmer must communicate about the form **method attribute** (get or post), form **action attribute** (URL of the server-side script), and any special form element control(s) expected by the server-side script.

The method attribute is used on the form tag to indicate the way in which the name and value pairs should be passed to the server. The method attribute value of get causes the form data to be appended to the URL, which is easily visible and not secure. The method attribute value of post does not pass the form information in the URL; it passes it in the entity body of the HTTP request, which makes it more private. The W3C recommends the method="post" method.

The action attribute is used on the <form> tag to invoke a server-side script. The name attribute and the value attribute associated with each form control are passed to the server-side script. The name attribute may be used as a variable name in the server-side processing.

Privacy and Forms

A **privacy policy** lists the guidelines that you develop to protect the privacy of your visitors' information. Websites either indicate this policy on the form page itself or create a separate page that describes the privacy policy (and other company policies).

If you browse popular sites such as Amazon.com or eBay.com, you'll find links to their privacy policies (sometimes called a privacy notice) in the page footer area. Include a privacy notice in your site to inform your visitors how you plan to use the information they share with you. The Better Business Bureau recommends that a privacy policy describes the type of information collected, the methods used to collect the information, the way that the information is used, the methods used to protect the information, and provisions for customers or visitors to control their personal information.[5]

 Sources of Free Remote-Hosted Form Processing

If your web host provider does not support server-side processing, free remotely hosted scripts may be an option. Check out the author's textbook companion website at https://webdevbasics.net/6e/chapter10.html for options to explore.

Practice with a Form

 Hands-On Practice 10.6 ─────────────────────────────

In this Hands-On Practice, you will modify the form page that you created earlier in this chapter, configuring the form so that it uses the post method to invoke a server-side script. Your computer must be connected to the Internet when you test your work. The post method is more secure than the get method because the post method does not pass the form information in the URL; it passes form information in the entity-body of the HTTP Request, which makes it more private.

When using a server-side script, you will need to obtain some information, or documentation, from the person or organization providing the script. You will need to know the location of the script, whether it requires the get or post method, whether it requires any specific names for the form controls, and whether it requires any hidden form elements. The `action` attribute is used on the `<form>` tag to invoke a server-side script. A server-side script has been created at https://webdevbasics.net/scripts/demo.php for students to use for this exercise. The documentation for the server-side script is listed in Table 10.10.

TABLE 10.10 *Server-Side Script Documentation*

Script URL	https://webdevbasics.net/scripts/demo.php
Form method	`post`
Script purpose	This script will accept form input and display the form control names and values in a web page. This is a sample script for student assignments. It demonstrates that server-side processing has been invoked. A script used by an actual website would perform a function such as sending an e-mail message or updating a database.

As you view Table 10.10, notice that the script's URL begins with https:// instead of http://. Coding https:// in the action value will cause the browser to use HTTPS, which stands for Hypertext Transfer Protocol Secure. HTTPS combines HTTP with a security and encryption protocol called Secure Sockets Layer (SSL)—see Chapter 12 for a brief introduction to SSL. Using HTTPS provides a more secure transaction because the browser encrypts the information entered in the form before sending it to the web server.

Launch a text editor and open the file you created in Hands-On Practice 10.5, also found in the student files (chapter10/10.5/contact2.html).

Modify the `<form>` tag by adding a `method` attribute with a value of post and an `action` attribute with a value of https://webdevbasics.net/scripts/demo.php. Coding https:// will cause the browser to use HTTPS, which combines HTTP with a security and encryption protocol called Secure Sockets Layer. The HTML for the revised `<form>` tag follows:

```
<form method="post" action="https://webdevbasics.net/scripts/demo.php">
```

Save your page with the name mycontact.html and test it in a browser. Your screen should look similar to Figure 10.22. Compare your work with the solution in the student files (chapter10/10.6/mycontact.html).

Now you are ready to test your form. You must be connected to the Internet to test your form successfully. Enter information in the form controls and click the submit button. You should see a confirmation page similar to the one shown in Figure 10.23.

FIGURE 10.22 *The contact form.*

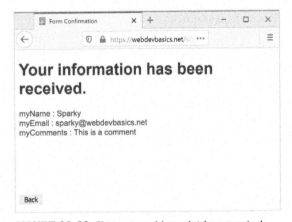

FIGURE 10.23 *The server-side script has created this page in response to the form.*

The demo.php script creates a web page that displays a message and the form information you entered. This confirmation page was created by the server-side script on the `action` attribute in the `<form>` tag. Writing scripts for server-side processing is beyond the scope of this textbook. If you are curious, the script's source code can be found at https://webdevbasics.net/6e/chapter10.html.

What should I do if nothing happened when I tested my form?
Try these troubleshooting hints:

▶ Verify that your computer is connected to the Internet.
▶ Verify the spelling of the script location in the `action` attribute.
▶ Recall that attention to detail is crucial!

Password, Hidden, and File Upload Controls

The Password Box

The **password box** form control is similar to the text box, but it is used to accept information that must be hidden as it is entered, such as a password.

Sample Password Box

Password: ●●●●●●

FIGURE 10.24 *The characters secret9 were typed, but the browser does not display them.*

The input element with `type="password"` configures a password box. A password box is a specialized text box. See Table 10.3 for a list of text box attributes. When the user types information in a password box, asterisks (or another symbol, depending on the browser) are displayed instead of the characters that have been typed, as shown in Figure 10.24. This hides the information from someone looking over the shoulder of the person typing. The actual characters typed are sent to the server, and the information is not really secret or hidden. The HTML follows:

```
Password: <input type="password" name="pword" id="pword">
```

The Hidden Input Control

The **hidden input control** stores text or numeric information, but it is not visible in the browser viewport. The information stored in the hidden input control should not be considered secure because a web page visitor can use the browser developer tools to change the value of the control even thought it is not displayed in the browser viewport. Hidden controls can be accessed by both client-side and server-side scripting.

The input element with `type="hidden"` configures a hidden field. Common hidden field attributes are listed in Table 10.11. The HTML to create a hidden form control with the `name` attribute set to "sendto" and the `value` attribute set to an e-mail address as follows:

```
<input type="hidden" name="sendto" id="sendto" value="order@site.com">
```

TABLE 10.11 *Common Hidden Input Control Attributes*

Attribute	Value	Usage
type	hidden	Configures the hidden form element
name	Alphanumeric, no spaces, begins with a letter	Uniquely names the form element so that it can be easily accessed by client-side scripting and server-side processing
id	Alphanumeric, no spaces, begins with a letter	Provides a unique identifier for the form element
value	Text or numeric characters	Assigns a value to the hidden control; this value can be accessed by client-side scripting languages and server-side processing

File Upload Control

The **file upload control** provides for files to be uploaded.

The input element with `type="file"` configures a file upload control. The form control display varies depending on the browser and operating system being used. The user is able to browse and choose one or more files to upload. Figure 10.25 shows the initial display of a file upload control with an option to browse for a file.

FIGURE 10.25 *The initial display of a file upload control.*

The HTML to create a form control that accepts a typical photograph file is

```
Profile Photo:
<input type="file" name="photo" id="photo" accept="image/*">
```

Common attributes for the file upload control are listed in Table 10.12. Note that `enctype="mutlipart/form-data"` must be coded on the form element when the form contains a file upload control.

TABLE 10.12 *Common file upload form control attributes.*

Attribute	Value	Purpose
type	file	Configure a file upload control
name	Alphanumeric, no spaces, begins with a letter	Names the form element so that it can be easily accessed by client-side scripting languages or by server-side processing; the name should be unique
id	Alphanumeric, no spaces, begins with a letter	Provides a unique identifier for the form element
accept	A comma delineated list of file extensions and/or MIME types	Indicates which file types are acceptable for upload. For example, the value `image/*` accepts any file with an image MIME type.[6]
capture	user	File is captured from the inward-facing camera/microphone
	environment	File is captured from the outward-facing camera/microphone
multiple		Boolean attribute; if present, more than one file can be uploaded

More Text Form Controls

The E-mail Address Input Form Control

The **e-mail address form control** is similar to the text box. Its purpose is to accept information that must be in e-mail format, such as DrMorris2010@gmail.com. The input element with `type="email"` configures an e-mail address form control. Figure 10.26 (chapter10/email.html in the student files) shows an error message displayed when text other than an e-mail address is entered. Note that the browser does not verify that the e-mail address actually exists—just that the text entered is in the correct format. The HTML follows:

```
<label for="myEmail">E-mail:</label>
<input type="email" name="myEmail" id="myEmail">
```

FIGURE 10.26 *The browser displays an error message.*

The URL Form Input Control

The **URL form control** is similar to the text box. It is intended to accept any type of URL or URI, such as https://webdevbasics.net. The input element with `type="url"` configures a URL form control. Figure 10.27 (chapter10/url.html in the student files) shows an error message displayed when text other than a URL is entered. Note that the browser does not verify that the URL actually exists—just that the text entered is in the correct format. The HTML follows:

```
<label for="myWebsite">Suggest a Website:</label>
<input type="url" name="myWebsite" id="myWebsite">
```

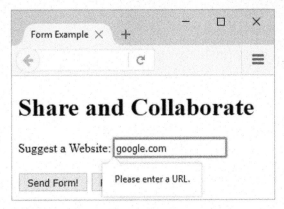

FIGURE 10.27 *The browser displays an error message.*

The Telephone Number Input Form Control

The **telephone number form control** is similar to the text box. Its purpose is to accept a telephone number. The input element with `type="tel"` configures a telephone number form control. An example is in the student files (chapter10/tel.html). Some mobile devices display a numeric keypad for entry into telephone number input form controls. The HTML follows:

```
<label for="mobile">Mobile Number:</label>
<input type="tel" name="mobile" id="mobile">
```

The Search Input Form Control

The **search form control** is similar to the text box and is used to accept a search term. The input element with `type="search"` configures a search input form control. An example is in the student files (chapter10/search.html). The HTML follows:

```
<label for="keyword">Search:</label>
<input type="search" name="keyword" id="keyword">
```

? FAQ How can I tell which browsers support these form control?

Current versions of modern browsers support most of the form controls introduced in this section. However, there's no substitute for testing. With that in mind, https://caniuse.com provides information about browser support for HTML and CSS.

Datalist Element

Figure 10.28 shows the **datalist form control** in action. Notice how a selection of choices is offered to the user along with a text box for entry. A datalist can be used to suggest predefined input values to the web page visitor. Configure a datalist using three components: an input element, the datalist element, and one or more option elements. Most modern browsers support the datalist element and will display and process the datalist items. Other browsers ignore the datalist element and render the form control as a text box.

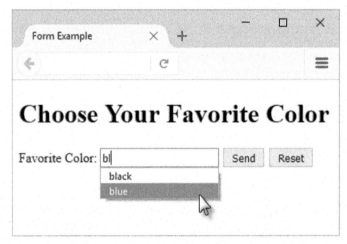

FIGURE 10.28 *Firefox displays the datalist form control.*

The source code for the datalist is available in the student files (chapter10/list.html). The HTML follows:

```
<label for="color">Favorite Color:</label>
<input type="text" name="color" id="color" list="colors">
  <datalist id="colors">
  <option>black</option>
  <option>red</option>
  <option>green</option>
  <option>blue</option>
  <option>yellow</option>
  <option>pink</option>
  <option>cyan</option>
  <option>purple</option>
  <option>gold</option>
  <option>silver</option>
</datalist>
```

Notice that the value of the **list attribute** on the input element is the same as the value of the id attribute on the datalist element. This creates the association between the text box and the datalist form control. One or more option elements can be used to offer predefined choices to your web page visitor. The text between each pair of option tags configures the text displayed in each list entry.

The list can be used to suggest input values to the user corresponding to the characters they type in the text box. When a web page visitor types characters in the text box, the option elements with text that match the characters typed are displayed. The web page visitor can choose an option from the displayed list (see Figure 10.28) or type directly in the text box, as shown in Figure 10.29. The datalist form control offers a convenient way to offer choices yet provide for flexibility on a form.

FIGURE 10.29 *The user can choose to type a value not on the list in the text box.*

 What happens in browsers that do not support a form control?

Browsers that do not support an input type will display it as a text box and ignore unsupported attributes or elements.

Slider and Spinner Controls

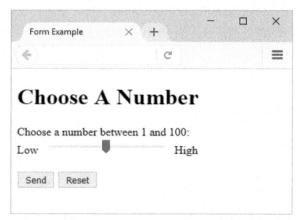

FIGURE 10.30 *The Firefox browser displays the range form control.*

The Slider Input Form Control

The **slider form control** provides a visual, interactive user interface that accepts numerical information. The input element with `type="range"` configures a slider control in which a number within a specified range is chosen. The default range is from 0 to 100. Most modern browsers support the `range` attribute value and will display the interactive slider control, shown in Figure 10.30 (chapter10/range.html in the student files). Note the position of the slider in Figure 10.30; this resulted in the value 80 being chosen. The nondisplay of the value to the user may be a disadvantage of the slider control. Nonsupporting browsers render this form control as a text box.

The slider control accepts attributes listed in Tables 10.3 and 10.13. The `min`, `max`, and `step` attributes are new. Use the **min attribute** to configure the minimum range value. Use the **max attribute** to configure the maximum range value. Use the **step attribute** to configure a value for the step between values to be other than 1.

The HTML for the slider control rendered in Figure 10.30 is shown below.

```
<label for="myChoice">Choose a number between 1 and 100:</label><br>
Low
<input type="range" min="1" max="100" name="myChoice" id="myChoice">
High
```

The Spinner Input Form Control

The **spinner form control** displays an interface that accepts numerical information and provides feedback to the user. The input element with `type="number"` configures a spinner control in which the user can either type a number into the text box or select a number from a specified range. Most modern browsers support the `number` attribute value and will display the interactive spinner control, shown in Figure 10.31 (chapter10/spinner.html in the student files). Other browsers render this form control as a text box.

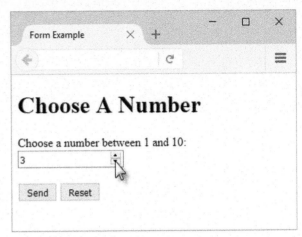

FIGURE 10.31 *A spinner control displayed in the Firefox browser.*

The spinner control accepts attributes listed in Tables 10.2 and 10.13. Use the `min` attribute to configure the minimum value. Use the `max` attribute to configure the maximum value. Use the `step` attribute to configure a value for the step between values to be other than 1. The HTML for the spinner control displayed in Figure 10.31 follows:

```
<label for="myChoice">Choose a number between 1 and 10:</label><br>
<input type="number" name="myChoice" id="myChoice" min="1" max="10">
```

TABLE 10.13 *Additional Attributes for Slider, Spinner, and Date/Time Form Controls*

Attribute	Value	Usage
max	Maximum value	Attribute for range, number, and date/time input controls; specifies a maximum value
min	Minimum value	Attribute for range, number, and date/time input controls; specifies a minimum value
step	Incremental step value	Attribute for range, number, and date/time input controls; specifies a value for incremental steps

Progressive Enhancement

Use form elements with the concept of progressive enhancement in mind. Nonsupporting browsers will display text boxes in place of form elements that are not recognized. Supporting browsers will display and process the new form controls. This is progressive enhancement in action—everyone sees a usable form, and those using modern browsers benefit from enhanced features.

Date and Color-Well Controls

The Date and Time Form Control

HTML5 provides a variety of **date and time form controls** to accept date- and time-related information. Use the input element and configure the `type` attribute to specify date and time controls. Table 10.14 lists the HTML5 date and time controls.

TABLE 10.14 *Date and Time Controls*

Type Attribute	Value	Purpose Format
`date`	A date	YYYY-MM-DD Example: January 2, 2022, is represented by "20220102"
`datetime`	A date and time with time zone information; note that the time zone is indicated by the offset from UTC time	YYYY-MM-DDTHH:MM:SS-##:##Z Example: January 2, 2022, at exactly 9:58 AM Chicago time (CST) is represented by "2022-01-02T09:58:00-06:00Z"
`datetime-local`	A date and time without time zone information	YYYY-MM-DDTHH:MM:SS Example: January 2, 2022, at exactly 9:58 AM is represented by "2022-01-02T09:58:00"
`time`	A time without time zone information	HH:MM:SS Example: 1:34 PM is represented by "13:34"
`month`	A year and month	YYYY-MM Example: January 2022, is represented by "2022-01"
`week`	A year and week	YYYY-W##, where ## represents the week in the year Example: The third week in 2022 is represented by "2022-W03"

The form in Figure 10.32 (chapter10/date.html in the student files) uses the input element with `type="date"` to configure a calendar control with which the user can select a date.

The HTML for the date control displayed in Figure 10.31 follows:

```
<label for="myDate">Choose a
Date</label>
<input type="date" name="myDate"
id="myDate">
```

The date and time controls accept attributes listed in Tables 10.3 and 10.14. The implementation of the date control is determined by the browser.

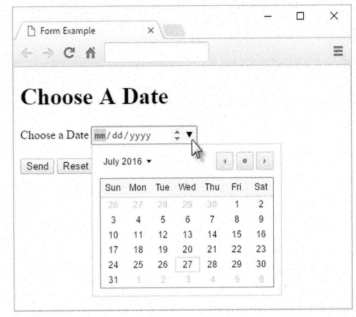

FIGURE 10.32 *A date form control displayed in the Google Chrome browser.*

The Color-Well Form Control

The **color-well form control** displays an interface that offers a color-picker interface to the user, as shown in Figure 10.33 (chapter10/color.html in the student files). The input element with `type="color"` configures a control with which the user can choose a color. The implementation of the color-well control is determined by the browser.

The HTML for the color-well form control rendered in Figure 10.33 follows:

```
<label for="myColor">Choose
a color:</label>
<input type="color"
name="myColor" id="myColor">
```

In the next section, you'll get some practice using form controls.

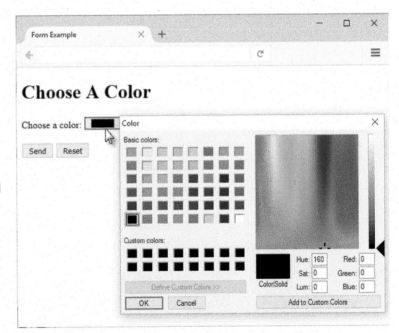

FIGURE 10.33 *The Google Chrome browser supports the color-well form control.*

More Form Practice

  Hands-On Practice 10.7 ————————————————————————————

FIGURE 10.34 *The form configured with Flexbox layout.*

In this Hands-On Practice, you will configure a form designed for browsers that support Flexbox layout and newer HTML5 form controls (although the form will also display adequately in nonsupporting browsers—demonstrating the concept of progressive enhancement). Shown in Figure 10.35, the form will accept a first name, a last name, an e-mail address, a rating value, and comments from a website visitor. Figure 10.34 displays the form in the Firefox browser, which supports Flexbox and the form controls used in the Hands-On Practice.

To get started, launch a text editor and open the file located at chapter10/template.html in the student files. Save the file with the name comment.html. You will modify the file to create a web page similar to the example in Figure 10.34.

1. Modify the title element to display the text: Comment Form. Configure the text within the h1 element to be: Comment Form. Add a paragraph to indicate: Required fields are marked with an asterisk *.

2. Configure the form element to submit the form information to the textbook's form processor at https://webdevbasics.net/scripts/demo.php.

```
<form method="post"
action="https://webdevbasics.net/scripts/demo.php">
```

3. Code the form labels and controls. Configure the first name, last name, e-mail, and comment information to be required. Use an asterisk to inform your web page visitor about the required fields. Use `type="email"` instead of `type="input"` for the e-mail address. Use the `placeholder` attribute to provide hints to the user in the name and e-mail form controls. Add a slider control (use `type="range"`) to generate a value from 1 to 10 for the rating. The HTML follows:

```
<form method="post"
action="https://webdevbasics.ent/scripts/demo.php">
    <label for="myFirstName">* First Name</label>
    <input type="text" name="myFirstName"
           id="myFirstName" required
           placeholder="your first name">
    <label for="myLastName">* Last Name</label>
```

```
<input type="text" name="myLastName"
       id="myLastName" required
       placeholder="your last name">
<label for="myEmail">* E-mail</label>
<input type="email" name="myEmail" id="myEmail"
       required
       placeholder="you@yourdomain.com">
<label for="myRating">Rating (1 &mdash; 10)
       </label>
<input type="range" name="myRating"
       id="myRating" min="1" max="10">
<label for="myComments">* Comments</label>
<textarea name="myComments" id="myComments"
          rows="2" cols="40" required
          placeholder="your comments here">
</textarea>
<input type="submit" value="Submit">
</form>
```

FIGURE 10.35 *The form displayed before configuring Flexbox layout.*

4. As shown in Figure 10.35, the layout of the form is a single column. We'll use CSS flexbox layout in this exercise. With progressive enhancement in mind, we will also provide for browsers that do not support flexbox. Code the CSS described below as embedded CSS in the head section:

   a. Configure styles for browsers that do not support flexbox. Code styles for the input element selector and textarea element selector. Set the display property to block and the bottom margin to .5em. The CSS follows:

   ```
   input, textarea { display: block;
                     margin-bottom: .5em; }
   ```

   Save your file. Test your page in a browser. It will look similar to Figure 10.35.

   b. Configure a one-column flexbox layout. Edit your file. Code styles for the form element selector. Set flex display, flex-direction set to column, flex-wrap to nowrap, and a maximum width of 25em. The CSS follows:

   ```
   form { display: flex;
          flex-direction: column; flex-wrap: nowrap;
          max-width: 25em; }
   ```

FIGURE 10.36 *The Firefox browser displays an error message.*

5. Save your file. Test your page in a browser. If you use a modern browser, your form should look similar to Figure 10.34.

6. Try submitting the form without entering any information. Figure 10.36 shows the error message displayed by Firefox.

See the student files (chapter10/10.7) for a suggested solution. You can view an example that uses CSS grid layout in the student files chapter10/10.7/grid.html). This Hands-On Practice demonstrates that it is possible to design forms with progressive enhancement in mind.

Review and Apply

Review Questions

Multiple Choice. Choose the best answer for each item.

1. What will happen when a browser encounters a form control that it does not support?
   a. The computer will shut down.
   b. The browser will display an error message.
   c. The browser will crash.
   d. The browser will display an input text box.

2. Which attribute of the `<form>` tag is used to specify the name and location of the script that will process the form field values?
   a. `action`   b. `process`
   c. `method`   d. `id`

3. Forms contain various types of _____, such as text boxes and buttons, that accept information from a web page visitor.
   a. hidden elements   b. labels
   c. form controls   d. legends

4. Choose the tag that would configure a text box with the name "city" and a width of 35 characters.
   a. `<input type="text" id="city" width="35">`
   b. `<input type="text" name="city" size="35">`
   c. `<input type="text" name="city" space="35">`
   d. `<input type="text" name="city" width="35">`

5. You would like to accept a number that is in a range from 1 to 50. The user needs visual verification of the number selected. Which of the following form controls is best to use for this purpose?
   a. spinner   b. radio button
   c. check box   d. slider

6. Which of the following form controls would be appropriate for an area that your visitors can use to type in their e-mail address?
   a. check box   b. select list
   c. text box   d. scrolling text box

7. You would like to conduct a survey and ask your web page visitors to vote for their favorite search engine. Which of the following form controls is best to use for this purpose?
   a. radio button   b. text box
   c. scrolling text box   d. check box

8. Which of the following form controls would be appropriate for an area that your visitors can use to type in comments about your website?
   a. text box   b. select list
   c. radio button   d. scrolling text box

9. Which tag would configure a scrolling text box with the name comments, four rows, and up to thirty characters in a row?
   a. `<textarea name="comments" width="30" rows="4"></textarea>`
   b. `<input type="textarea" name="comments" size="30" rows="4">`
   c. `<textarea name="comments" rows="4" cols="30"></textarea>`
   d. `<textarea name="comments" width="30" rows="4">`

10. Choose the item that would associate a label displaying the text E-mail: with the text box named `email`.

    a. `E-mail: <input type="textbox" name="email" id="email">`

    b. `<label>E-mail: <input type="text" name="email" id="email"></label>`

    c. `<label for="email">E-mail:</label> <input type="text" name="email" id="email">`

    d. both b and c

Review Answers

1. d   2. a   3. c   4. b   5. a   6. c   7. a   8. d   9. c   10. d

Hands-On Exercises

1. Write the code to create the following:
   a. A text box named username that will accept the user name of web page visitors. The text box should allow a maximum of 30 characters to be entered.
   b. A group of radio buttons that website visitors can check to vote for their favorite day of the week.
   c. A select list that asks website visitors to select their favorite social networking website.
   d. A fieldset and legend with the text "Billing Address" around the form controls for the following fields: AddressLine1, AddressLine2, City, State, Zip Code.
   e. A hidden form control with the name of userid.
   f. A password form control with the name of password.

2. Create a web page with a form that accepts requests for a brochure to be sent in the mail. Use the `required` attribute to configure the browser to verify that all fields have been entered by the user. Sketch out the form on paper before you begin.

3. Create a web page with a form that accepts feedback from website visitors. Use the input `type="email"` along with the `required` attribute to configure the browser to verify the data entered. Also configure the browser to require user comments with a maximum length of 1200 characters accepted. Sketch out the form on paper before you begin.

4. Create a web page with a form that accepts a website visitor's name, e-mail, and birthdate. Use the input `type="date"` to configure a calendar control on browsers that support the attribute value.

Focus on Web Design

The design of a form, such as the justification of the labels, the use of background colors, and even the order of the form elements can either increase or decrease the usability of a form. Visit some of the resources listed below to begin exploring form design. Feel free to search on your own and locate additional resources. Create a web page that lists the URLs of at least two useful resources along with a brief description of the information you found most interesting or valuable. Design a form on the web page that applies what you've just learned in your exploration of form design. Place your name in an e-mail link on the web page.

- Designing Efficient Web Forms: https://www.smashingmagazine.com/2017/06/designing-efficient-web-forms/
- Form Design Best Practices: https://blog.hubspot.com/marketing/form-design
- Best Practices for Mobile Form Design: https://www.smashingmagazine.com/2018/08/best-practices-for-mobile-form-design/

Case Study

You will continue the case studies from Chapter 9 as you configure a web page with a form.

Pacific Trails Resort Case Study

In this chapter's case study, you will use the existing Pacific Trails Resort website (Chapter 9) as a starting point. You will add a new page to the Pacific Trails website—the Reservations page. Refer back to the site map for the Pacific Trails website in Chapter 2, Figure 2.28. The Reservations page will use the same two-column layout as the other Pacific Trails web pages. You'll apply your new skills from this chapter and code a form in the content area of the Reservations page.

You have three tasks in this case study:

1. Create a folder for the Pacific Trails website.
2. Modify the CSS to configure style rules needed for the Reservations page, shown in Figure 10.37.

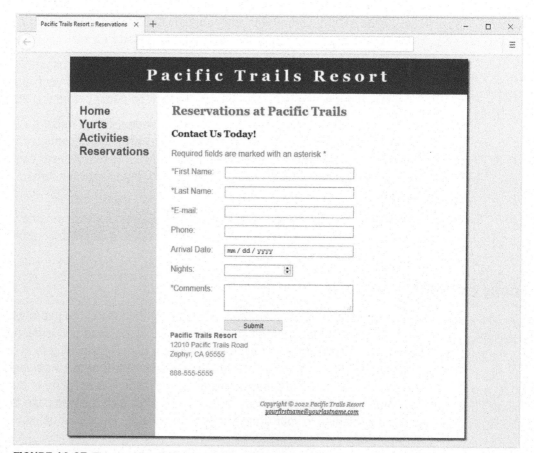

FIGURE 10.37 *The new Pacific Trails Reservations page.*

3. Create the Reservations page. Your new page (reservations.html) will be similar to Figure 10.37 when you have completed this step.

Task 1: Create a folder called ch10pacific to contain your Pacific Trails Resort website files. Copy the files from the Chapter 9 Case Study ch9pacific folder to your new ch10pacific folder.

Task 2: Configure the CSS. Review Figure 10.37 and the grid layout sketch in Figure 10.38. Notice how the text labels for the form controls are on the left side of the content area. Notice the empty vertical space between each form control. When displaying on a narrow viewport, the display will be more pleasing if there is only one column, as shown in Figure 10.39. Open pacific.css in a text editor. Configure the CSS as follows:

a. Configure the single column display for narrow viewports using flexbox. Add CSS above the media queries to configure a flex layout for the form.

1. Configure a form element selector. Set flex display, `flex-direction` to column, and `flex-wrap` to nowrap.

2. Configure the input and textarea element selectors with .5em bottom margin.

b. Configure the two-column display with grid layout. Add CSS to the first media query to accomplish this.

1. Configure a form element selector. Set 60% width, 30em maximum width, grid display with 1em grid gap, and two columns (6em width and 1fr width).

2. Configure an attribute selector for the submit button. Use the `grid-column` property to place this in the second column. Set width to 9em.

Save the pacific.css file.

Task 3: Create the Reservations Page. A productivity technique is to create new pages based on existing pages so you can benefit from your previous work. Your new Reservations page will use the index.html page as a starting point. Open the index.html page for the Pacific Trails Resort website in a text editor. Select File > Save As and save the file with the new name of reservations.html in the ch10pacific folder.

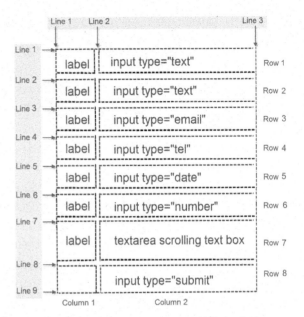

FIGURE 10.38 *The grid layout sketch of the form.*

FIGURE 10.39 *Single-column form in a narrow viewport.*

Launch a text editor and edit the reservations.html file.

1. Modify the page title. Change the text between the `<title>` and `</title>` tags to: Pacific Trails Resort :: Reservations.

2. The Reservations page will not feature a large image. Remove the div element assigned to the `homehero` id.

3. Replace the text within the `<h2>` tags with: Reservations at Pacific Trails.

4. Delete the paragraph and the unordered list. Do not delete the logo, navigation, contact information, or footer areas of the page.

5. Position your cursor on a blank line below the h2 element. Configure an h3 element with the following text: Contact Us Today!

6. Add a paragraph below the h3 element to indicate:
Required fields are marked with an asterisk *.

7. Position your cursor on a blank line under the h3 element. You are ready to configure the form. Begin with a `<form>` tag that uses the post method and the action attribute to invoke server-side processing. Unless directed otherwise by your instructor, use https://webdevbasics.net/scripts/pacific.php as the value of the `action` attribute.

8. Configure the form control for the First Name information. Create a `<label>` element that contains the text "*First Name:". Create a text box configured with "myFName" as the value of the id and `name` attributes. Configure the `required` attribute. Use the `for` attribute to associate the label element with the form control.

9. In a similar way, configure the following form controls and labels:
   a. the form control and label for the Last Name
   b. the e-mail address form control and label for the E-Mail Address
   c. the telephone number form control and label for the Phone Number (which is not required to be entered by the user); set the maxlength of the telephone form control to 12
   d. the calendar form control and label for the Arrival Date (which is not required to be entered by the user)
   e. the spinner form control and label for the Nights (which is not required to be entered by the user); configure the spinner form control to accept a value from 1 to 14 (inclusive)

10. Configure the Comments area on the form. Create a label element that contains the text "*Comments:". Create a textarea element configured with "myComments" as the value of the id and `name` attributes, `rows` set to 2, and `cols` set to 30. Use the `for` attribute to associate the label element with the form control.

11. Configure the submit button on the form.

12. Code an ending `</form>` tag on a blank line after the submit button.

Save your file. Display your web page in browser. It should be similar to Figure 10.37. If you resize the browser viewport to be narrower, the display should be similar to Figure 10.39. Submit the form with missing information or only a partial e-mail address. Depending on the browser's level of HTML5 support, the browser may perform form validation and display an error message. Figure 10.40 shows the Reservations page rendered in a browser with an incorrectly formatted e-mail address.

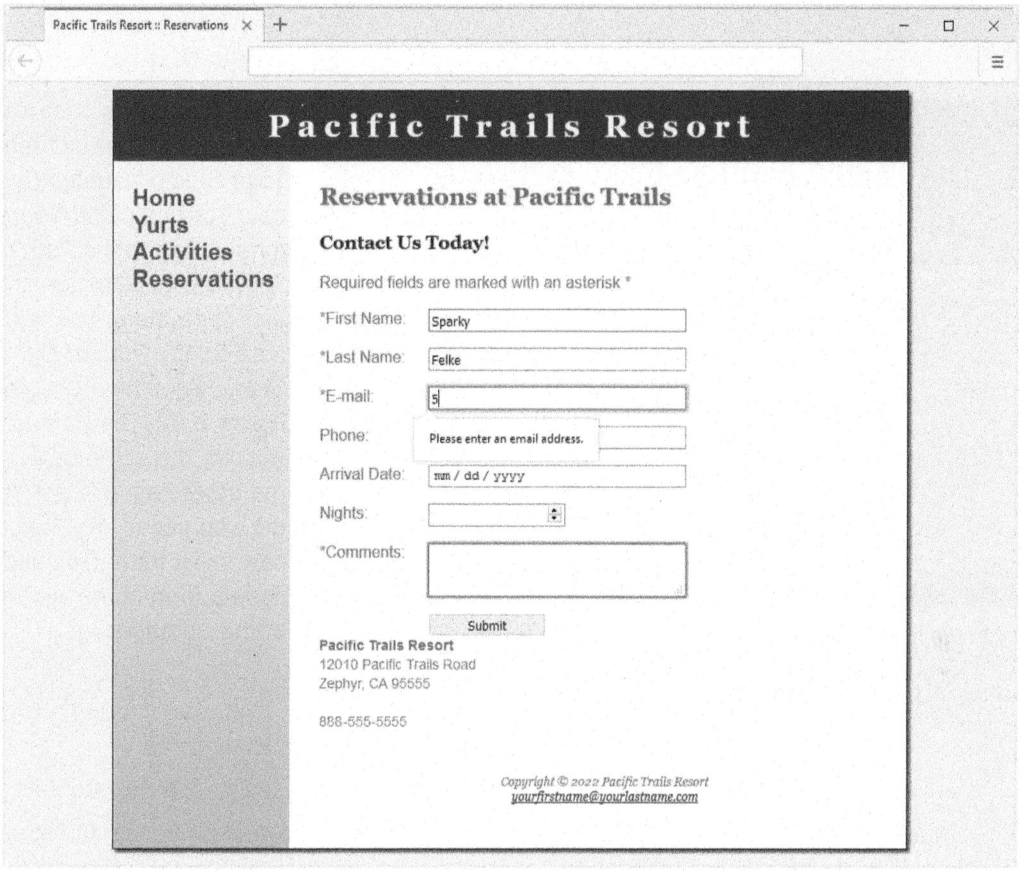

FIGURE 10.40 *The browser checks for required information.*

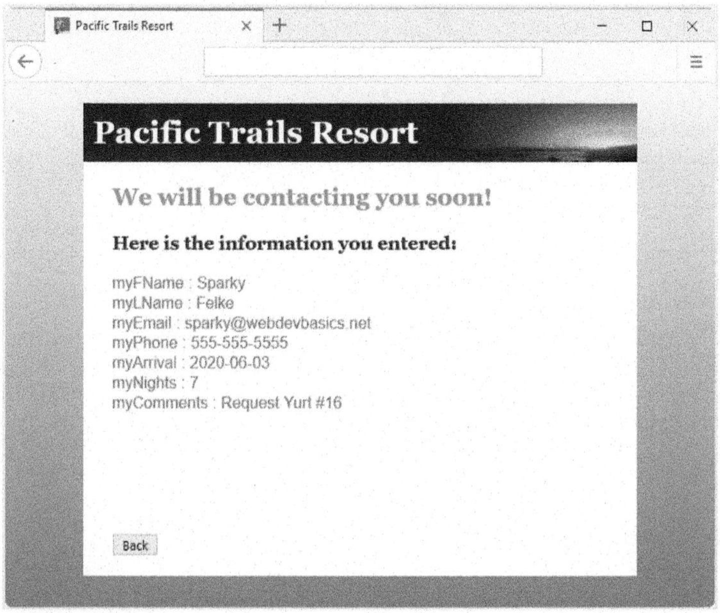

FIGURE 10.41 *The form confirmation page.*

Provide information for all the form controls and click the submit button to submit the form. If you are connected to the Internet, this will send your form information to the server-side script configured in the `<form>` tag. A confirmation page similar to Figure 10.41 will be displayed that lists the form control names and the values you entered.

In this case study, you have coded and styled a form, configured form processing, and completed the final page in the Pacific Trails Resort website.

FIGURE 10.42 *The new Contact page.*

Path of Light Yoga Studio Case Study

In this chapter's case study, you will use the existing Path of Light Yoga Studio website (Chapter 9) as a starting point. You will add a new page to the Path of Light Yoga Studio website—the Contact page. Refer back to the site map for the Path of Light Yoga Studio website in Chapter 2, Figure 2.32. The Contact page will use the same page layout as the other Path of Light Yoga Studio web pages. You'll apply your new skills from this chapter and code a form in the content area of the Contact page.

You have three tasks in this case study:

1. Create a new folder for this Path of Light Yoga Studio case study.
2. Modify the style sheet (yoga.css) to configure style rules for the new Contact page.
3. Create the Contact page: contact.html. Your new page will be similar to Figure 10.42 when you have completed this step.

Task 1: Create a folder called ch10yoga to contain your Path of Light Yoga Studio website files. Copy the files from the Chapter 9 Case Study ch9yoga folder to your new ch10yoga folder.

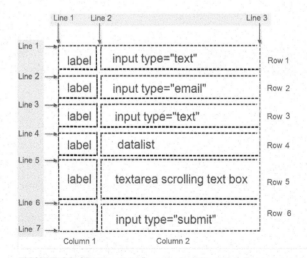

FIGURE 10.43 *The grid layout sketch of the form.*

Task 2: Configure the CSS. Review Figure 10.42 and the wireframe in Figure 10.43. Notice how the text labels for the form controls are on the left side of the content area but contain right-aligned text. Notice the empty vertical space between each form control. When displaying on a narrow viewport, the display will be more pleasing if there is only one column, as shown in Figure 10.44. Open yoga.css in a text editor. Configure the CSS as follows:

a. Configure the single column display for narrow viewports using flex-box. Add CSS above the media queries to configure the a flex layout for the form.

  1. Configure a form element selector. Set flex display, `flex-direction` to column, and `flex-wrap` to nowrap.

  2. Configure the input, datalist, and textarea element selectors with .5em bottom margin.

b. Configure the two-column display with grid layout. Add CSS to the first media query to accomplish this.

  1. Configure a form element selector. Set 60% width, maximum width 40em, grid display with 1em grid gap, and two columns (9em width and 1fr width).

  2. Configure an attribute selector for the submit button. Use the `grid-column` property to place this in the second column. Set width to 9em and bottom margin to .5em.

Save the yoga.css file.

FIGURE 10.44 *Single-column form in a narrow viewport.*

Task 3: **Create the Contact Page.** Use the Classes page as the starting point for the Contact page. Launch a text editor and open classes.html. Save the file as contact.html. Modify your contact.html file to look similar to the Contact page (shown in Figure 10.42) as follows:

1. Change the page title to an appropriate phrase.

2. The Contact page will display a form in the main element. Delete all HTML and content within the main element except for the `<h2>` element and its text.

3. Change the text in the `<h2>` element to "Contact Path of Light Yoga Studio".

4. Add a paragraph below the h2 element to indicate: Required fields are marked with an asterisk * .

5. Prepare to code the HTML for the form area. Begin with a form element that uses the post method and the `action` attribute to invoke server-side processing. Unless directed otherwise by your instructor, configure the `action` attribute to send the form data to https://webdevbasics.net/scripts/yoga.php.

6. Configure the form control for the Name information. Create a label element that contains the text "* Name:". Create a text box configured with "myName" as the value of the `id` and `name` attributes. Use the `for` attribute to associate the label element with the form control. Configure the required attribute.

7. Configure the e-mail form control for the E-mail information (use `type="email"`). Create a label element that contains the text "* E-mail:". Create a text box configured with "myEmail" as the value of the `id` and `name` attributes. Use the `for` attribute to associate the label element with the form control. Configure the required attribute.

8. Code a label element containing the text "Referred by?" that is associated with a textbox and datalist form control with the following options configured: Google, Bing, Facebook, Friend, Radio Ad.

9. Configure the Comments area on the form. Create a label element that contains the text "* Comments:". Create a textarea element configured with "myComments" as the value of the `id` and `name` attributes, `rows` set to 2, and `cols` set to 20. Use the `for` attribute to associate the label element with the form control. Configure the required attribute.

10. Configure the submit button to display "Send Now".

11. Code an ending `</form>` tag on a blank line after the submit button.

Save your file and test your web page in a browser. It should look similar to the pages shown in Figures 10.42 and 10.44, depending on the size of your browser viewport. Submit the form with missing information or only a partial e-mail address. Depending on the browser's level of HTML5 support, the browser may perform form validation and display an error message. Figure 10.45 shows the Contact page rendered in a browser with an incorrectly formatted e-mail address.

FIGURE 10.45 *The browser checks for required information.*

Provide information for all the form controls and click the submit button to submit the form. If you are connected to the Internet, this will send your form information to the server-side script configured in the `<form>` tag. A confirmation page similar to Figure 10.46 will be displayed that lists the form control names and the values you entered.

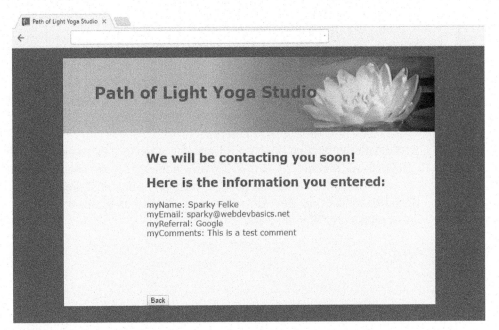

FIGURE 10.46 *The form confirmation page.*

In this case study you have coded and styled a form, configured form processing, and completed the final page in the Path of Light Yoga Studio website.

Endnotes

1. "HTML Living Standard: 4.10.18.7.1 Autofilling Form Controls: the Auto-complete Attribute", *HTML Standard*, WHATWG, html.spec.whatwg.org/#autofilling-form-controls:-the-autocomplete-attribute.

2. "H4: Creating a Logical Tab Order through Links, Form Controls, and Objects." Edited by Alastair Campbell et al., *Techniques for WCAG 2.1*, W3C, 2018, www.w3.org/WAI/WCAG21/Techniques/html/H4.html.

3. "Techniques for Providing a Mechanism to Allow Users to Remap or Turn off Character Key Shortcuts." Edited by Alastair Campbell et al., *Techniques for WCAG 2.1*, W3C, 2018, www.w3.org/WAI/WCAG21/Techniques/general/G217.

4. "Cascading Style Sheets Level 2 Revision 2 (CSS 2.2) Specification: 5.8 Attribute Selectors." Edited by Bert Box et al., *W3C*, W3C, 21 Apr. 2016, www.w3.org/TR/CSS22/selector.html#attribute-selectors.

5. "BBB Tip: Writing an Effective Privacy Policy for Your Business' Website." *BBB*, Better Business Bureau, 28 Jan. 2020, www.bbb.org/article/news-releases/21390-bbb-tip-writing-an-effective-privacy-policy-for-your-small-business-website.

6. "HTML Media Capture." Edited by Anssi Kostiainen et al., *W3C*, W3C, 1 Feb. 2018, www.w3.org/TR/html-media-capture/.

Media and Interactivity Basics

Videos and sounds on your web pages can make them more interesting and informative. This chapter introduces you to working with multimedia and interactive elements on web pages. Methods to add audio and video to your web pages are introduced. Sources of these media types, the HTML code needed to place the media on a web page, and suggested uses of the media are discussed. You'll explore more **CSS** properties as you create an interactive photo gallery and a drop-down menu with **CSS**. Adding the right touch of multimedia and interactivity to a web page can make it engaging and compelling for your visitors.

You'll learn how to...

▸ Describe types of multimedia files used on the Web

▸ Configure hyperlinks to multimedia files

▸ Configure audio and video on a web page

▸ Use the CSS transform and transition properties

▸ Configure an interactive drop-down menu

▸ Configure an interactive widget with the details and summary elements

▸ Describe features and common uses of JavaScript and jQuery

▸ Describe the purpose of HTML5 APIs such as geolocation, web storage, manifest, service workers, and canvas

Getting Started with Audio and Video

The easiest way to give your website visitors access to an audio or a video file is to create a simple hyperlink to the file. For example, the code to hyperlink to a sound file named WDFpodcast.mp3 follows:

```
<a href="WDFpodcast.mp3">Podcast Episode 1</a> (MP3)
```

When your website visitor clicks the link, the player for .mp3 files that is installed on the device will typically display embedded in a new browser window or tab. Your web page visitor can then play the sound. If your website visitor right-clicks on the hyperlink, the media file can be downloaded and saved.

 Hands-On Practice 11.1

In this Hands-On Practice, you will create a web page similar to Figure 11.1 that contains an h1 element and a hyperlink to an MP3 file. The web page will also provide a hyperlink to a text transcript of that file to provide for accessibility. It can be useful to your web page visitors if you indicate the type of file (such as an MP3) and, optionally, the size of the file to be accessed.

FIGURE 11.1 *Providing hyperlinks to an audio file and text transcript.*

Copy the podcast.mp3 and podcast.txt files from the chapter11/starters folder in the student files and save them to a folder named podcast. Use the chapter11/template.html file as a starting point and create a web page containing a page title of Podcast, an h1 element with the text Web Design Podcast, a hyperlink to the MP3 file, and a hyperlink to the text transcript. Save your page as podcast.html. Display the file in a browser. Try to test your page in different browsers and browser versions. When you click on the MP3 hyperlink, an audio player (whichever player is configured on the device) will display in a new browser window or tab. When you click on the hyperlink for the text transcript, the text will display in the browser. Compare your work to the sample in the student files (chapter11/11.1/podcast.html).

Multimedia and Browser Compatibility Issues

Providing your website visitor a hyperlink to download and save a multimedia file is the most basic method to allow access to your media, although your visitor will need an application installed on his or her device to play the file after download. In this chapter, you will work with the HTML5 audio and video elements which are native to the browser and do not require browser plug-ins or players.

Containers and Codecs

When working with native HTML5 video and audio, it's useful to be aware of the container and the codec. The **container** is designated by the file extension, such as .mpa or .aac. The **codec** is the algorithm used to compress and decompress the media, such as the Advanced Audio Coding (AAC) codec.

Explore Tables 11.1 and 11.2, which list common media file extensions, including a description with codec information (if applicable for HTML5) and the MIME type. Note that there is not always a one-to-one correspondence between file extensions and codecs. In some cases, multiple codecs may use the same files extension as their container.

TABLE 11.1 *Commonly Used Audio File Types*

File Extension	Description	MIME Type
.aac, .mp4	MPEG-4 Audio;[1] audio-only MPEG-4 format that uses the Advanced Audio Coding (AAC) codec	audio/mp4
.mp3	MPEG-1 Audio Layer-3[2]	audio/mpeg
.ogg	Open-source audio file format; Vorbis and Opus codecs[3]	audio/ogg
.wav	Waveform Audio File Format	audio/wav

TABLE 11.2 *Commonly Used Video File Types*

File Extension	Description	MIME Type
.av1, .mp4	Open-source, royalty-free AV1 video codec; Alliance for Open Media[4]	video/avi
.m4v, .mp4	MPEG-4 (MP4) codec; H.264 codec	video/mp4
.mov	Developed by Apple; originally indicated a video with a MPEG-4 codec used in Apple's Quicktime application	video/quicktime
.mov, .mp4, .hevc	Newer High-efficiency Video Coding HEVC codec (also called H.265 codec);[5] Apple has begun using the .mov file extension for HEVC video	video/mp4
.ogv, .ogg	Open-source video file format;[6] Theora codec	video/ogg
.webm	Open media file format;[7] VP8 and VP9 video codecs	video/webm

There is no single codec that is supported by all popular browsers. Table 11.3 shows a partial list of browser support of video codecs at the time this was written.

TABLE 11.3 *Current Browser Support of Video Codecs*

Codecs	Chrome	Edge	Firefox	Opera	Safari
AV1[8]	Yes	Yes	Yes	Yes	No
H.264[9]	Yes	Yes	Yes	Yes, but not Opera Mini	Yes
H.265[10]	No	No	No	No	Yes
Theora[11]	Yes	Yes	Yes	Yes	No
VP9[12]	Yes	Yes	Yes	Yes	Partial

For the most current information about browser support for HTML video and codecs, visit https://caniuse.com.

Audio Element and Source Element

The Audio Element

The **audio element** supports native play of audio files in the browser—without the need for plug-ins or players. The audio element begins with the `<audio>` tag and ends with the `</audio>` tag. Table 11.4 lists the attributes of the audio element.

TABLE 11.4 *Audio Element Attributes*

Attribute	Value	Description
src	file name	Optional; audio file name
type	MIME Type	Optional; the MIME type of the audio file, such as audio/mpeg or audio/ogg
autoplay		Optional; boolean attribute; if present, indicates whether audio should start playing automatically; use with caution
controls		Optional; boolean attribute; if present, indicates whether controls should be displayed; recommended
loop		Optional; boolean attribute; if present, indicates whether audio should be played over and over
preload	none, auto, metadata	Optional; values: none (no preload), metadata (only download media file metadata), and auto (download the media file)
title		Optional; specifies a brief text description that may be displayed by browsers or assistive technologies

You'll need to supply multiple versions of the audio due to browser support of different codecs. Plan to supply audio files in at least two different containers, including ogg and mp3. It is typical to omit the src and type attributes from the audio tag and, instead, configure multiple versions of the audio file with the source element.

The Source Element

The **source element** is a self-contained, or void, tag that specifies a multimedia file and a MIME type. The src attribute identifies the file name of the media file. The type attribute indicates the MIME type of the file. Code `type="audio/mpeg"` for an MP3 file. Code `type="audio/ogg"` for audio files using the Vorbis codec. Configure a source element for each version of the audio file. Place the source elements before the closing audio tag.

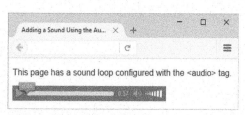

FIGURE 11.2 *The Firefox browser supports the HTML5 audio element.*

The following code sample configures the web page shown in Figure 11.2 (also in the student files chapter11/audio.html) to display a controller for an audio file:

```
<audio controls>
  <source src="soundloop.mp3" type="audio/mpeg">
  <source src="soundloop.ogg" type="audio/ogg">
  <a href="soundloop.mp3">Download the Audio File</a> (MP3)
</audio>
```

Current versions of modern browsers support the HTML5 audio element. The controls displayed by each browser are different. Review the code sample just given and note the hyperlink placed between the second source element and the closing audio tag. Any HTML elements or text placed in this area is rendered by browsers that do not support the HTML5 audio element. This is referred to as fallback content—if the audio element is not supported, the MP3 version of the file is made available for download.

 ## Hands-On Practice 11.2

In this Hands-On Practice, you will launch a text editor and create a web page (see Figure 11.3) that displays an audio control to play a podcast.

Copy the podcast.mp3, podcast.ogg, and podcast.txt files from the chapter11/starters folder in the student files and save them to a folder named audio. Use the chapter11/template.html file as a starting point and create a web page containing a page title and an h1 element with the text Web Design Podcast, an audio control (use the audio element and two source elements), and a hyperlink to the text transcript. Configure a hyperlink to the MP3 file as the fallback content. The code for the audio element follows:

```
<audio controls>
  <source src="podcast.mp3" type="audio/mpeg">
  <source src="podcast.ogg" type="audio/ogg"><br>
  <a href="podcast.mp3">Download the Podcast</a> (MP3)
</audio>
```

Save your page as index.html in the audio folder. Display the file in a browser. Try to test your page in different browsers and browser versions. When you click on the hyperlink for the text transcript, the text will display in the browser. Compare your work to the sample in the student files (chapter11/11.2/audio.html).

FIGURE 11.3 *Using the audio element to provide access to a podcast.*

Audio and Accessibility

Provide alternate content for the audio files you use on your website. For example, create a text transcript for audio files such as podcasts. Often you can use the podcast script as the basis of the text transcript file that you create as a PDF and upload to your website.

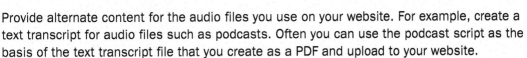

How can I convert an audio file to the Ogg Vorbis codec?

The open-source Audacity application supports Ogg Vorbis. For download information, see https://www.audacityteam.org. If you are looking for a free Web-based converter, you can upload and share an audio file at the Internet Archive (https://archive.org), and an .ogg format file will automatically be generated.

Video Element and Source Element

VideoNote

HTML5 Video

The Video Element

The **video element** supports native play of video files in the browser—without the need for plug-ins or players. The video element begins with the `<video>` tag and ends with the `</video>` tag. Table 11.5 lists the attributes of the video element.

TABLE 11.5 *Video Element Attributes*

Attribute	Value	Description
src	file name	Optional; video file name
type	MIME Type	Optional; the MIME type of the video file, such as video/mp4 or video/ogg
autoplay		Optional; boolean attribute; if present, indicates whether video should start playing automatically; use with caution
controls		Optional; boolean attribute; if present, indicates whether controls should be displayed
height	number	Optional; video height in pixels
loop		Optional; boolean attribute; if present, indicates whether video plays continuously
poster	file name	Optional; specifies an image to display while the video is downloading and before the browser plays the video
preload	none, metadata, auto	Optional; values: none (no preload), metadata (only download media file metadata), and auto (download the media file)
title		Optional; specifies a brief text description that may be displayed by browsers or assistive technologies
width	number	Optional; video width in pixels

Due to browser support of different codecs, plan to supply video files in at least two different containers, including mp4 and ogg (or ogv). It is typical to omit the `src` and `type` attributes from the video tag and, instead, configure multiple versions of the audio file with the source element.

The Source Element

The **source element** is a self-contained, or void, tag that specifies a multimedia file and a MIME type. The `src` attribute identifies the file name of the media file. The `type` attribute indicates the MIME type of the file. Code `type="video/mp4"` for video files using the MP4 codec. Code `type="video/ogg"` for video files using the Theora codec. Configure a source element for each version of the video file. Place the source elements before the closing video tag.

The following code sample configures the web page shown in Figure 11.4 (see the student files chapter11/sparky.html) with the native HTML5 browser controls to display and play a video.

FIGURE 11.4 *Video on a web page.*

```
<video controls poster="sparky.jpg"
       width="160" height="150">
  <source src="sparky.m4v" type="video/mp4">
  <source src="sparky.ogv" type="video/ogg">
  <a href="sparky.mov">Sparky the Dog</a> (.mov)
</video>
```

Current versions of modern browsers support the HTML5 video element. The controls displayed by each browser are different. Internet Explorer versions 9 and later support the video element, but earlier versions do not.

Review the code sample just given and note the anchor element placed between the second source element and the closing video tag. Any HTML elements or text placed in this area is rendered by browsers that do not support the HTML5 video element. This is referred to as fallback content. In this case, a hyperlink to a .mov version of the file is supplied for the user to download.

Video and Accessibility

Provide alternate content for the video files you use on your website. For example, create captions for video files. When you upload a video to YouTube, captions can be automatically generated (although you'll probably want to make some corrections). You can also create a transcript or text captions for an existing YouTube video.[13]

Practice with Video

 Hands-On Practice 11.3

In this Hands-On Practice, you will launch a text editor and create the web page in Figure 11.5, which displays a video control to play a movie. Copy the sedona.m4v, sedona.ogv, sedona.mov, and sedona.jpg files from the chapter11/starters folder in the student files and save them to a new folder named video.

FIGURE 11.5 *Video element.*

Open the chapter11/template.html file in a text editor. Save the file with the name index.html in the video folder.

Edit the index.html file:

1. Modify the title element and configure the h1 element with the text: Sedona Scenes.

2. Configure a video control (use the video element and two source elements) to display the video.

   a. Configure a hyperlink for the sedona.mov file as fallback content.

   b. Configure the sedona.jpg file as a poster image, which will display as the video downloads and before the browser plays the video.

c. The code for the video element follows:

```
<video controls poster="sedona.jpg">
  <source src="sedona.m4v" type="video/mp4">
  <source src="sedona.ogv" type="video/ogg">
  <a href="sedona.mov">Photos of Sedona, Arizona</a> (.mov)
</video>
```

3. Notice that the video element does not contain `height` and `width` attributes. Recall the method to configure flexible images in Chapter 8 and configure the HTML5 video to be flexible with CSS. Place your cursor in the head section and code a style element. Configure the following style rule to set 100% width, auto height, and a maximum width of 1280 pixels (which is the actual width of the video).

```
video { width: 100%; height: auto; max-width: 1280px; }
```

Save your page as index.html in the video folder. Display the file in a browser. Try to test in different browsers and browser versions. Compare your work with Figure 11.5 and the sample in the student files (chapter11/11.3/video.html).

How can I convert a video file to the new codecs?

There are many free online video converters. Online-Convert offers free conversion from MP4 to OGV format (https://video.online-convert.com/convert-to-ogv) and free conversion to WebM format (https://video.online-convert.com/convert-to-webm).

Can I use anything I find on the Web?

Be careful to use only imedia that you have personally created or have obtained the rights or license to use. If another individual has created an image, sound, video, or document that you think would be useful on your own website, ask permission to use the material instead of simply taking it. All work (web pages, images, sounds, videos, and so on) is copyrighted, even if there is no copyright symbol and date on the material. Be aware that there are times when students and educators can use portions of another's work with attribution and not be in violation of copyright law. This is called fair use. **Fair use** is the use of a copyrighted work for purposes such as criticism, reporting, teaching, scholarship, or research. The criteria used to determine fair use are as follows:

▶ The use must be educational rather than commercial.

▶ The nature of the work copied should be factual rather than creative.

▶ The amount copied must be as small a portion of the work as possible.

▶ The copy does not impede the marketability of the original work.

Some individuals may want to retain ownership of their work, but make it easy for others to use or adapt it. Creative Commons (https://creativecommons.org) provides a free service that allows authors and artists to register a type of copyright license called a Creative Commons license, which informs others exactly what others can and cannot do with the creative work.

Iframe Element

Inline frames are widely used on the Web for a variety of marketing and promotional purposes, including displaying banner ads, playing multimedia that may be hosted on an external web server, and serving content for associate and partner sites to display. An advantage of inline frames is separation of control. The dynamic content—such as an ad banner or multimedia clip—can be modified by the partner site at any time, just as YouTube dynamically configures the format of the video display in this section.

The iframe Element

The **iframe element** configures an **inline frame** that displays the contents of another web page within your web page document, referred to as *nested browsing*. The iframe element begins with the `<iframe>` tag and ends with the `</iframe>` tag. Figure 11.6 shows a web page that displays a YouTube video within an iframe element. See Table 11.6 for a list of commonly used iframe element attributes.[14]

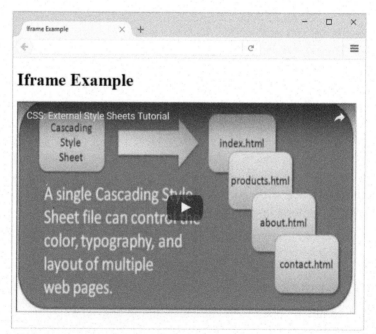

FIGURE 11.6 *The iframe element in action.*

TABLE 11.6 *Commonly Used iframe Element Attributes*

Attribute	Description and Value
src	URL of the web page to display in the inline frame
height	Optional; inline frame height in pixels or percentage
width	Optional; inline frame width in pixels or percentage
id	Optional; text name, alphanumeric, beginning with a letter, no spaces—the value must be unique and not used for other id values on the same web page document
loading	Optional; eager (default, display immediately), lazy (defer display until the user scrolls near it)
name	Optional; text name, alphanumeric, beginning with a letter, and no spaces
sandbox	Optional; disallow/disable features such as plug-ins, scripts, and forms
title	Optional; specifies a brief text description
allow	Optional; feature policy applied to the iframe
allowfullscreen	Optional; boolean attribute; if present, the iframe has the option to display the video to fill the device's screen

 Hands-On Practice 11.4 ———————————————————————

VideoNote
Configure an
Inline Frame

In this Hands-On Practice, you will create a web page that displays a YouTube video (located at https://www.youtube.com/watch?v=2CuOug8KDWI) within an iframe element. You can choose to embed this video or select a different video. The process is to display the YouTube page for the video and copy the video identifier, which is the text after the "=" in the URL. In this example, the video identifier is 2CuOug8KDWI.

Use the chapter11/template.html file as a starting point and configure a web page containing a page title and an h1 element with the text "Iframe Example" and an iframe element. Code the src attribute with https://www.youtube.com/embed/ followed by the video identifier. In this example, set the src attribute to the value https://www.youtube.com/embed/2CuOug8KDWI. The code to display the video shown in Figure 11.6 follows:

```
<iframe src="https://www.youtube.com/embed/2CuOug8KDWI"
  width="640" height="385" allowfullscreen>
</iframe>
```

Save your page as myiframe.html and display it in a browser. Compare your work with Figure 11.6 and the sample in the student files (chapter11/11.4).

 A sample web page using CSS to configure a responsive iframe is in the student files (chapter11/11.4/ri.html). Explore the following resources about this technique:

▶ https://css-tricks.com/responsive-iframes/

▶ https://davidwalsh.name/responsive-iframes

▶ https://benmarshall.me/responsive-iframes/

CSS Transform Property

The CSS **transform property** allows you to change the display of an element and provide functions to rotate, scale, skew, and reposition an element. Both two-dimensional (2D) and three-dimensional (3D) transforms are possible. Table 11.7 lists commonly used 2D transform property function values and their purpose.[15] We'll focus on the rotate and scale transforms in this section.

FIGURE 11.7 *The transform property in action.*

CSS Rotate Transform

The **rotate() transform** function takes a value in degrees (like an angle in geometry). Rotate to the right with a positive value. Rotate to the left with a negative value. The rotation is around the origin, which, by default, is the middle of the element. The web page in Figure 11.7 demonstrates the use of the CSS transform property to slightly rotate the figure.

TABLE 11.7 *Values of the Transform Property*

Value	Purpose
rotate (*degree*)	Rotates the element by the angle
scale (*number, number*)	Scales or resizes the element along the X- and Y-axes (X,Y); if only one value is provided, it configures the horizontal and vertical scale amount
scaleX (*number*)	Scales or resizes the element along the X-axis
scaleY (*number*)	Scales or resizes the element along the Y-axis
skewX (*number*)	Distorts the display of the element along the X-axis
skewY (*number*)	Distorts the display of the element along the Y-axis
translate (*number, number*)	Repositions the element along the X- and Y-axes (X,Y)
translateX (*number*)	Repositions the element along the X-axis
translateY (*number*)	Repositions the element along the Y-axis

CSS Scale Transform

The **scale() transform** function resizes an element in three different ways: along the X-axis, along the Y-axis, and along both the X- and Y-axes. Specify the amount of resizing using a number without units. For example, `scale(1)` does not change the element's size, `scale(2)` indicates the element should render two times as large, `scale(3)` indicates the element should render three times as large, and `scale(0)` indicates the element should not display.

 Hands-On Practice 11.5 —————————————

In this Hands-On Practice, you will configure the rotate and scale transforms shown in Figure 11.7. Create a new folder named transform. Copy the light.gif and lighthouse.jpg images from the chapter11/starters folder in the student files to your transform folder. Launch a text editor and open the starter.html file in the chapter11 folder. Save the file as index.html in your transform folder. Launch the file in a browser, and it will look similar to Figure 11.8.

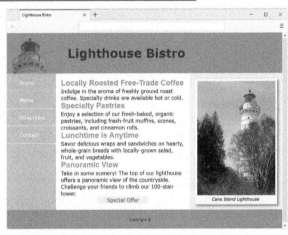

FIGURE 11.8 *Before the transform property.*

Open index.html in a text editor and view the embedded CSS.

1. Locate the figure element selector. You will add a style declaration to the figure element selector that will configure a three-degree rotation transform. The CSS follows:

```
figure { margin: auto; padding: 8px; width: 265px;
        background-color: #FFF; border: 1px solid #CCC;
        box-shadow: 5px 5px 5px #828282;
        transform: rotate(3deg);   }
```

2. Locate the #offer selector. This configures the "Special Offer" div displayed above the page footer. You will add a style declaration to the selector that configures the browser to display the element two times larger. The CSS follows:

```
#offer { background-color: #EAEAEA;
        width: 10em;
        margin: 2em auto 0 auto;
        text-align: center;
        transform: scale(2);   }
```

Save the file and display it in a browser. You should see two changes: the figure displayed on a slight angle and larger "Special Offer" text. Compare your work to Figure 11.7 and the sample in the student files (chapter11/11.5/index.html).

 This section provided a brief introduction to the transform property, but there is much more to explore. Visit https://html-css-js.com/css/generator/transform/ to generate the CSS for rotate, scale, translate, and skew transforms. Find out more about transforms at https://css-tricks.com/almanac/properties/t/transform/ and https://developer.mozilla.org/en/CSS/Using_CSS_transforms.

CSS Transition Property

The CSS **transition property** provides for changes in property values to display in a smoother manner over a specified time. You can apply a transition to a variety of CSS properties including `color`, `background-color`, `border`, `font-size`, `font-weight`, `margin`, `padding`, `opacity`, and `text-shadow`. When you configure a transition for a property, you need to configure values for the `transition-property`, `transition-duration`, `transition-timing-function`, and `transition-delay` properties. These can be combined in a single transition shorthand property. Table 11.8 lists the transition properties and their purpose. Table 11.9 lists commonly used `transition-timing-function` values and their purpose.[16]

TABLE 11.8 *CSS Transition Properties*

Property	Description
`transition-property`	Indicates the CSS property to which the transition applies
`transition-duration`	Indicates the length of time to apply the transition; default value 0 configures an immediate transition; a numeric value specifies time (usually in seconds); W3C WCAG 2.1 Accessibility Guidelines recommend limiting the duration to five seconds or less[17]
`transition-timing-function`	Configures changes in the speed of the transition by describing how intermediate property values are calculated; common values include `ease` (default), `linear`, `ease-in`, `ease-out`, `ease-in-out`
`transition-delay`	Indicates the beginning of the transition; default value 0 configures no delay; a numeric value specifies time (usually in seconds)
`transition`	Shorthand property; list the value for `transition-property`, `transition-duration`, `transition-timing-function`, and `transition-delay` separated by spaces; default values can be omitted, but the first time unit applies to `transition-duration`

TABLE 11.9 *Commonly Used* `transition-timing-function` *Values*

Value	Purpose
`ease`	Default; transition effect begins slowly, speeds up, and ends slowly
`linear`	Transition effect has a constant speed
`ease-in`	Transition effect begins slowly and speeds up to a constant speed
`ease-out`	Transition effect begins at a constant speed and slows down
`ease-in-out`	Transition effect is slightly slower; begins slowly, speeds up, and slows down

Hands-On Practice 11.6

Recall that the CSS `:hover` pseudo-class provides a way to configure styles to display when the web page visitor moves the mouse over an element. The change in display happens somewhat abruptly. Web designers can use a CSS transition to create a more gradual change to the hover state. You'll try this out in this Hands-On Practice when you configure a transition for the navigation hyperlinks on a web page.

Create a new folder named transition. Copy the light.gif and lighthouse.jpg images from the chapter11/starters folder in the student files to your transition folder. Launch a text editor and open the starter.html file in the chapter11 folder. Save the file as index.html in your transition folder. Open index.html in a browser, and the background color of each navigation hyperlink area is blue. Place your mouse pointer over one of the navigation hyperlinks and notice that the background color and text color change immediately.

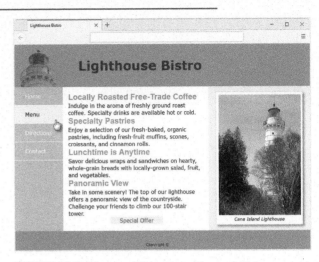

FIGURE 11.9 *The transition causes the hyperlink background color to change in a more gradual manner.*

Open index.html in a text editor and view the embedded CSS. Locate the `nav a:hover` selector and notice that the color and background-color properties are configured. You will add a style declaration to the `nav a` selector to cause a more gradual change in the background color when the user places the mouse over the hyperlink. The CSS follows:

```
nav a { text-decoration: none;
        display: block;
        padding: 1em 2em;
        transition: background-color 2s linear;  }
```

Save the file and display it in a browser. Place your mouse over one of the navigation hyperlinks and notice that while the text color changes immediately, the background color changes in a more gradual manner—the transition is working! Compare your work to Figure 11.9 and the student files (chapter11/11.6/index.html).

 If you'd like more control over the transition than what is provided by the values listed in Table 11.9, explore using the cubic-bezier value for the transition-timing-function. A Bezier curve is a mathematically defined curve often used in graphic applications to describe motion. Explore the following resources:

▶ https://www.the-art-of-web.com/css/timing-function

▶ https://cubic-bezier.com

Practice with Interactivity

 Hands-On Practice 11.7

FIGURE 11.10 *The initial display of the page with the new CSS applied.*

In this Hands-On Practice you will use the CSS `transition`, `transform`, `opacity`, `position`, and `z-index` properties to configure an interactive photo gallery web page.

The web page will initially display six semi-transparent images (shown in Figure 11.10) that a web page visitor can hover over or click on.

When the visitor hovers over an image, the photo increases in size, becomes opaque and displays a caption, as shown in Figure 11.11. The visitor can also click on an image to see a larger version of the photo.

Create a new folder named ch11gallery. Copy all the files in the student files chapter11/starters/gallery folder into your ch11gallery folder.

Display the index.html file in browser. It should be similar to Figure 11.12, with six photos and captions.

FIGURE 11.11 *Hover over a photo to see a larger version and a caption.*

FIGURE 11.12 *Part of the initial web page.*

Launch a text editor and view the HTML. Locate the div element assigned to a class named `gallery`. This div contains six figure elements, which each contain an image hyperlink and a figcaption element. You will configure the div element as a flexbox container for the figure elements. You will also configure styles for the figure and `figure:hover` selectors that handle the initial display of the photos and the transition that occurs when the visitor hovers over each image.

Configure the following CSS in the head section between the style tags.

1. Code styles for a class named `gallery`. Set flex display, `flex-wrap` to `wrap`, and `justify-content` to `space-around`. Also set 70% width and center the div with margin set to `auto`. The CSS follows:

```
.gallery { display: flex; flex-wrap: wrap;
           justify-content: space-around;
           width: 70%; margin: auto; }
```

2. Configure the initial display of the figure elements. Because the larger, opaque photo should overlap the rest of the gallery, the z-index property is required. Recall from Chapter 7 that for the z-index property to take effect, the element must also have a position property set to either absolute, relative, fixed, or sticky. Code styles for the figure element selector that set relative positioning, z-index to 1, opacity to .75, margin to 1em, white background color, transparent black text (use RGBA color), .5em font size, centered text, and a three second `ease-out transition`.
The CSS follows:

```
figure { position: relative;
         z-index: 1; opacity: .75;
         background-color: #FFF; color: rgba(0,0,0,0);
         margin: 1em; font-size: .5em, text-align: center;
         transition: all ease-out 3s; }
```

3. Configure the result of the transition. Code styles for the `figure:hover` selector that set opacity to 1, z-index to 999, a 3px dark grey shadow, black opaque text and create a transform that is three times the original size. The CSS follows:

```
figure:hover { opacity: 1; z-index: 999;
               box-shadow: 3px 3px 3px #333;
               color: rgba(0,0,0,1);
               transform: scale(3); }
```

Save your page and display it in a browser. The initial display of your page should be similar to Figure 11.10 with semi-transparent photos. If you resize your browser viewport, the gallery is responsive as the rows and columns shift.

Hover your mouse over one of the images and it should become similar to Figure 11.11. The photo will become larger, opaque, and the figure caption will display. Compare your work to the sample in the student files (chapter11/11.7).

CSS Drop-Down Menu

 Hands-On Practice 11.8

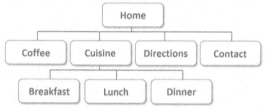

FIGURE 11.13 *Site map.*

In this Hands-On Practice, you will configure a navigation menu that is interactive and displays a drop-down menu. Figure 11.13 displays a site map for the website. Notice how the Cuisine page has three subpages: Breakfast, Lunch, and Dinner. You will configure a drop-down menu that displays when a visitor hovers over the Cuisine navigation hyperlink as shown in Figure 11.14.

Create a folder named mybistro. Copy the files from the chapter11/bistro folder in the student files into your mybistro folder. Notice the main menu has hyperlinks for Home, Coffee, Cuisine, Directions, and Contact. You will edit the CSS and edit each page to configure a Cuisine submenu that provides hyperlinks to three pages (Breakfast, Lunch, and Dinner).

Task 1: Configure the HTML.

Launch a text editor and open the index.html file. You will modify the nav area to contain a new unordered list with hyperlinks to the Breakfast, Lunch, and Dinner pages. You will configure a new ul element that is contained *within* the Cuisine li element. The new ul element will contain an li element for each room. The HTML follows:

```
<nav>
<ul>
  <li><a href="index.html">Home</a></li>
  <li><a href="coffee.html">Coffee</a></li>
  <li><a href="cuisine.html">Cuisine</a>
    <ul>
      <li><a href="breakfast.html">Breakfast</a></li>
      <li><a href="lunch.html">Lunch</a></li>
      <li><a href="dinner.html">Dinner</a></li>
    </ul>
  </li>
  <li><a href="directions.html">Directions</a></li>
  <li><a href="contact.html">Contact</a></li>
</ul>
</nav>
```

Save the file and display it in a browser. Don't worry if the navigation area seems a bit garbled—you'll configure the submenu CSS in Step 2. Next, edit the nav area in each page (coffee.html, cuisine.html, breakfast.html, lunch.html, dinner.html, directions.html, and contact.html) as you did in the index.html file.

Task 2: Configure the CSS.

Launch a text editor and open the bistro.css file.

a. Configure the submenu with absolute positioning. Recall from Chapter 7 that absolute positioning precisely specifies

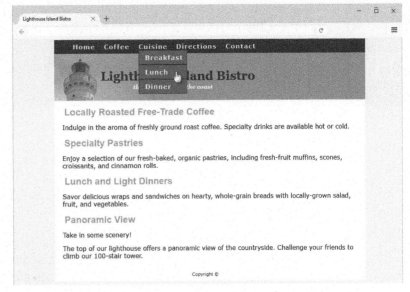

FIGURE 11.14 *The drop-down menu displays.*

the location of an element outside of normal flow in relation to its first parent non-static element. The nav element's position is static by default so add the following declaration to the styles for the nav element selector: `position: relative;`

b. The submenu that displays the hyperlinks for the Breakfast, Lunch, and Dinner pages is configured using a new ul element that is contained within the existing ul element in the nav area. Configure a descendent `nav ul ul` selector and code style declarations to use absolute positioning, #5564A0 background color, 0 padding, display set to none. The CSS follows:

```
nav ul ul { position: absolute; background-color: #5564A0;
            padding: 0; display: none; }
```

c. To style each li element within the submenu, use a descendent `nav ul ul li` selector and configure the li elements in the submenu with a border, block display, 8em width, 1em left padding, and 0 left margin. The CSS follows:

```
nav ul ul li { border: 1px solid #00005D;
               display: block; width: 8em;
               padding-left: 1em; margin-left: 0; }
```

d. Configure the submenu ul to display when the `:hover` is triggered for the li elements in the nav area. The CSS follows:

```
nav li:hover ul { display: block; }
```

Test your pages in a browser. The drop-down menu should look similar to Figure 11.14. You can compare your work to the sample in the student files (chapter11/11.8/horizontal). An example of a web page with a vertical fly-out menu is available in the student files (chapter11/11.8/vertical).

Details Element and Summary Element

The details element and summary element are used together to configure an interactive widget that will hide and show information.

Details Element

The purpose of the **details element** is to configure the browser to render an interactive widget, which contains one summary element and detailed information (which can be a combination of text and HTML tags). The details element begins with the `<details>` tag and ends with the `</details>` tag.

Summary Element

The **summary element** is coded within the details element. The purpose of the summary element is to contain the text summary (typically some type of term or heading) shown in the interactive widget. The summary element begins with the `<summary>` tag and ends with the `</summary>` tag.

Details and Summary Widget

Figures 11.15 and 11.16 show the details and summary elements in action using the Chrome browser. Figure 11.15 shows the initial display of the web page with each summary item (in this case the terms Repetition, Contrast, Proximity, and Alignment) visible and displayed next to a triangle rendered automatically by the Chrome browser.

In Figure 11.16, the visitor has selected the first summary item (Repetition), which caused browser to display the detailed information for that item. The visitor can select the same summary item again to hide the details or can select another summary item to also show its corresponding detailed information.

FIGURE 11.15 *Initial browser display.*

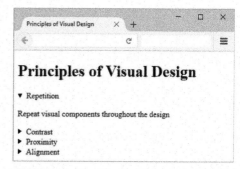

FIGURE 11.16 *Detailed information displays.*

Current versions of modern browsers support the details and summary elements. Browsers that do not support the details and summary elements display all the information immediately and do not provide interactivity.

 Hands-On Practice 11.9 ─────────────────────────────

In this Hands-On Practice, you will configure an interactive widget with the details and summary elements as you create the page shown in Figures 11.15 and 11.16. Create a new folder named ch11details. Launch a text editor and open the template file located at chapter11/template.html in the student files. Save the file as index.html in your ch11details folder. Modify the file to configure a web page as indicated:

1. Configure the text, Principles of Visual Design, within an h1 element and within the title element.

2. Code the following in the body of the web page:

```
<details>
   <summary>Repetition</summary>
   <p>Repeat visual components throughout the design</p>
</details>
<details>
   <summary>Contrast</summary>
   <p>Add visual excitement and draw attention</p>
</details>
<details>
   <summary>Proximity</summary>
   <p>Group related items</p>
</details>
<details>
   <summary>Alignment</summary>
   <p>Align elements to create visual unity</p>
</details>
```

Save your file and test your page in a browser. The initial display should be similar to Figure 11.15. Try selecting or clicking on one of the terms or arrows to display the information you coded within the details element. If you select the term Repetition, your browser should be similar to Figure 11.16.

If you are using a browser that does not support the details and summary elements, your display will be similar to Figure 11.17.

A suggested solution is in the student files chapter11/11.9 folder.

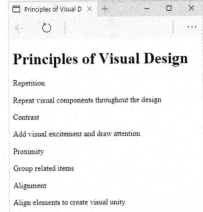

FIGURE 11.17 *Display in a nonsupporting browser.*

JavaScript

Although some interactivity on web pages can be achieved with CSS, JavaScript powers much of the interactivity on the Web. **JavaScript**, developed initially by Brendan Eich at Netscape, is an object-based, client-side scripting language interpreted by a web browser. JavaScript is considered to be **object-based** because it's used to work with the objects associated with a web page document: the browser window, the document itself, and elements such as forms, images, and hyperlinks.

JavaScript statements can be placed in a separate file (with a .js extension) that is accessed by a web browser or within an HTML script element. The purpose of the **script element** is to either contain scripting statements or indicate a file that contains scripting statements. Some JavaScript also can be coded within the HTML. In all cases, the web browser interprets the JavaScript statements. Because JavaScript is interpreted by a browser, it is considered to be a **client-side scripting** language.

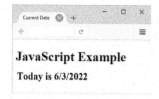

FIGURE 11.18 *JavaScript in action.*

JavaScript can be used to respond to events such as moving the mouse, clicking a button, and loading a web page. This technology is also often utilized to edit and verify information on HTML form controls such as text boxes, check boxes, and radio buttons. Other uses for JavaScript include pop-up windows, image slideshows, animation, date manipulation, and calculations. Figure 11.18 shows a web page (found in the student files at chapter11/date.html) that uses JavaScript to determine and display the current date. The JavaScript statements are enclosed within an HTML **script element** and coded directly in the .html file. The code sample follows:

```
<h2>Today is
<script>
var myDate = new Date()
var month = myDate.getMonth() + 1
var day = myDate.getDate()
var year = myDate.getFullYear()
document.write(month + "/" + day + "/" + year)
</script>
</h2>
```

JavaScript is a powerful scripting language and is a good choice to learn as you continue your studies. There are many free resources for JavaScript code and JavaScript tutorials on the Web. Here are a few sites that offer free tutorials or free scripts:

- Mozilla Developer Network JavaScript Guide: https://developer.mozilla.org/en-US/docs/Web/JavaScript/Guide
- JavaScript Tutorial: https://javascript.info
- TutorialsPoint: https://www.tutorialspoint.com/javascript

jQuery

Web developers often need to configure the same type of common interactive features (such as slideshows, form validation, and animation) on web pages. One approach is to write one's own JavaScript code and test it in a wide variety of browsers and operating systems. As you might guess, this can become quite time-consuming. The free, open-source **jQuery** JavaScript library was developed by John Resig in 2006 to simplify client-side scripting. The jQuery Foundation is a volunteer organization that contributes to the continued development of jQuery and provides jQuery documentation at https://api.jquery.com.

An **application programming interface** (API) is a protocol that allows software components to communicate—interacting and sharing data. The jQuery API can be used to configure many interactive features, including the following:

▶ image slideshows

▶ animation (moving, hiding, and fading)

▶ event handling (mouse movements and mouse clicking)

▶ document manipulation

Many web developers and designers have found that jQuery is easier to learn and work with than writing their own JavaScript, although a basic understanding of JavaScript is needed to be efficient when using jQuery. An advantage of the jQuery library is its compatibility with all current browsers.

jQuery is often used on popular websites, such as Amazon and Twitter. Because jQuery is an open-source library, anyone can extend the jQuery library by writing a new **jQuery plugin** that provides a new or enhanced interactive feature. For example, the jQuery Cycle plugin[18] supports a variety of transition effects. Figure 11.19 (see https://webdevfoundations.net/jquery) shows an example of using jQuery and the Cycle plugin to create an image slideshow. There are many jQuery plugins available, providing interactions and functionality such as slideshows, tooltips, and form validation.

FIGURE 11.19 *jQuery plugin slideshow.*

 There are many free tutorials and resources that can help you learn about jQuery. You may wish to do further research using some of the following resources:

▶ How jQuery Works: https://learn.jquery.com/about-jquery/how-jquery-works/

▶ Tutorial Republic: https://www.tutorialrepublic.com/jquery-tutorial

▶ jQuery Tutorial: https://jquery-tutorial.net/

HTML5 APIs

You've already been introduced to the term application programming interface (API), which is a protocol that allows software components to communicate—interacting and sharing data. A variety of APIs that are intended to work with HTML5, CSS, and JavaScript are currently under development and in the W3C approval process. We'll explore some of the new APIs in this section that provide for:

▶ geolocation
▶ web storage
▶ progressive web applications
▶ two-dimensional drawing

Geolocation

The **Geolocation API**[19] allows your web page visitors to share their geographic location. The browser will first confirm that your visitors want to share their location. Then, their location may be determined by the IP address, wireless network connection, local cell tower, or Global Positioning System (GPS) hardware depending on the type of device and browser. JavaScript is used to work with the latitude and longitude coordinates provided by the browser. Visit https://developers.google.com/maps/documentation/javascript/examples/map-geolocation for an example of geolocation in action.

Web Storage

Web developers have traditionally used the JavaScript cookie object to store information in key-value pairs on the client (the website visitor's computer). The **Web Storage API**[20] provides two new ways to store information on the client side: local storage and session storage. An advantage to using web storage is the increase in the amount of data that can be stored (5MB per domain). The **localStorage** object stores data without an expiration date. The **sessionStorage** object stores data only for the duration of the current browser session. JavaScript is used to work with the values stored in the localStorage and sessionStorage objects. Visit https://webdevfoundations.net/storage and https://html5demos.com/storage for examples of web storage.

Progressive Web Application

You've most likely used native applications (apps) for mobile phones. A native app must be built and distributed specifically for the platform it will be used on. If your client would like a native mobile app for both an iPhone and an Android, you'll need to create two different apps! In contrast, a web application can be written with HTML, CSS, and JavaScript and can run in any browser—as long as you are online. A **progressive web application (PWA)** offers a rich experience similar to a native app on a mobile device—the user can choose to add the website's icon to the home screen, and the website has some level of functionality even when the device is not connected to the Internet.

An early approach to progressive web applications[21] utilized an application cache that informed the browser about files to automatically download and update, fallback files to display when a resource has not been cached, and files that are only available when online. However, there were issues with this approach and the W3C is developing a combination of new APIs to power PWAs including Manifest and Service Workers.

The **Manifest API**[22] contains information about the PWA; including the data needed for the PWA's icon to be added to the home screen of a device. The **Service Workers API**[23] provides a way for websites to perform persistent background processing such as push notifications and background data syncing. A **service worker** is JavaScript that runs in the background, separate from a web page, and listens for events such as install, activate, message, fetch, sync, and push. To provide more security, service workers must run over HTTPS.

For more information about PWAs, visit the following resources:

▶ https://developer.mozilla.org/en-US/docs/Web/Apps/Progressive/Introduction

▶ https://web.dev/progressive-web-apps/

▶ https://medium.com/samsung-internet-dev/a-beginners-guide-to-making-progressive-web-apps-beb56224948e

Drawing with the Canvas Element

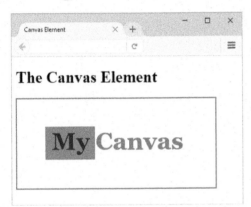

FIGURE 11.20 *The canvas element.*

The HTML5 **canvas element** is a container for dynamic graphics. The canvas element begins with the `<canvas>` tag and ends with the `</canvas>` tag. The canvas element is configured with the **Canvas 2D Context API,**[24] which provides a way to dynamically draw and transform lines, shapes, images, and text on web pages. If that wasn't enough, the canvas API also provides for interaction with actions taken by the user, like moving the mouse.

The Canvas API offers methods for two-dimensional (2D) bitmap drawing, including lines, strokes, arcs, fills, gradients, images, and text. However, instead of drawing visually using a graphics application, you draw programmatically by writing JavaScript statements. A very basic example of using JavaScript to draw within the canvas element is shown in Figure 11.20 (see chapter11/canvas.html in the student files). Experience examples of the canvas element in action at https://codepen.io/CraneWing/pen/egaBze.

?FAQ

What is SVG?

SVG (Scalable Vector Graphics) is a markup language that describes vector-based, two-dimensional graphics in XML[25]. Vector graphic shapes, images, and text objects can be included in an SVG, which can scale to increase or decrease in size without losing clarity. Advantages of using SVG images include scalability and small file size. SVG content is stored in the .svg file extension and can be interactive and animated. You can write the XML code for an SVG yourself, but it's common to use a vector graphics editor, such as the open-source Inkscape to generate an SVG file. There are several methods commonly used to display an SVG on a web page: an img element with an .svg file as the src attribute value, a CSS background image, and an svg element that contains the XML code for the SVG graphic. See the chapter11/svg folder in the student files for examples of SVG in use. To learn more about SVG, visit the following resources:

▶ https://developer.mozilla.org/en-US/docs/Web/SVG/Tutorial

▶ https://css-tricks.com/using-svg

CHAPTER 11

Review and Apply

Review Questions

Multiple Choice. Choose the best answer for each item.

1. Which property provides a way for you to rotate, scale, skew, or move an element?
   a. `position`
   b. `transition`
   c. `transform`
   d. `z-index`

2. What type of files are .webm, .ogv, and .m4v?
   a. audio files
   b. video files
   c. image files
   d. none of the above

3. What happens when a browser does not support the `<video>` or `<audio>` element?
   a. The computer crashes.
   b. The web page does not display.
   c. The fallback content, if it exists, will display.
   d. None of the above.

4. Which property enables changes in property values to display in gradual manner over a specified time?
   a. `transition`
   b. `transform`
   c. `position`
   d. `z-index`

5. Which of the following is an open-source video codec?
   a. Theora
   b. MP3
   c. Vorbis
   d. API

6. Which of the following is an object-based, client-side scripting language?
   a. HTML
   b. CSS
   c. JavaScript
   d. API

7. Which of the following is an HTML API that stores information on the client?
   a. geolocation
   b. web storage
   c. client storage
   d. canvas

8. Which elements can be used to configure an interactive widget?
   a. hide and show
   b. details and summary
   c. display and hidden
   d. title and summary

9. Which of the following should you do to provide for usability and accessibility?
   a. Use video and sound whenever possible.
   b. Supply text descriptions of audio files and caption video files that appear in your web pages.
   c. Never use audio and video files.
   d. None of the above.

10. Which of the following elements displays the contents of another web page document?
    a. `iframe`
    b. `script`
    c. `document`
    d. `video`

Review Answers

1. c   2. b   3. c   4. a   5. a   6. c   7. b   8. b   9. b   10. a

Hands-On Exercises

1. Write the HTML for a hyperlink to a video called sparky.mov on a web page.

2. Write the HTML to embed an audio file called soundloop.mp3 on a web page that can be controlled by the visitor.

3. Write the HTML to display a video on a web page. The video files are named prime.m4v, prime.webm, and prime.ogv. The width is 213 pixels. The height is 163 pixels. The poster image is prime.jpg.

4. Write the HTML to display a details and summary widget with three items a web page.

5. Write the HTML to configure an inline frame to display the home page of https://webdevbasics.net in your web page.

 Although you can configure an inline frame to display another website, it is an ethical practice to only do so when you have obtained permission or have an arrangement with the other website.

6. Create a web page about your favorite movie that contains an audio file with your review of the movie. Use an application of your choice to record your review. Place an e-mail hyperlink to yourself on the web page. Save the page as review.html.

7. Create a web page about your favorite music group that contains either a brief audio file with your review or an audio clip of the group. Use an application of your choice to record your review. Place an e-mail hyperlink to yourself on the web page. Save the page as music.html.

8. Add new transitions to the Lighthouse Bistro home page (found in the student files at chapter11/11.6/index.html). Configure the opacity property to display the lighthouse figure initially at 50% opacity and slowly change the opacity to 100% when the visitor places their mouse over the figure area.

9. Create a web page that uses the iframe element to embed the video from the Internet Archive found at https://archive.org/details/lake-tahoe-memories. Place an e-mail hyperlink to yourself on the web page. Save the page as iframe.html.

Focus on Web Design

There are web design usability and accessibility issues associated with HTML5 video. Visit the following sites to become aware of these issues:

- https:/developer.mozilla.org/en-US/docs/Learn/Accessibility/Multimedia
- https://developer.mozilla.org/en-US/docs/Learn/Tools_and_testing/ Cross_browser_Testing/Accessibility
- https://www.digitala11y.com/accessible-jquery-html5-media-players

Write a one-page report that describes HTML5 video usability issues that web designers should be mindful of. Cite the URLs of the resources you used.

Case Study

You will continue the case studies from Chapter 10 as you add media and and an interactive feature.

Pacific Trails Resort Case Study

In this chapter's case study, you will use the existing Pacific Trails Resort website (Chapter 10) as a starting point to create a new version of the website that incorporates multimedia and interactivity. You have three tasks in this case study:

1. Create a new folder for this Pacific Trails case study.
2. Modify the style sheet (pacific.css) to configure a transition for the navigation hyperlink color.
3. Add a video to the home page (index.html) and update the external CSS file.

Task 1: Create a folder called ch11pacific to contain your Pacific Trails Resort website files. Copy the files from the Chapter 10 Case Study ch10pacific folder. Copy the following files from the chapter11/casestudystarters folder in the student files and save them in your ch11pacific folder: pacifictrailsresort.mp4, pacifictrailsresort.ogv, pacifictrailsresort.mov, and pacifictrailsresort.jpg.

Task 2: Configure a Navigation Transition with CSS. Open pacific.css in a text editor. Locate the `nav a` selector. Code additional style declarations to configure a three-second ease-out transition in the `color` property. Save the file. Display any of the web pages in a browser (see Figure 11.21) that supports transitions and place your mouse pointer over a navigation link. You should see a gradual change in the color of the text in the navigation link.

FIGURE 11.21 *Navigation links with a transition.*

Task 3: Configure the Video. Launch a text editor and open the home page (index.html). Code an HTML5 video control below the paragraph element. Configure the video, source, and anchor elements to work with the following files: pacifictrailsresort.mp4, pacifictrailsresort.ogv, pacifictrailsresort.mov, and pacifictrailsresort.jpg. The dimensions of the video are 320 pixels wide by 240 pixels high. Use the anchor element to configure a hyperlink to the .mov file as fallback content. Save the file. Check your HTML syntax using the W3C validator (https://validator.w3.org). Correct and retest if necessary.

Next, configure the CSS. Launch a text editor and open pacific.css.

Code CSS above the media queries to configure a style rule for the video element selector that configures the video to float on the right with a 1em margin.

Save the pacific.css file. Launch a browser and test your new index.html page. It should look similar to Figure 11.22. You have enhanced the website with interactivity and video.

FIGURE 11.22 *Pacific Trails Resort Home page.*

Path of Light Yoga Studio Case Study

In this chapter's case study, you will use the existing Path of Light Yoga Studio website (Chapter 10) as a starting point to create a new version of the website that incorporates multimedia and interactivity. You have three tasks in this case study:

1. Create a new folder for this Path of Light Yoga Studio case study.
2. Modify the style sheet (yoga.css) to configure a transition for the navigation background color.
3. Configure the Classes page (classes.html) to display an audio control and update the external CSS file.

Task 1: Create a folder called ch11yoga to contain your Path of Light Yoga Studio website files. Copy the files from the Chapter 10 Case Study ch10yoga folder to your new ch11yoga folder. Copy the savasana.mp3 and savasana.ogg files from the chapter11/casestudystarters folder in the student files and save them in your ch11yoga folder.

Task 2: Configure a Header Hyperlink Transition with CSS. Open yoga.css in a text editor. Locate the `header a` selector. Change the text color for the `:hover` pseudo-class to #8F92B2. Code a style declaration to configure a ten-second text color ease-out transition. Save the file. Display any of the web pages in a browser that supports transitions and place your mouse pointer over the text in the header area. You should see a gradual change in the text color.

Task 3: Configure the Audio. Open the Classes page (classes.html) in a text editor. Modify classes.html so that a heading, a paragraph, and an HTML5 audio control display between the div assigned to the id `flow` and the div assigned to the id `mathero` (see Figure 11.23). Use an h2 element to display the text "Relax Anytime with Savasana." Add a paragraph that contains the following text:

"Prepare yourself for savasana. Lie down on your yoga mat with your arms at your side with palms up. Close your eyes and breathe slowly but deeply. Sink into the mat and let your worries slip away. When you are ready, roll on your side and use your arms to push yourself to a sitting position with crossed legs. Place your hands in a prayer position. Be grateful for all that you have in life. Namaste."

Refer to Hands-On Practice 11.2 when you create the audio control. Configure the audio and source elements to work with the savasana.mp3 and savasana.ogg files. Configure a

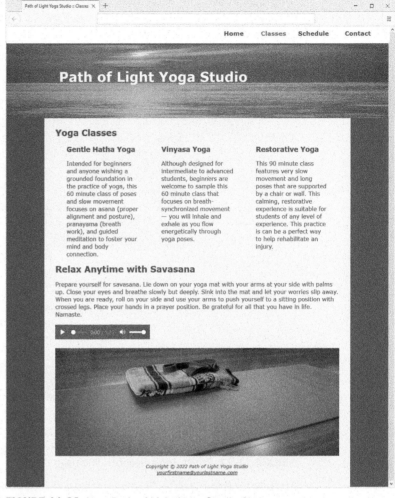

FIGURE 11.23 *New Path of Light Yoga Studio Classes page.*

hyperlink to the savasana.mp3 file as a fallback if the audio element is not supported. Save the file. Check your HTML syntax using the W3C validator (https://validator.w3.org). Correct and retest if necessary.

Next, configure the CSS. Launch a text editor. Open yoga.css. Configure a style above the media queries for the audio element selector that sets a 1em bottom margin. Save the yoga.css file.

Launch a browser to test your classes.html page. It should look similar to Figure 11.23. You have enhanced the website with interactivity and audio.

Endnotes

1. Quackenbush, S. "Audio Standard: MPEG-4." *MPEG*, The Moving Picture Experts Group, Oct. 2005, mpeg.chiariglione.org/standards/mpeg-4/audio.
2. Quackenbush, S. "Audio Standard: MPEG-1." *MPEG*, The Moving Picture Experts Group, Oct. 2005, mpeg.chiariglione.org/standards/mpeg-1/audio.
3. "Vorbis audio compression." Xiph.org, xiph.org/vorbis/.
4. "Home." Alliance for Open Media, 18 July 2019, aomedia.org/.
5. Harding, Scharon. "What Are H.265 and H.264? Video Codecs and Content Streaming Explained." Tom's Hardware, Tom's Hardware, 31 Dec. 2019, www.tomshardware.com/reference/h264-h265-hevc-codec-definition.
6. "Theora video compression." Xiph.org, xiph.org/theora/.
7. "WebM: an Open Web Media Project." The WebM Project, www.webmproject.org/.
8. Deveria, Alexis. "Can I use AV1 video format?" Can I Use... Support Tables for HTML5, CSS3, Etc, Apr. 2020, caniuse.com/#feat=av1
9. Deveria, Alexis. "Can I use MPEG-4/H.264 video format?" Can I Use... Support Tables for HTML5, CSS3, Etc, Apr. 2020, caniuse.com/#feat=mpeg4.
10. Deveria, Alexis. "Can I use HEVC/H.265 video format?" Can I Use... Support Tables for HTML5, CSS3, Etc, Apr. 2020, caniuse.com/#feat=hevc.
11. Deveria, Alexis. "Can I use Ogg/Theora Video Format?" Can I Use... Support Tables for HTML5, CSS3, Etc, Apr. 2020, caniuse.com/#feat=ogv.
12. Deveria, Alexis. "Can I use WebM video format?" Can I Use... Support Tables for HTML5, CSS3, Etc, Apr. 2020, caniuse.com/#feat=webm.
13. "Translate Videos & Captions - YouTube Help." YouTube Help, *Google,* support.google.com/youtube/topic/9257536?visit_id=637051226249550044-246125716&rd=1.
14. "HTML: 4.8.5 The Iframe Element." *HTML Living Standard*, WHATWG, 21 May 2020, html.spec.whatwg.org/multipage/iframe-embed-object.html#the-iframe-element.
15. "CSS Transforms Module Level 1." Edited by Simon Fraser et al., *W3C*, 14 Feb. 2019, www.w3.org/TR/css-transforms-1/.
16. "CSS Transitions." Edited by L. David Baron et al., *W3C*, 11 Oct. 2018, www.w3.org/TR/css-transitions-1/.
17. "Web Content Accessibility Guidelines (WCAG) 2.1." Edited by Andrew Kirkpatrick et al., *W3C*, 5 June 2018, www.w3.org/TR/WCAG21/#pause-stop-hide.
18. Alsup, Mike. "JQuery Cycle Plugin." *JQuery Cycle Plugin*, jquery.malsup.com/cycle/.
19. "Geolocation API Specification 2nd Edition." Edited by Andrei Popescu, *W3C*, 8 Nov. 2016, www.w3.org/TR/geolocation-API/.
20. "Web Storage (Second Edition)." Edited by Ian Hickson, *W3C*, 16 Apr. 2016, www.w3.org/TR/webstorage/.
21. "HTML5: 5.6 Offline Web Applications." Edited by Ian Hickson, *W3C*, 25 June 2011, www.w3.org/TR/2011/WD-html5-20110525/offline.html.
22. "Web App Manifest." Edited by Marcos Caceres et al., *W3C*, 26 Sept. 2019, www.w3.org/TR/appmanifest/.
23. "Service Workers 1." Edited by Alex Russell et al., *W3C*, 13 Aug. 2019, www.w3.org/TR/service-workers-1/.
24. "HTML Canvas 2D Context, Level 2." Edited by Rik Cabanier et al., *HTML Canvas 2D Context, Level 2*, 29 Sept. 2015, www.w3.org/TR/2dcontext2/.
25. "Scalable Vector Graphics (SVG)." *W3C SVG Working Group, W3C*, www.w3.org/Graphics/SVG/.23

Web Publishing Basics

Well, you've designed and built a website, but there is still much more to do. You need to obtain a domain name, select a web host, publish your files to the Web, and submit your site to search engines. In addition to discussing these tasks, this chapter introduces you to evaluating the accessibility and usability of your website.

You'll learn how to...

▶ Describe criteria to consider when you're selecting a web host

▶ Obtain a domain name for your website

▶ Describe best practices for website file organization

▶ Code relative hyperlinks to files in folders within a website

▶ Publish a website using FTP

▶ Design web pages that are friendly to search engines

▶ Submit a website for inclusion in a search engine

▶ Determine whether a website meets accessibility requirements

▶ Evaluate the usability of a website

File Organization

An unorganized website often contains a long list of files, which can become difficult to maintain over time. It's common practice to create a separate folder for images on a website. It's also a good idea to organize your web pages into folders by purpose or subject. This section introduces you to coding relative hyperlinks for a website with multiple folders.

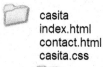

casita
index.html
contact.html
casita.css

images
logo.gif
scenery.jpg

rooms
canyon.html
javelina.html

events
weekend.html
festival.html

FIGURE 12.1 *The web page files are organized in folders.*

As discussed in Chapter 2, a relative hyperlink is used to link to web pages within your site. You've been coding relative links to display web pages that are all inside the same folder. Let's consider a website for a bed and breakfast that features rooms and events. The folder and file listing is shown in Figure 12.1. The main folder for this website is called casita, and the web developer has created separate subfolders—named images, rooms, and events—to organize the site.

Relative Link Examples

Recall that when linking to a file located in the same folder or directory, the value of the `href` attribute is the name of the file. For example, to link from the home page (index.html) to the contact.html page, code the anchor element as follows:

```
<a href="contact.html">Contact</a>
```

When linking to a file that is inside a folder within the current directory, use both the folder name and the file name in the relative link. For example, to link from the home page (index.html) to the canyon.html page (located in the rooms folder), code the anchor element as follows:

```
<a href="rooms/canyon.html">Canyon</a>
```

As shown in Figure 12.1, the canyon.html page is located in the rooms subfolder of the casita folder. The home page for the site (index.html) is located in the casita folder. When linking to a file that is up one directory level from the current page, use the ". ./" notation. To link to the home page for the site from the canyon.html page, code the anchor element as follows:

```
<a href="../index.html">Home</a>
```

When linking to a file that is in a folder on the same level as the current folder, the href value will use the ". ./" notation to indicate moving up one level; then specify the desired folder. For example, to link to the weekend.html page in the events folder from the canyon.html page in the rooms folder, code the anchor element as follows:

```
<a href="../events/weekend.html">Weekend Events</a>
```

Don't worry if the use of ". ./" notation and linking to files in different folders seems new and different. You can explore the example of the bed and breakfast website located in the student files (see chapter12/CasitaExample) to become more familiar with coding references to files in different folders.

Hands-On Practice 12.1

This Hands-On Practice provides an opportunity to practice coding hyperlinks to files in different folders. The website you'll be working with has pages in prototype form—the navigation and layout of the pages are configured, but the specific content has not yet been added. You'll focus on the navigation area in this Hands-On Practice. Figure 12.2 shows a partial screen shot of the bed and breakfast's home page with a navigation area on the left side of the page.

Examine Figure 12.3 and notice the new juniper.html file listed within the rooms folder. You will create a new web page (Juniper Room) named juniper.html and save it in the rooms folder. Then, you will update the navigation area on each existing web page to link to the new Juniper Room page.

FIGURE 12.2 *The navigation area.*

1. Copy the CasitaExample folder (chapter12/CasitaExample) from the student files. Rename the folder casita.

2. Display the index.html file in a browser and click through the navigation links. View the source code of the pages and notice how the `href` values of the anchor tags are configured to link to and from files within different folders.

3. Launch a text editor and open the canyon.html file. You'll use this file as a starting point for your new Juniper Room page. Save the file as juniper.html in the rooms folder.

   a. Edit the page title and h2 text: change "Canyon" to "Juniper".

   b. Add a new li element in the navigation area that contains a hyperlink to the juniper.html file.

   ```
   <li><a href="juniper.html">Juniper Room</a></li>
   ```

   Place this hyperlink between the Javelina Room and Weekend Events navigation hyperlinks as shown in Figure 12.4. Save the file.

4. Use the coding for the Canyon and Javelina hyperlinks as a guide as you add the Juniper Room link to the navigation area on each of the following pages:

   index.html
   contact.html
   rooms/canyon.html
   rooms/javelina.html
   events/weekend.html
   events/festival.html

FIGURE 12.3 *New juniper.html file is in the rooms folder.*

FIGURE 12.4 *The new navigation area.*

Save all the .html files and test your pages in a browser. The navigation hyperlink to the new Juniper Room page should work from every other page. The hyperlinks on the new Juniper Room page should function well and open other pages as expected. A solution is in the student files (chapter12/12.1 folder).

Register a Domain Name

VideoNote

Choosing a Domain Name

A crucial part of establishing an effective web presence is choosing a **domain name**; it serves to locate your website on the Internet (Figure 12.5). If your business or organization is brand new, then it's often convenient to select a domain name while you are deciding on a company name. If your organization is well established, you should choose a domain name that relates to your existing business presence. Although many domain names have already been purchased, there are still many options available.

FIGURE 12.5 *Your domain name establishes your presence on the Web.*

Describe Your Business. Although there is a long-standing trend to use "fun" words as domain names (such as yahoo.com, google.com, bing.com, woofoo.com, and so on), think carefully before doing so. Domain names for traditional businesses and organizations are the foundation of the organization's web presence and should include the business name or purpose.

Be Brief, If Possible. Although most people find new websites with search engines, some of your visitors will type your domain name in a browser. A shorter domain name is preferable to a longer one—it's easier for your web visitors to remember.

Avoid Hyphens in Your Domain Name. Using the hyphen character (commonly called a dash) in a domain name makes it difficult to pronounce the name. Also, someone typing your domain name may forget the dash and end up at a competitor's site! If you can, avoid the use of dashes in a domain name.

There's More Than .com. While the .com top-level domain name (TLD) is still the most popular for commercial and personal websites, consider also registering your domain name with other TLDs, such as .biz, .net, .us, and so on. Commercial businesses should avoid the .org TLD, which is the first choice for nonprofit organizations. You don't have to create a website for each domain name that you register. You can arrange with your domain name registrar for the "extra" domain names to point visitors to the domain name where your website is located. This is called **domain name redirection**.

Brainstorm Potential Keywords. Think about words that a potential visitor might type into a search engine when looking for your type of business or organization. This is the starting point for your list of **keywords.** If possible, work one or more keywords into your domain name (but still keep it as short as possible).

Avoid Trademarked Words or Phrases. The U.S. Patent and Trademark Office (USPTO)[1] defines a **trademark** as a word, phrase, symbol, or design, or a combination of words, phrases, symbols, or designs, that identifies and distinguishes the source of the goods of one party from those of others. A starting point in researching trademarks is the USPTO Trademark Electronic Search System (TESS) at http://tess2.uspto.gov.

Know the Territory. Explore the way your potential domain name and keywords are already used on the Web. It's a good idea to type your potential domain names (and related words) into a search engine to see what may already exist.

Verify Availability. Check with one of the many **domain name registrars** to determine whether your domain name choices are available. A few of the many sites that offer domain name registration services are as follows:

- https://register.com
- https://networksolutions.com
- https://godaddy.com

Each of these sites offers a search feature that provides a way to determine whether a potential domain name is available and, if it is owned, who owns it. Often the domain name is already taken. If that's the case, the sites listed here will provide alternate suggestions that may be appropriate. Don't give up; a domain name is out there waiting for your business.

Registering a Domain Name

Once you've found your perfect domain name, don't waste any time in registering it. The cost of registering a domain name varies but is quite reasonable. It's perfectly OK to register a domain name even if you are not ready to publish your website immediately. Some web hosting services offer discounted domain name registration, but you will have the most control over your domain name if you register it directly with a domain name registrar. When you register a domain name, your contact information (such as name, phone number, mailing address, and e-mail address) will be entered into the WHOIS database and available to anyone unless you choose the option for private registration. While there is usually a small annual fee for **private registration**, it shields your personal information from unwanted spam and curiosity seekers.

Obtaining a domain name is just one part of establishing a web presence—you also need to host your website somewhere. The next section introduces you to factors involved in choosing a web host.

Choose a Web Host

A **web host provider** is an organization that offers storage for your website files along with the service of making them available on the Internet. Your domain name, such as webdevbasics.net, is associated with an IP address that points to your website on the web server at the web host provider.

It is common for web host providers to charge a setup fee in addition to the monthly hosting fee. Hosting fees vary widely. The cheapest hosting company is not necessarily the one to use. Never consider using a "free" web host provider for a business website. These free sites are great for kids, college students, and hobbyists, but they are unprofessional. The last thing you or your client wants is to be perceived as unprofessional or not serious about the business at hand. As you consider different web host providers, try contacting their support phone numbers and e-mail addresses to determine just how responsive they really are. Word of mouth, web searches, and online directories[2] are all resources in your quest for the perfect web host provider.

Types of Web Hosting

▶ **Virtual Hosting,** or shared hosting, is a popular choice for small websites. The web host provider's physical web server is divided into a number of virtual domains, and multiple websites are setup on the same computer. You have the authority to update files in your own website space, while the web host provider maintains the web server computer and Internet connectivity.

▶ **Dedicated Hosting** is the rental and exclusive use of a computer and connection to the Internet that is housed on the web hosting company's premises. A dedicated server is usually needed for a website that could have a considerable amount of traffic, such as tens of millions of hits a day. The server can usually be configured and operated remotely from the client company, or you can pay the web host provider to administer it for you.

▶ **Colocated Hosting** uses a computer that your organization has purchased and configured. Your web server is housed and connected to the Internet at the web host's physical location, but your organization administers this computer.

▶ **Cloud Hosting** is different from other types of hosting in that your website is stored on multiple web servers, usually in different physical locations. A major benefit of cloud hosting is reliability. Instead of depending on a single web server, there are multiple servers that can serve your website resources at any time. Cloud hosting is also scalable and can flexibly handle unexpected increases in website traffic more easily than a website hosted on a single web sever.

Choosing a Virtual Host

There are a number of factors to consider when choosing a virtual web host. Table 12.1 provides a checklist.

TABLE 12.1 *Virtual Web Host Checklist*

Operating system	☐ UNIX ☐ Linux ☐ Windows	Some web hosts offer a choice of these platforms. If you need to integrate your web host with your business systems, choose the same operating system for both.
Web server	☐ Apache ☐ IIS	These two web server applications are the most popular. Apache usually runs on a UNIX or Linux operating system. Internet Information Services (IIS) is bundled with selected versions of Microsoft Windows.
Bandwidth	☐ ___ GB per month ☐ ___ Charge for overage	Some web hosts carefully monitor your data transfer bandwidth and charge you for overages. While unlimited bandwidth is great, it is not always available. A typical low-traffic website varies between 100GB and 200GB per month. A medium-traffic site should be OK with about 500GB of data transfer bandwidth per month.
Technical support	☐ E-mail ☐ Chat ☐ Forum ☐ Phone	Review the description of technical support on the web host's site. Is it available 24 hours a day, 7 days a week? E-mail or phone a question to test it. If the organization is not responsive to you as a prospective customer, be leery about the availability of its technical support later.
Service agreement	☐ Uptime guarantee ☐ Automatic monitoring	A web host that offers a Service Level Agreement (SLA) with an uptime guarantee shows that service and reliability are valued. The use of automatic monitoring will inform the web host technical support staff when a server is not functioning.
Disk space	☐ ___ GB	Many virtual hosts routinely offer 100GB+ disk storage space. If you have a small site that is not graphic intensive, you may never even use more than 50MB of disk storage space.
E-mail	☐ ___ Mailboxes	Most virtual hosts offer multiple e-mail mailboxes per site. These can be used to filter messages—customer service, technical support, general inquiries, and so on.
Uploading files	☐ FTP access ☐ Web-based file manager	A web host that offers FTP access will allow the most flexibility. Others only allow updates through a web-based file manager application. Some web hosts offer both options.
Scripts	☐ Form processing	Many web hosts supply canned, prewritten scripts to process form information.
Scripting support	☐ PHP ☐ .NET	If you plan to use server-side scripting (refer back to Chapter 10) on your site, determine which, if any, scripting is supported by your web host.
Database support	☐ MySQL ☐ SQL Server	If you plan to access a database with your scripting, determine which, if any, database is supported by your web host.
E-commerce packages	☐ _____	If you plan to enter into e-commerce, it may be easier if your web host offers a shopping cart package. Check to see if one is available.
SSL	☐ $___ setup fee ☐ $___ per month	Determine if your web host offers SSL—you may want to use https in the future to prevent security issues.
Scalability	☐ Scripting ☐ Database ☐ E-commerce	You probably will choose a basic (low-end) plan for your first website. Note the scalability of your web host—are there other plans with scripting, database, e-commerce packages, and additional bandwidth or disk space available as your site grows?
Backups	☐ Daily ☐ Periodic ☐ No backups	Most web hosts will back up your files regularly. Check to see how often the backups are made and if they are accessible to you. Be sure to make your own site backups as well.
Site statistics	☐ Raw log file ☐ Log reports ☐ No log	The website log contains useful information about your visitors, how they find your site, and what pages they visit. Check to see if the log is available to you. Some web hosts provide reports about the log.
Domain name	☐ Included ☐ OK to register on your own	Some web hosts offer a package that includes registering your domain name. You may prefer to register your domain name yourself and retain control of your domain name account.
Price	☐ $___ setup fee ☐ $___ per month	Price is last in this list for a reason. Do not choose a web host based on price alone—the old adage "you get what you pay for" is definitely true here. It is not unusual to pay a one-time setup fee and then a periodic fee—monthly, quarterly, or annually.

Secure Sockets Layer (SSL)

Secure Sockets Layer (SSL) is a protocol that allows data to be privately exchanged over public networks. It was initially developed by Netscape in 1994 to encrypt data sent between a client (usually a web browser) and a web server.

Transport Layer Security (TLS) was later developed and implemented as an improvement and replacement for Secure Sockets Layer. However, the acronym SSL is commonly used to indicate encrypted secure communication between a web browser and a web server.

SSL provides secure communication between a client and a server by using the following:

- Server and (optionally) client digital certificates for authentication
- Symmetric-key cryptography with a "session key" for bulk encryption
- Public-key cryptography for transfer of the session key
- Message digests (hash functions) to verify the integrity of the transmission.

You can tell that a website is using SSL by the protocol in the web browser address text box—it shows https instead of http. When a URL begins with https://, it indicates that the browser is using **HTTPS**, which stands for **Hypertext Transfer Protocol Secure**. HTTPS combines HTTP with SSL. In addition to displaying the https protocol, browsers typically display a lock icon or other indicator of SSL, as shown in Figure 12.6.

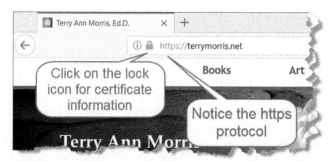

FIGURE 12.6 *The browser indicates that SSL is being used.*

? FAQ When some websites are displayed in a browser, there is a color bar in the address area. What's up?

If a website displays a color bar in the address area of the browser in addition to the lock icon in the status bar, you know that it is using **Extended Validation SSL (EV SSL)**. EV SSL signifies that the business has undergone more rigorous background checks to obtain its digital certificate, including verification of the following:

- The applicant owns the domain.
- The applicant works for the organization.
- The applicant has the authority to update the website.
- The organization is a valid, recognized place of business.

Digital Certificate

SSL enables two computers to communicate securely by posting a digital certificate for authentication. A **digital certificate** is a form of an asymmetric key that also contains information about the certificate, the holder of the certificate, and the issuer of the certificate. The contents of a digital certificate include the following:

▶ The public key

▶ The effective date of the certificate

▶ The expiration date of the certificate

▶ Details about the certificate authority (the issuer of the certificate)

▶ Details about the certificate holder

▶ A digest of the certificate content

VeriSign, Thawte, and Entrust are well-known certificate authorities.

To obtain your own digital certificate, you will need to generate a certificate signing request (CSR) and a private/public key pair.[3] Next, you request a certificate from a certificate authority, pay the application fee, and provide your CSR and public key. The certificate authority verifies your identity. There may be a waiting period, and you will need to pay an annual fee. After verification, the certificate authority signs and issues your certificate. You store the certificate in your software, such as a web server, a web browser, or an e-mail application. When linking to your secure web pages, use "https" instead of "http" on your absolute hyperlinks.

? FAQ Do I have to apply for a certificate?

If you are accepting any personal information on your website such as credit card numbers, you should be using SSL. Using SSL not only improves the security of your website, it may also help with marketing your website. Google's PageRank algorithm ranks secure pages higher than non-secure pages. However, you may not need to apply for your own certificate; other options exist. Cloudflare offers an online content delivery network (CDN) service that will route copies of your web pages through their servers and encrypt via SSL.[4] There are several levels of service plans at varying costs, including a free starter plan. Also, many web hosts offer basic SSL with their web hosting packages. Check with your web host provider to determine if they offer this feature.

Publish with File Transfer Protocol

Once you obtain your web hosting, you'll need to upload your files. While your web host may offer a web-based file manager application for client use, a common method of transferring files is to use **File Transfer Protocol** (FTP). A **protocol** is a convention or standard that enables computers to speak to one another. **FTP** is used to copy and manage files and folders over the Internet. FTP uses two ports to communicate over a network—one for the data (typically port 20) and one for control commands (typically port 21).

FTP Applications

There are many FTP applications available for download or purchase on the Web; several are listed in Table 12.2.

TABLE 12.2 *FTP Applications*

Application	Platform	URL	Cost
FileZilla	Windows, Mac, Linux	https://filezilla-project.org	Free download
SmartFTP	Windows	https://www.smartftp.com	Free download
CuteFTP	Windows, Mac	https://www.cuteftp.com	Free trial download, academic pricing available
WS_FTP	Windows	https://www.ipswitch.com	Free trial download

Connecting with FTP

Your web host will provide you with the following information for connecting to your FTP server, along with any other specifications, such as whether the FTP server requires the use of active mode or passive mode:

FTP Host: ftp://*yourhostaddress*

Username: *your_account_username*

Password: *your_account_password*

Overview of Using an FTP Application

This section focuses on FileZilla, a free FTP application with versions for the Windows, Mac, and Linux platforms. A free download of FileZilla is available at https://filezilla-project.org. After you download an FTP application of your choice, install the program on your computer using the instructions provided.

Launch and Login. Launch Filezilla or another FTP application. Enter the information required by your web host (such as FTP host, username, and password) and initiate the connection. An example screenshot of FileZilla after a connection is shown in Figure 12.7.

As you examine Figure 12.7, notice the text boxes near the top of the application for the Host, Username, and Password information. Under this area is a display of messages from the FTP server. Review this area to confirm a successful connection and the results of file transfers. Next, notice that the applica-

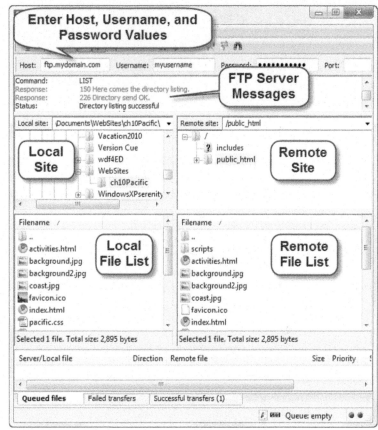

FIGURE 12.7 *The FileZilla FTP application.*

tion is divided into a left panel and a right panel. The left panel is the local site—it displays information about your local computer and allows you to navigate to your drives, folders, and files. The right panel is the remote site—it displays information about your website and provides a way to navigate to its folders and files.

Uploading a File. It's really easy to transfer a file from your local computer to your remote website—just select the file with your mouse in the left panel (local site list) and drag it to the right panel (remote site list).

Downloading a File. If you need to download a file from your website to your local computer, just drag the file from the right panel (remote site list) to the left panel (local site list).

Deleting a File. To delete a file on your website, right-click (Ctrl-click if using a Mac) on the file name (in the right panel) and select Delete from the context-sensitive menu.

And There's More! Explore other functions offered by FileZilla (and most FTP applications)—right-click (Ctrl-click if using a Mac) on a file in the remote site list to display a context-sensitive menu with several options, including renaming a file, creating a new directory (also known as a folder), and viewing a file.

Search Engine Submission

Using a search engine is a popular way to navigate the Web and find websites. Search engine listings can be an excellent marketing tool for your business. To harness the power of search engines, it helps to know how they work.

According to a Global Stats StatCounter report, Google was the most popular search engine during a recent month. Google was reported to have an overwhelming market share of 91.89%, while the closest competitors were Bing (2.79%), Yahoo! (1.87%), and Baidu (1.1%).[5] Google's popularity has continued to grow since it was founded in the late 1990s. The simple and whimsical interface, combined with quick-loading and useful results, has made it a favorite of web users.

Components of a Search Engine

Search engines have the following components:

- ▶ Robot
- ▶ Database
- ▶ Search form

The components of a search engine work together to obtain information about web pages, store information about web pages, and provide a graphical user interface to facilitate searching for and displaying a list of web pages relevant to given keywords.

Robot. A **robot** (sometimes called a spider or bot) is a program that automatically traverses the hypertext structure of the Web by retrieving a web page document and following the hyperlinks on the page. It moves like a robot spider on the Web, accessing and storing information about the web pages in a database. Visit The Web Robots Pages at https://www.robotstxt.org if you'd like more details about web robots.

Database. A **database** is a collection of information organized so that its contents can easily be accessed, managed, and updated. Database management systems (DBMSs) such as Oracle, Microsoft SQL Server, MySQL, or IBM DB2 are used to configure and manage the database. The web page that displays the results of your search, called the **Search Engine Results Page (SERP)**, lists information from the database accessed by the search engine.

Search Form. The search form is the graphical user interface that allows a user to type in the word or phrase he or she is searching for. It is typically a text box and a submit button.

The visitor to the search engine types words (called keywords) related to his or her search into the text box. When the form is submitted, the keywords are sent to a server-side script that searches the database for matches. The results are displayed on a SERP and formatted with a hyperlink to each page along with additional information that might include the page title, a brief description, the first few lines of text, or the size of the page. The order in which the results are displayed may depend on paid advertisements, alphabetical order, and link popularity. The **link popularity** of a website is a rating determined by a search engine based on the quantity and quality of incoming hyperlinks. Each search engine has its own policy for ordering the search results. Be aware that these policies can change over time.

Submitting Your Site to a Search Engine

Search engine spiders routinely traverse the web and should eventually visit your website, but it can take some time. You can potentially speed up the process by submitting your website manually to a search engine. Back in the day, all you needed to do to submit your website to Google or Bing was to anonymously fill out a form on their website. Now, the process is more complex.

To work with Google, visit https://search.google.com. Sign in with your Google account and look for the "add property" option. You'll provide the URL of your website and follow instructions provided by Google to verify that you are the actual site owner. Once verified, visit https://www.google.com/webmasters/tools/submit-url and follow the instructions to either submit a Sitemap or use the Fetch as Google tool.

To work with Bing, visit https://www.bing.com/toolbox/webmaster. Sign in with an account of your choice (Microsoft, Google, or Facebook). You'll provide the URL of your website and follow instructions provided by Bing to verify that you are the actual site owner. Once verified, you will be able to submit URLs to Bing (in the Dashboard select "Configure My Site" and "Submit URLs") by filling out a form. You can also submit a Sitemap to Bing using the Dashboard.

After you submit your URL, it's time to wait for the search engine's spider or bot to visit your website. This may take several weeks. Try to be patient.

Several weeks after you submit your website, check the search engine or search directory to see if your site is listed. If it is not listed, review your pages and check whether they are optimized for search engines (see the next section) and display in common browsers.

FAQ

Is advertising on a search engine worth the cost?

It depends. How much is it worth to your client to appear on the first page of the search engine results? You select the keywords that will trigger the display of your ad. You also set your monthly budget and the maximum amount to pay for each click. While costs and charges vary by search engine, at this time Google charges are based on cost per click—you'll be charged each time a visitor to Google clicks on your advertisement. Visit https://ads.google.com/ for more information about their program.

Search Engine Optimization

If you have followed recommended web design practices, you've already designed your website so that the pages are appealing and compelling to your target audience. How can you also make your site work well with search engines? Here are some suggestions and hints on designing your pages for optimal ranking by search engines—a process called **Search Engine Optimization (SEO)**.

Keywords

Spend some time brainstorming about terms and phrases that people may use when searching for your site. These terms or phrases that describe your website or business are your **keywords**.

Page Titles

A descriptive page title (the text between the `<title>` tag and `</title>` tag) that includes your company and/or website name will help your site market itself. It's common for search engines to display the text in the page title in the SERP. The page title is also saved by default when a visitor bookmarks your site and is often included when a visitor prints a page of your site. Avoid using the exact same title for every page; include keywords in the page title that are appropriate for the page.

Heading Tags

Use structural tags such as `<h1>`, `<h2>`, and so on to organize your page content. If it is appropriate for the web page content, also include some keywords in the text contained within heading tags. Some search engines will give a higher list position if keywords are included in a page title or headings. Also include keywords as appropriate within the page text content.

Description

What is special about your website that would make someone want to visit? With this in mind, write a few sentences about your website or business. This description should be inviting and interesting so that a person searching the Web will choose your site from the list provided by a search engine or search directory. Some search engines will display your description on the SERP. You can configure a description for your web page by coding a meta tag in the page header area.

The Meta Tag

A **meta tag** is a self-contained tag that is placed in the header section of a web page. You've been using a meta tag to indicate character encoding. There are a number of other uses for a meta tag. We'll focus here on providing a description of a website for use by search engines. The description meta tag content is displayed on the SERP by some search engines, such as Google. The **name** attribute indicates the purpose of the meta tag. The **content** attribute provides the value needed for the specific purpose. For example, the description meta tag for a website about a web development consulting firm called Acme Design could be configured as follows:

```
<meta name="description" content="Acme Design, a web consulting group that
specializes in e-commerce, website design, development, and redesign.">
```

> **? FAQ** **What if I don't want a search engine to index a page?**
> To indicate to a search engine robot that a page should not be indexed and the links should not be followed, do not place keywords and description meta tags in the page. Instead, add a "robots" meta tag to the page as follows:
>
> ```
> <meta name="robots" content="noindex,nofollow">
> ```

Linking

Verify that all hyperlinks are working and not broken. Each page on your website should be reachable by a text hyperlink. The text should be descriptive—avoid phrases like "more info" and "click here"—and should include keywords as appropriate. Inbound links (sometimes called incoming links) are also a factor in SEO. All these linking issues can affect your website's link popularity, and its link popularity can determine its order in the search engine results page.

File Names

Google's Search Console Help[6] suggests that web developers use file names that contain readable words and consider separating the words with hyphens.

HTTPS Protocol

As part of Google's support for security on the Web, the search engine considers the HTTPS protocol in its page rank algorithms.[7] If possible, select a web host that will provide HTTPS for your website.

Images and Multimedia

Be mindful that search engine robots do not "see" the text embedded within your images and multimedia. Configure meaningful alternate text for images. Include relevant keywords in the alternate text.

Valid Code

Search engines do not require that your HTML and CSS code pass validation tests. However, code that is valid and well structured is likely to be more easily processed by search engine robots. This may help with your placement in the search engine results.

Content of Value

Probably the most basic, but often overlooked, component of SEO is providing content of value that follows web design best practices (see Chapter 3). Your website should contain high-quality, well-organized content that is of value to your visitors.

Accessibility Testing

Universal Design and Accessibility

Recall from Chapter 1 that universal design is a "strategy for making products, environments, operational systems, and services welcoming and usable to the most diverse range of people possible".[8] Web pages that follow the principle of universal design are **accessible** to all individuals, including those with visual, hearing, mobility, and cognitive challenges. As you've worked through this book, accessibility has been an integral part of your web page design and coding rather than an afterthought. You've configured headings and subheadings, navigation within unordered lists, images with alternate text, alternate text for multimedia, and associations between text and form controls. These techniques all increase the accessibility of a web page.

Web Accessibility Standards

The accessibility recommendations presented in this text are intended to satisfy Section 508 of the Rehabilitation Act and the W3C's Web Content Accessibility Guidelines.

Section 508 of the Rehabilitation Act. Section 508[9] requires electronic and information technology, including web pages, that are used by federal agencies to be accessible to people with disabilities. The U.S. Access Board released a Section 508 Refresh in 2017 that updated the standard to harmonize with WCAG 2.0 Success Criteria.

Web Content Accessibility Guidelines (WCAG 2.1). WCAG 2.1[10] considers an accessible web page to be perceivable, operable, and understandable for people with a wide range of abilities. The page should be robust to work with a variety of browsers and other user agents, such as screen readers and mobile devices. WCAG 2.1 includes and extends WCAG 2.0 accessibility Success Criteria. The guiding principles of WCAG 2.0 and WCAG 2.1 are as follows:

1. Content must be **P**erceivable.
2. Interface components in the content must be **O**perable.
3. Content and controls must be **U**nderstandable.
4. Content should be **R**obust enough to work with current and future user agents, including assistive technologies.

?FAQ What is assistive technology and what is a screen reader?

Assistive technology is a term that describes any tool that a person can use to help him or her to overcome a disability and use a computer. Examples of assistive technologies include screen readers, head- and mouth-wands, and specialized keyboards, such as a single-hand keyboard. A screen reader is a software application that can be controlled by the user to read aloud what is displayed on the computer screen.

Testing for Accessibility Compliance

No single testing tool can automatically test for all web standards. The first step in testing the accessibility of a web page is to verify that it is coded according to W3C standards with the Markup Validation Service (https://validator.w3.org) and the CSS Validation Service (https://jigsaw.w3.org/css-validator).

Automated Accessibility Testing. An automated accessibility evaluation tool is no substitute for your own manual evaluation but can be useful to quickly identify potential issues with a web page. WebAim Wave[11] and ATRC AChecker[12] are two popular free online accessibility evaluation tools. The online applications typically require the URL of a web page and reply with an accessibility report. The Web Developer Extension[13] is a browser toolbar that can be used to assess accessibility. The Web Developer Extension is multifunctional, with options to validate HTML, validate CSS, disable images, view alt text, outline block-level elements, resize the browser viewport, disable styles, and more. Figure 12.8 shows the Web Developer Extension toolbar in action.

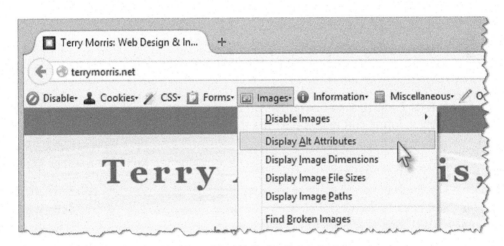

FIGURE 12.8 *Selecting the Images > Display Alt Attributes feature.*

Manual Accessibility Testing. It's important not to rely completely on automated tests—you'll want to review the pages yourself. For example, while an automated test can check for the presence of an `alt` attribute, it takes a human to critically think and decide whether the text of the `alt` attribute is an appropriate description for a person who cannot view the image.

Usability Testing

In addition to accessibility, another aspect of universal design is the usability of the website. **Usability** is the measure of the quality of a user's experience when interacting with a website.[14] It's about making a website that is easy, efficient, and pleasant for your visitors. Usability.gov describes factors that affect the user's experience:

▶ **Intuitive Design**—How easy is it for a new visitor to understand the organization of the site? Is the navigation intuitive for a new user?

▶ **Ease of Learning**—How easy is it to learn to use the website? Does a new visitor consider it easy to learn to perform basic tasks on the website or is he or she frustrated?

▶ **Efficiency of Use**—How do experienced users perceive the website? Once they are comfortable, are they able to complete tasks efficiently and quickly or are they frustrated?

▶ **Memorability**—When a visitor returns to a website, does he or she remember enough to use it productively or is the visitor back at the beginning of the learning curve (and frustrated)?

▶ **Error Frequency and Severity**—Do website visitors make errors when navigating or filling in forms on the website? Are they serious errors? Is it easy to recover from errors or are visitors frustrated?

▶ **Subjective Satisfaction**—Do users "like" using the website? Are they satisfied? Why or why not?

Conducting a Usability Test

Testing how people use a website is called **usability testing**. Usability testing can be conducted at any phase of a website's development and is often performed more than once. A usability test is conducted by asking users to complete tasks on a website, such as placing an order, looking up the phone number of a company, or finding a product. The exact tasks will vary depending on the website being evaluated. The users are monitored while they try to perform these tasks. They are asked to think out loud about their doubts and hesitations. The results are recorded (often on video) and discussed with the web design team. Often changes are made to the navigation and page layouts based on these tests.

If usability testing is done early in the development phase of a website, it may use the printed page layouts and site map. If the web development team is struggling with a design issue, sometimes a usability test can help to determine which design idea is the better choice. When usability is done during a later phase after the pages have been built, the actual website is tested. This can lead to confirmation that the site is easy to use and well designed, to last-minute changes in the website, or to a plan for website enhancements in the near future.

Hands-On Practice 12.2

Perform a small-scale usability test with a group of other students. Decide who will be the "typical users," the tester, and the observer. You will perform a usability test on your school's website.

- The "typical users" are the test subjects.
- The tester oversees the usability test and emphasizes that the users are not being tested—the website is being tested.
- The observer takes notes on the user's reactions and comments.

Step 1:   The tester welcomes the users and introduces them to the website they will be testing.

Step 2:   For each of the following scenarios, the tester introduces the scenario and questions to the users as they work through the task. The tester should ask the users to indicate when they are in doubt, confused, or frustrated. The observer takes notes.

- Scenario 1: Find the phone number of the contact person for the Web Development program at your school.
- Scenario 2: Determine when to register for the next semester.
- Scenario 3: Find the requirements to earn a degree or certificate in Web Development or a related area.

Step 3:   The tester and observer organize the results and write a brief report. If this were a usability test for an actual website, the development team would meet to review the results and discuss necessary improvements to the site.

Step 4:   Hand in a report with your group's usability test results. Complete the report using a word processor. Write no more than one page about each scenario. Write one page of recommendations for improving your school's website.

 Continue to explore the topic of usability testing at the following resources:

- Advanced Common Sense—the website of usability expert Steve Krug: https://www.sensible.com
- Usability Evaluation Basics: https://www.usability.gov/what-and-why/usability-evaluation.html
- Usability Testing Materials: https://www.infodesign.com.au/usabilityresources/ usabilitytestingmaterials

Review and Apply

Review Questions

Multiple Choice. Choose the best answer for each item.

1. Which of the following is a protocol commonly used to transfer files over the Internet?
   a. HTML
   b. HTTP
   c. FTP
   d. SMTP

2. In which of the following sections of a web page should meta tags be placed?
   a. head
   b. body
   c. comment
   d. none of the above

3. Which of the following statements is true?
   a. No single testing tool can automatically test for all web standards.
   b. Include as many people as possible when you conduct usability tests.
   c. Search engine listings are effective immediately after submission.
   d. None of the above statements are true.

4. Which are the four principles of the Web Content Accessibility Guidelines?
   a. contrast, repetition, alignment, and proximity
   b. perceivable, operable, understandable, and robust
   c. accessible, readable, maintainable, and reliable
   d. hierarchical, linear, random, and sequential

5. What is the measure of the quality of a user's experience when interacting with a website?
   a. accessibility
   b. usability
   c. universal design
   d. assistive technology

6. Which web hosting option is appropriate for the initial web presence of an organization?
   a. dedicated hosting
   b. free web hosting
   c. virtual hosting
   d. colocated hosting

7. What is the purpose of private registration for a domain name?
   a. It protects the privacy of your web host.
   b. It is the cheapest form of domain name registration.
   c. It protects the privacy of your contact information.
   d. None of the above.

8. Which of the following is true about domain names?
   a. It is recommended to register multiple domain names that are redirected to your website.
   b. It is recommended to use long, descriptive domain names.
   c. It is recommended to use hyphens in domain names.
   d. There is no reason to check for trademarks when you are choosing a domain name.

9. Which of the following is a rating determined by a search engine based on the number and quality of hyperlinks to a website?
   a. link checking
   b. link rating
   c. link popularity
   d. search engine optimization

10. Which of the following is the design of products and environments to be usable by all people, to the greatest extent possible, without the need for adaptation or specialized design?
    a. accessibility
    b. usability
    c. universal design
    d. functionality

Review Answers

1. c   2. a   3. a   4. b   5. b   6. c   7. c   8. a   9. c   10. c

Hands-On Exercises

1. Practice writing description meta tags. For each scenario described here, write the HTML to create an appropriate meta tag that includes keywords which may be used by visitors to search for the business.

   a. Lanwell Publishing is a small independent publisher of English as a second language (ESL) books used for secondary school and adult continuing education learners. The website offers textbooks and teacher's manuals.

   b. RevGear is a small specialty truck and auto repair shop in Schaumburg, Illinois. The company sponsors a local drag racing team.

   c. Morris Accounting is a small accounting firm that specializes in tax return preparation and accounting for small businesses. The owner, Greg Morris, is a CPA and Certified Financial Planner.

2. Run an automated accessibility test on the home page of your school website. Use both the WebAim Wave (https://wave.webaim.org) and ATRC AChecker (https://www.achecker.ca/checker) automated tests. Describe the differences in the way these tools report the results of the test. Did both tests find similar errors? Write a one-page report that describes the results of the tests. Include your recommendations for improving the website.

3. Search for web host providers and report on three that provide at least 1GB space, support PHP, support MySQL, and offer e-commerce capabilities.

Use your favorite search engine to find web host providers or visit web host directories such as https://www.hosting-review.com and http://www.hostindex.com. The web host survey results provided by http://uptime.netcraft.com/perf/reports/Hosters may also be useful. Create a web page that presents your findings. Include links to the three web host providers you selected. Your web page should include a table of information such as setup fees, monthly fees, domain name registration costs, amount of hard disk space, type of e-commerce package, and cost of e-commerce package. Use color and graphics appropriately on your web page. Place your name and e-mail address at the bottom of your web page.

Focus on Web Design

1. Explore how to design your website so that it is optimized for search engines (Search Engine Optimization or SEO). Visit the following resources as a starting point as you search for three SEO tips or hints:

   - https://backlinko.com/seo-this-year
   - https://www.seomoz.org/beginners-guide-to-seo
   - https://www.bruceclay.com/seo/search-engine-optimization.htm

Write a one-page report that describes three tips that you found interesting or potentially useful. Cite the URLs of the resources you used.

2. Using a single page website for a small business or an event is trendy but offers challenges for search engine optimization. Visit the following resources as a starting point as you search for three recommendations for single page website SEO:

   - https://www.awwwards.com/seo-tricks-for-one-page-websites.html
   - https://yoast.com/one-page-website-seo/
   - https://www.99signals.com/single-page-websites-seo/
   - https://www.popwebdesign.net/popart_blog/en/2018/05/ how-to-optimize-a-one-page-website/

   Create a web page that describes your findings. Provide URLs of the websites you used as resources. Place your name in an e-mail link on the web page.

3. Explore how to reach out to your current and potential website visitors with **Social Media Optimization** (SMO), which is described by Rohit Bhargava as optimizing a website so that it is "more easily linked to, more highly visible in social media searches on custom search engines (such as Technorati), and more frequently included in relevant posts on blogs, podcasts, and vlogs." Benefits of SMO include increased awareness of your brand and/or site along with an increase in the number of inbound links (which can help with SEO). Visit the following resources as a starting point as you search for three SMO tips or hints:

   - The Beginners Guide to Social Media: https://moz.com/beginners-guide-to-social-media
   - http://www.rohitbhargava.com/2010/08/ the-5-new-rules-of-social-media-optimization-smo.html
   - https://www.searchenginejournal.com/2020-social-media-trends/342851/

Write a one-page report that describes three tips that you found interesting or potentially useful. Cite the URLs of the resources you used.

Case Study

You will continue the case studies from Chapter 11 as you configure description meta tags.

Pacific Trails Resort Case Study

In this chapter's case study, you will use the existing Pacific Trails Resort website (Chapter 11) as a starting point to create a new version of the website that implements the description meta tag on each page. You have three tasks in this case study:

1. Create a new folder for this Pacific Trails case study.
2. Write a description of the Pacific Trails Resort business.
3. Code a description meta tag on each page in the website.

Task 1: Create a folder called ch12pacific to contain your Pacific Trails Resort website files. Copy the files from the Chapter 11 Case Study ch11pacific folder.

Task 2: Write a Description. Review the Pacific Trails Resort pages that you created in earlier chapters. Write a brief paragraph that describes the Pacific Trails Resort site. Edit the paragraph down to a description that is only a few sentences and less than 25 words in length.

Task 3: Update Each Page. Open each page in a text editor and add a description meta tag to the head section. Save the files and test them in a browser. They will not look different, but they are much friendlier to search engines!

Path of Light Yoga Studio Case Study

In this chapter's case study, you will use the existing Path of Light Yoga Studio website (Chapter 11) as a starting point to create a new version of the website that implements the description meta tag on each page. You have three tasks in this case study:

1. Create a new folder for this Path of Light Yoga Studio case study.
2. Write a description of the Path of Light Yoga Studio business.
3. Code a description meta tag on each page in the website.

Task 1: Create a folder called ch12yoga to contain your Path of Light Yoga Studio website files. Copy the files from the Chapter 11 Case Study ch11yoga folder.

Task 2: Write a Description. Review the Path of Light Yoga Studio pages that you created in earlier chapters. Write a brief paragraph that describes the Path of Light Yoga Studio site. Edit the paragraph down to a description that is only a few sentences and less than 25 words in length.

Task 3: Update Each Page. Open each page in a text editor and add a description meta tag to the head section. Save the files and test them in a browser. They will not look different, but they are much friendlier to search engines!

Endnotes

1. "Trademarks > Trademark Electronic Search System (TESS)." *United States Patent and Trademark Office*, tess2.uspto.gov/.
2. "TOP 10 WEB HOSTING PROVIDERS." *Hosting Review: Compares, Rates & Reviews Best Web Hosting Companies.*, www.hosting-review.com/.
3. Anicas, Mitchell. "How To Install an SSL Certificate from a Commercial Certificate Authority." *DigitalOcean*, 24 Nov. 2014, www.digitalocean.com/community/tutorials/how-to-install-an-ssl-certificate-from-a-commercial-certificate-authority.
4. "Cloudflare Free SSL/TLS: Get SSL Certificates." *Cloudflare*, 2020, www.cloudflare.com/ssl/.
5. "Search Engine Market Share Worldwide." *StatCounter Global Stats*, Global Stats, Mar. 2020, gs.statcounter.com/search-engine-market-share.
6. "Keep a Simple URL Structure - Search Console Help." *Search Console Help*, Google, 2020, support.google.com/webmasters/answer/76329?hl=en.
7. Bahiajii, Zineb, and Gary Illyes. "HTTPS as a Ranking Signal." *Official Google Webmaster Central Blog*, Google, 6 Aug. 2014, webmasters.googleblog.com/2014/08/https-as-ranking-signal.html.
8. "Universal Design." *Universal Design - Office of Disability Employment Policy - United States Department of Labor*, www.dol.gov/odep/topics/UniversalDesign.htm.
9. "GSA Government-Wide IT Accessibility Program." *Section508.Gov | GSA Government-Wide IT Accessibility Program*, www.section508.gov/.
10. "Web Content Accessibility Guidelines (WCAG) 2.1." Edited by Kirkpatrick, Andrew, et al., *W3C*, 5 June 2018, www.w3.org/TR/WCAG21/.
11. "WAVE Web Accessibility Evaluation Tool." *WAVE Web Accessibility Evaluation Tool*, WebAIM, Utah State University, wave.webaim.org/.
12. "Web Accessibility Checker." *IDI Web Accessibility Checker : Web Accessibility Checker*, University of Toronto, 2011, achecker.ca/checker/.
13. Pederick, Chris. "Web Developer Extension." *Blog on Chrispederickcom*, Chris Pederick, chrispederick.com/work/web-developer/.
14. "Usability Evaluation Basics." *Usability.gov*, Department of Health and Human Services, 8 Oct. 2013, www.usability.gov/what-and-why/usability-evaluation.html.

HTML5 Cheat Sheet

Commonly Used HTML5 Tags

Tag	Purpose	Commonly Used Attributes
`<!-- -->`	Comment	
`<a>`	Anchor tag: configures hyperlinks	accesskey, class, href, id, name, rel, style, tabindex, target, title
`<abbr>`	Configures an abbreviation	class, id, style
`<address>`	Configures contact information	class, id, style
`<area>`	Configures an area in an image map	accesskey, alt, class, href, hreflang, id, media, rel, shape, style, tabindex, target, type
`<article>`	Configures an independent section of a document as an article	class, id, style
`<aside>`	Configures tangential content	class, id, style
`<audio>`	Configures an audio control native to the browser	autoplay, class, controls, id, loop, preload, src, style, title
`<b>`	Configures bold text with no implied importance	class, id, style
`<bdi>`	Configures text used in bi-directional text formatting (bi-directional isolation)	class, id, style
`<bdo>`	Specifies a bi-directional override	class, id, style
`<blockquote>`	Configures a long quotation	class, id, style
`<body>`	Configures the body section	class, id, style
` `	Configures a line break	class, id, style
`<button>`	Configures a button	accesskey, autofocus, class, disabled, format, formaction, formenctype, formmethod, formtarget, formnovalidate, id, name, type, style, value
`<canvas>`	Configures dynamic graphics	class, height, id, style, title, width
`<caption>`	Configures a caption for a table	class, id, style
`<cite>`	Configures the title of a cited work	class, height, id, style, title

Tag	Purpose	Commonly Used Attributes
`<code>`	Configures a fragment of computer code	`class, id, style`
`<col>`	Configures a table column	`class, id, span, style`
`<colgroup>`	Configures a group of one or more columns in a table	`class, id, span, style`
`<command>`	Configures an area to represent commands	`class, id, style, type`
`<datalist>`	Configures a control that contains one or more option elements	`class, id, style`
`<dd>`	Configures a description area in a description list	`class, id, style`
`<del>`	Configures deleted text (with strikethrough)	`cite, class, datetime, id, style`
`<details>`	Configures a control to provide additional information to the user on demand	`class, id, open, style`
`<dfn>`	Configures the definition of a term	`class, id, style`
`<dialog>`	Configures an interactive component such as a dialog box	`class, id, open, style`
`<div>`	Configures a generic section or division in a document	`class, id, style`
`<dl>`	Configures a description list (formerly called a definition list)	`class, id, style`
`<dt>`	Configures a term in a description list	`class, id, style`
`<em>`	Configures emphasized text (usually displays in italics)	`class, id, style`
`<embed>`	Plug-in integration (such as Adobe Flash Player)	`class, id, height, src, style, type, width`
`<fieldset>`	Configures a grouping of form elements with a border	`class, id, style`
`<figcaption>`	Configures a caption for a figure	`class, id, style`
`<figure>`	Configures a figure	`class, id, style`
`<footer>`	Configures a footer area	`class, id, style`
`<form>`	Configures a form	`accept-charset, action, autocomplete, class, enctype, id, method, name, novalidate, style, target`
`<h1>...<h6>`	Configures headings	`class, id, style`
`<head>`	Configures the head section	
`<header>`	Configures a header area	`class, id, style`

(Continued)

Tag	Purpose	Commonly Used Attributes
`<hr>`	Configures a horizontal line; indicates a thematic break in HTML5	`class, id, style`
`<html>`	Configures the root element of a web page document	`lang, manifest`
`<i>`	Configures italic text	`class, id, style`
`<iframe>`	Configures an inline frame	`allow, allowfullscreen, class, height, id, loading, name, sandbox, src, style, width`
`<img>`	Configures an image	`alt, class, height, id, ismap, longdesc, name, sizes, src, srcset, style, usemap, width, loading`
`<input>`	Configures an input control; such as a text box, email text box, URL text box, search text box, telephone number text box, scrolling text box, submit button, reset button, password box, calendar control, slider control, spinner control, color picker control, or hidden field form control	`accept, accesskey, autocomplete, autofocus, capture, class, checked, disabled, form, id, list, max, maxlength, min, multiple, name, pattern, placeholder, readonly, required, size, step, style, tabindex, type, value`
`<ins>`	Configures text that has been inserted with an underline	`cite, class, datetime, id, style`
`<kbd>`	Configures a representation of user input	`class, id, style`
`<label>`	Configures a label for a form control	`class, for, form, id, style`
`<legend>`	Configures a caption for a fieldset element	`class, id, style`
`<li>`	Configures a list item in an unordered or ordered list	`class, id, style, value`
`<link>`	Associates a web page document with an external resource	`class, href, hreflang, id, rel, media, sizes, style, type`
`<main>`	Configures the main content area of a web page	`class, id, style`
`<map>`	Configures an image map	`class, id, name, style`
`<mark>`	Configures text as marked (or highlighted) for easy reference	`class, id, style`
`<menu>`	Configures a list of commands	`class, id, label, style, type`
`<meta>`	Configures meta data	`charset, content, http-equiv, name`
`<meter>`	Configures visual gauge of a value	`class, id, high, low, max, min, optimum, style, value`
`<nav>`	Configures an area with navigation hyperlinks	`class, id, style`

Tag	Purpose	Commonly Used Attributes
`<noscript>`	Configures content for browsers that do not support client-side scripting	
`<object>`	Configures a generic embedded object	classid, codebase, data, form, height, name, id, style, title, tabindex, type, width
`<ol>`	Configures an ordered list	class, id, reversed, start, style, type
`<optgroup>`	Configures a group of related options in a select list	class, disabled, id, label, style
`<option>`	Configures an option in a select list	class, disabled, id, selected, style, value
`<output>`	Configures result of processing in a form	class, for, form, id, style
`<p>`	Configures a paragraph	class, id, style
`<param>`	Configures a parameter for plug-ins	name, value
`<picture>`	Configures images for responsive display with media queries	class, id, style
`<pre>`	Configures preformatted text	class, id, style
`<progress>`	Configures a visual progress indicator	class, id, max, style, value
`<q>`	Configures quoted text	cite, class, id, style
`<rp>`	Configures a ruby parentheses	class, id, style
`<rt>`	Configures ruby text component of a ruby annotation	class, id, style
`<ruby>`	Configures a ruby annotation	class, id, style
`<samp>`	Configures sample output from a computer program or system	class, id, style
`<script>`	Configures a client-side script (typically JavaScript)	async, charset, defer, src, type
`<section>`	Configures a section of a document	class, id, style
`<select>`	Configures a select list form control	class, disabled, form, id, multiple, name, size, style, tabindex
`<small>`	Configures a disclaimer in small text size	class, id, style
`<source>`	Configures a media file and MIME type	class, id, media, sizes, src, srcset, style, type
`<span>`	Configures a generic section of a document with inline display	class, id, style

Tag	Purpose	Commonly Used Attributes
`<strong>`	Configures text with strong importance (typically displayed as bold)	`class, id, style`
`<style>`	Configures embedded styles in a web page document	`media, scoped, type`
`<sub>`	Configures subscript text	`class, id, style`
`<summary>`	Configures text as a summary, caption, or legend for a details control	`class, id, style`
`<sup>`	Configures superscript text	`class, id, style`
`<table>`	Configures a table	`class, id, style, border` (obsolete in HTML5)
`<tbody>`	Configures the body section of a table	`class, id, style`
`<td>`	Configures a table data cell in a table	`class, colspan, id, headers, rowspan`
`<textarea>`	Configures a scrolling text box form control	`accesskey, autofocus, class, cols, disabled, id, maxlength, name, placeholder, readonly, required, rows, style, tabindex, wrap`
`<tfoot>`	Configures the footer section of a table	`class, id, style`
`<th>`	Configures a table header cell in a table	`class, colspan, id, headers, rowspan, scope, style`
`<thead>`	Configures the head section of a table	`class, id, style`
`<time>`	Configures a date and/or time	`class, datetime, id, pubdate, style`
`<title>`	Configures the title of a web page document	
`<tr>`	Configures a row in a table	`class, id, style`
`<track>`	Configures a subtitle or caption track for media	`class, default, id, kind, label, src, srclang, style`
`<u>`	Configures text displayed with an underline	`class, id, style`
`<ul>`	Configures an unordered list	`class, id, style`
`<var>`	Configures text as a variable or placeholder text	`class, id, style`
`<video>`	Configures a video control native to the browser	`autoplay, class, controls, height, id, loop, poster, preload, src, style, width`
`<wbr>`	Configures a line-break opportunity	`class, id, style`

CSS Cheat Sheet

Commonly Used CSS Properties

Property	Description
align-items	Configures the extra space along the cross-axis of a flex container Value: `flex-start`, `flex-end`, `center`, `baseline`, or `stretch`
background	Shorthand to configure all the background properties of an element Value: `background-color`, `background-image`, `background-repeat`, `background-position`
background-attachment	Configures a background image as fixed-in-place or scrolling Value: `scroll` (default) or `fixed`
background-clip	Configures the area to display the background Value: `border-box`, `padding-box`, or `content-box`
background-color	Configures the background color of an element Value: Valid color value
background-image	Configures a background image for an element Value: `url` (*file name or path to the image*), `none` (default) Optional CSS3 functions: `linear-gradient()` and `radial-gradient()`
background-origin	Configures the background positioning area Value: `padding-box`, `border-box`, or `content-box`
background-position	Configures the position of a background image Value: Two percentages, pixel values, or position values (`left`, `top`, `center`, `bottom`, `right`)
background-repeat	Configures how the background image will be repeated Value: `repeat` (default), `repeat-y`, `repeat-x`, or `no-repeat`
background-size	Configures the size of the background images Value: Numeric value (px, em, rem, vh, and vw), percentage, `contain`, `cover`
border	Shorthand to configure the border of an element Value: `border-width`, `border-style`, `border-color`
border-bottom	Configures the bottom border of an element Value: `border-width`, `border-style`, `border-color`
border-collapse	Configures the display of borders in a table Value: `separate` (default) or `collapse`
border-color	Configures the border color of an element Value: Valid color value

(Continued)

Property	Description
`border-image`	Configures an image in the border of an element See https://www.w3.org/TR/css3-background/#the-border-image
`border-left`	Configures the left border of an element Value: `border-width`, `border-style`, `border-color`
`border-radius`	Configures rounded corners Value: One or two numeric values (px or em) or percentages that configure horizontal and vertical radius of the corner. If one value is provided, it applies to both horizontal and vertical radius. Related properties: `border-top-left-radius`, `border-top-right-radius`, `border-bottom-left-radius`, and `border-bottom-right-radius`
`border-right`	Configures the right border of an element Value: `border-width`, `border-style`, `border-color`
`border-spacing`	Configures the space between table cells in a table Value: Numeric value (px or em)
`border-style`	Configures the style of the borders around an element Value: `none` (default), `inset`, `outset`, `double`, `groove`, `ridge`, `solid`, `dashed`, or `dotted`
`border-top`	Configures the top border of an element Value: `border-width`, `border-style`, `border-color`
`border-width`	Configures the width of an element's border Value: Numeric pixel value (such as 1 px), `thin`, `medium`, or `thick`
`bottom`	Configures the offset position from the bottom of a containing element Value: Numeric value (px or em), percentage, or `auto` (default)
`box-shadow`	Configures a drop shadow on an element Value: Three or four numerical values (px or em) to indicate horizontal offset, vertical offset, blur radius, (optional) spread distance, and a valid color value. Use the `inset` keyword to configure an inner shadow.
`box-sizing`	Alters the default CSS box model that calculates widths and heights of elements Value: `content-box` (default), `padding-box`, or `border-box`
`caption-side`	Configures the placement of a table caption Value: `top` (default) or `bottom`
`clear`	Configures the display of an element in relation to floating elements Value: `none` (default), `left`, `right`, or `both`
`color`	Configures the color of text within an element Value: Valid color value
`column-gap`	Configures empty space or gutters between grid columns Value: Numeric length or percentage

Property	Description
display	Configures how and if an element will display Value: `inline`, `inline-block`, `none`, `block`, `flex`, `grid`, `inline-flex`, `list-item`, `table`, `table-row`, or `table-cell`
flex	Configures the size of each flex item proportionate to the whole Value: Numeric integer value, `auto`, `flex-grow`, `flex-shrink`, or `flex-basis`
flex-basis	Configures the initial dimension along the main axis of a flex item Value: `auto` (default), `content`, or a positive numeric value
flex-direction	Configures the direction of flex items Value: `row`, `column`, `row-reverse`, or `column-reverse`
flex-flow	Shorthand to configure direction and wrap of a flex container Value: `flex-direction` and `flex-wrap`
flex-grow	Determines the growth of a flex item relative to other items in the flex container Value: 0 (default) or a positive numeric value
flex-shrink	Configures the shrinkage of a flex item relative to others in the flex container Value: 1 (default) or another positive numeric value
flex-wrap	Configures whether flex items are displayed on multiple lines Value: `nowrap (default)`, `wrap`, or `wrap-reverse`
float	Configures the horizontal placement (left or right) of an element Value: `none` (default), `left`, or `right`
font	Shorthand to configure the font properties of an element Value: `font-style`, `font-variant`, `font-weight`, `font-size/line-height`, `font-family`
font-family	Configures the font typeface of text Value: List of valid font names or generic font family names
font-size	Configures the font size of text Value: Numeric value (px, pt, em), percentage value, `xx-small`, `x-small`, `small`, `medium` (default), `large`, `x-large`, `xx-large`, `smaller`, or `larger`
font-stretch	Configures a normal, condensed, or expanded face from a font family Value: `normal`, `wider`, `narrower`, `condensed`, `semi-condensed`, `expanded`, or `ultra-expanded`
font-style	Configures the font style of text Value: `normal` (default), `italic`, or `oblique`
font-variant	Configures whether text is displayed in small-caps font Value: `normal` (default) or `small-caps`

(Continued)

Property	Description
`font-weight`	Configures the weight (boldness) of text Value: `normal` (default), `bold`, `bolder`, `lighter`, `100`, `200`, `300`, `400`, `500`, `600`, `700`, `800`, or `900`
`gap`	Shorthand to configure empty space or gutters between grid tracks Value: `row-gap` and `column-gap`
`grid-area`	Associates a grid item with a named area of the grid Value: Name of a grid area
`grid-column`	Configures one or more columns in the grid for an item or area Value: See https://www.w3.org/TR/css-grid-1/#typedef-grid-row-start-grid-line
`grid-column-gap`	Configures empty space or gutters between grid columns Value: Numeric length or percentage
`grid-gap`	Shorthand to configure empty space or gutters between grid tracks Value: `grid-row-gap` and `grid-column-gap`
`grid-row`	Configures one or more rows in a grid for an item or area Value: See https://www.w3.org/TR/css-grid-1/#typedef-grid-row-start-grid-line
`grid-row-gap`	Configures empty space or gutters between grid rows Value: Numeric length or percentage
`grid-template`	Shorthand property that combines the grid-template-areas, grid-template-rows, and grid-template-columns properties Value: A series of strings for each row that indicate placement of named grid areas; the last row describes the columns widths See https://www.w3.org/TR/css-grid-1/#propdef-grid-template
`grid-template-areas`	Visually indicates the placement of the named grid areas on the grid Value: A series of strings for each row that indicate placement of named grid areas See https://www.w3.org/TR/css-grid-1/#grid-template-areas-property
`grid-template-columns`	Specifies how much space to reserve for each column in the grid Value: Numeric pixels, percentage, auto, and flex factor units; See https://www.w3.org/TR/css-grid-1/#propdef-grid-template-columns
`grid-template-rows`	Specifies how much space to reserve for each row in the grid Value: Numeric pixels, percentage, auto, and flex factor units; see https://www.w3.org/TR/css-grid-1/#propdef-grid-template-columns
`height`	Configures the height of an element Value: Numeric value (px, em, rem, vh, vmin, and vmax), percentage, or `auto` (default)
`justify-content`	Configures how the browser should display any extra space Value: `center`, `space-between`, `space-around`, `flex-start` (default), or `flex-end`

Property	Description
`left`	Configures the offset position from the left of a containing element Value: Numeric value (px or em), percentage, or `auto` (default)
`letter-spacing`	Configures the space between text characters Value: Numeric value (px or em) or `normal` (default)
`line-height`	Configures the line height of text Value: Numeric value (px or em), percentage, multiplier numeric value, or `normal` (default)
`list-style`	Shorthand to configure the properties of a list `list-style-type`, `list-style-position`, `list-style-image`
`list-style-image`	Configures an image as a list marker Value: `url` (*file name or path to the image*) or `none` (default)
`list-style-position`	Configures the position of the list markers Value: `inside` or `outside` (default)
`list-style-type`	Configures the type of list marker displayed Value: `none`, `circle`, `disc` (default), `square`, `decimal`, `decimal-leading-zero`, `Georgian`, `lower-alpha`, `lower-roman`, `upper-alpha`, or `upper-roman`
`margin`	Shorthand to configure the margin of an element Value: One to four numeric values (px or em), percentages, `auto` or 0
`margin-bottom`	Configures the bottom margin of an element Value: Numeric value (px or em), percentage, `auto` or 0
`margin-left`	Configures the left margin of an element Value: Numeric value (px or em), percentage, `auto` or 0
`margin-right`	Configures the right margin of an element Value: Numeric value (px or em), percentage, `auto` or 0
`margin-top`	Configures the top margin of an element Value: Numeric value (px or em), percentage, `auto` or 0
`max-height`	Configures the maximum height of an element Value: Numeric value (px, em, rem, and vh), percentage, `none` (default), `max-content`, `min-content`, `fit-content`, or `fill-available`
`max-width`	Configures the maximum width of an element Value: Numeric value (px, em, rem, and vw), percentage, or `none` (default), `max-content`, `min-content`, `fit-content`, or `fill-available`
`min-height`	Configures the minimum height of an element Value: Numeric value (px, em, rem, and vh), percentage, `max-content`, `min-content`, `fit-content`, `fill-available`, or `auto`
`min-width`	Configures the minimum width of an element Value: Numeric value (px, em, rem, and vw), or percentage, `max-content`, `min-content`, `fit-content`, or `fill-available`

(Continued)

Property	Description
opacity	Configures the transparency of an element Value: Numeric value between 1 (fully opaque) and 0 (completely transparent)
order	Configures the display of flex items and grid items in a different order than coded Value: Numeric integer
outline	Shorthand to configure an outline of an element Value: `outline-width, outline-style, outline-color`
outline-color	Configures the outline color of an element Value: Valid color value
outline-style	Configures the style of the outline around an element Value: `none` (default), `inset, outset, double, groove, ridge, solid, dashed,` or `dotted`
outline-width	Configures the width of an element's outline Value: Numeric pixel value (such as 1 px), `thin, medium,` or `thick`
overflow	Configures how content should display if it is too large for the area allocated Value: `visible` (default), `hidden, auto,` or `scroll`
padding	Shorthand to configure the padding of an element Value: One to four numeric values (px or em), percentages, or 0
padding-bottom	Configures the bottom padding of an element Value: Numeric value (px or em), percentage, or 0
padding-left	Configures the left padding of an element Value: Numeric value (px or em), percentage, or 0
padding-right	Configures the right padding of an element Value: Numeric value (px or em), percentage, or 0
padding-top	Configures the top padding of an element Value: Numeric value (px or em), percentage, or 0
page-break-after	Configures the page break after an element Value: `auto` (default), `always, avoid, left,` or `right`
page-break-before	Configures the page break before an element Value: `auto` (default), `always, avoid, left,` or `right`
page-break-inside	Configures the page break inside an element Value: `auto` (default) or `avoid`
position	Configures the type of positioning used to display an element Value: `static` (default), `absolute, fixed, relative,` or `sticky`
right	Configures the offset position from the right of a containing element Value: Numeric value (px or em), percentage, or `auto` (default)
row-gap	Configures empty space or gutters between grid rows Value: Numeric length or percentage

Property	Description
text-align	Configures the horizontal alignment of text Value: `left` (default), `right`, `center`, or `justify`
text-decoration	Configures the decoration added to text Value: `none` (default), `underline`, `overline`, `line-through`, or `blink`
text-indent	Configures the indentation of the first line of text Value: Numeric value (px or em) or percentage
text-outline	Configures an outline around text displayed within an element Value: One or two numerical values (px or em) to indicate thickness and (optionally) blur radius, and a valid color value
text-overflow	Configures how the browser indicates content that has overflowed the container and is not visible Value: `clip` (default), `ellipsis`, or string value
text-shadow	Configures a drop shadow on the text displayed within an element Value: Three or four numerical values (px or em) to indicate horizontal offset, vertical offset, blur radius, (optional) spread distance, and a valid color value
text-transform	Configures the capitalization of text Value: `none` (default), `capitalize`, `uppercase`, or `lowercase`
top	Configures the offset position from the top of a containing element Value: Numeric value (px or em), percentage, or `auto` (default)
transform	Configures change or transformation in the display of an element Value: A transform function such as `scale()`, `translate()`, `matrix()`, `rotate()`, `skew()`, or `perspective()`
transition	Shorthand property to configure the presentational transition of a CSS property value Value: List the value for `transition-property`, `transition-duration`, `transition-timing-function`, and `transition-delay` separated by spaces; default values can be omitted, but the first time unit applies to `transition-duration`
transition-delay	Indicates the beginning of the transition Value: Default value 0 configures no delay, otherwise use a numeric value to specify time (usually in seconds)
transition-duration	Indicates the length of time to apply the transition Value: Default value 0 configures an immediate transition, otherwise use a numeric value to specify time (usually in seconds)
transition-property	Indicates the CSS property that the transition applies to; a list of applicable properties is available at https://www.w3.org/TR/css3-transitions
transition-timing-function	Configures changes in the speed of the transition by describing how intermediate property values are calculated; common values include `ease` (default), `linear`, `ease-in`, `ease-out`, or `ease-in-out`

(Continued)

Property	Description
vertical-align	Configures the vertical alignment of an element Value: Numeric value (px or em), percentage, baseline (default), sub, super, top, text-top, middle, bottom, or text-bottom
visibility	Configures the visibility of an element Value: visible (default), hidden, or collapse
white-space	Configures white space inside an element Value: normal (default), nowrap, pre, pre-line, or pre-wrap
width	Configures the width of an element Value: Numeric value (px, em, rem, vw, vmin, and vmax), percentage, or auto (default)
word-spacing	Configures the space between words within text Value: Numeric value (px or em) or auto (default)
z-index	Configures the stacking order of an element Value: A numeric value or auto (default)

Commonly Used CSS Pseudo-Classes and Pseudo-Elements

Name	Purpose
:active	Configures an element that is being clicked
:after	Inserts and configures content after an element
:before	Inserts and configures content before an element
:first-child	Configures an element that is the first child of another element
:first-letter	Configures the first character of text
:first-line	Configures the first line of text
:first-of-type	Configures the first element of the specified type
:focus	Configures an element that has keyboard focus
:hover	Configures an element that has a mouse placed over it
:last-child	Configures the last child of an element
:last-of-type	Configures the last element of the specified type
:link	Configures a hyperlink that has not been visited
:nth-of-type(n)	Configures the "nth" element of the specified type Value: A number, odd, or even
:visited	Configures a hyperlink that has been visited

WCAG 2.1 Quick Reference

Web Content Accessibility Guidelines (WCAG) 2.1 reached W3C Recommendation status in June 2018. WCAG 2.1 includes all WCAG 2.0 success criteria. WCAG 2.1 also introduces new success criteria.

Perceivable

▶ **1.1 Text Alternatives:** Provide text alternatives for any nontext content so that it can be changed into other forms people need, such as large print, Braille, speech, symbols, or simpler language. *You configure images (Chapter 5) and multimedia (Chapter 11) on web pages and provide for alternate text content.*

▶ **1.2 Time-Based Media:** Provide alternatives for time-based media. *We don't create time-based media in this textbook, but keep this in mind for the future if you create animation or use client-side scripting for features such as interactive slide shows.*

▶ **1.3 Adaptable:** Create content that can be presented in different ways (for example, simpler layout) without losing information or structure. *In Chapter 2, you use block elements (such as headings, paragraphs, and lists) to create single-column web pages. You create multicolumn web pages in Chapter 7. You use media queries and apply principles of responsive web design in Chapter 8. You associate meaningful labels with form controls in Chapter 10.*

▶ **1.4 Distinguishable:** Make it easier for users to see and hear content, including separating foreground from background. *You are aware of the importance of good contrast between text and background.*

Operable

▶ **2.1 Keyboard Accessible:** Make all functionality available from a keyboard. *In Chapter 7, you configure hyperlinks to named fragment identifiers on a web page. The label element is introduced in Chapter 10.*

▶ **2.2 Enough Time:** Provide users enough time to read and use content. Provide a way to pause, stop and/or hide moving, blinking, or scrolling information if it begins automatically and lasts more than five seconds. *In Chapter 11, you limit the duration of a transition to five seconds or less.*

▶ **2.3 Seizures:** Do not design content in a way that is known to cause seizures. *Be careful when you use animation created by others; web pages should not contain elements that flash more than three times in a one-second period.*

▶ **2.4 Navigable:** Provide ways to help users navigate, find content, and determine where they are. *In Chapter 2, you use block elements (such as headings and lists) to organize web page content. In Chapter 7, you configure hyperlinks to named fragment identifiers on a web page.*

- ▶ **2.5 Input Modalities:** Design to support input via other devices than a keyboard. *In Chapter 8, you create responsive web pages that work well on both desktop and mobile devices.*

Understandable

- ▶ **3.1 Readable:** Make text content readable and understandable. *You explore techniques used when writing for the Web in Chapter 3.*
- ▶ **3.2 Predictable:** Make web pages appear and operate in predictable ways. *The websites you create in the case studies have a consistent design, with clearly labeled and functioning hyperlinks.*
- ▶ **3.3 Input Assistance:** Help users avoid and correct mistakes. *In Chapter 10, you used form controls and attributes that cause a supporting browser to validate basic form information and display error messages.*

Robust

- ▶ **4.1 Compatible:** *Maximize compatibility with current and future user agents, including assistive technologies. You provide for future compatibility by writing code that follows W3C Recommendations (standards).*

The How to Meet WCAG 2 (Quick Reference) (https://www.w3.org/WAI/WCAG21/quickref/) entries are copyright © 2019 World Wide Web Consortium, (MIT, ERCIM, Keio, Beihang). https://www.w3.org/Consortium/Legal/2002/ipr-notice-20021231.

Web Content Accessibility Guidelines (WCAG) 2.1 reached W3C Recommendation status in June 2018. WCAG 2.1 includes all WCAG 2.0 success criteria. WCAG 2.1 also introduces new success criteria, described at https://www.w3.org/WAI/standards-guidelines/wcag/new-in-21/. You'll find the most up-to-date information about WCAG 2.1 at the following resources:

- ▶ Web Content Accessibility Guidelines (WCAG) 2.1
  https://www.w3.org/TR/WCAG21
- ▶ Understanding WCAG 2.1
  https://www.w3.org/WAI/WCAG21/Understanding/
- ▶ How to Meet WCAG 2.1
  https://www.w3.org/WAI/WCAG21/quickref

Landmark Roles with ARIA

The W3C's Web Accessibility Initiative (WAI) has developed a standard to provide for additional accessibility, called **Accessible Rich Internet Applications (ARIA)**. ARIA provides methods intended to increase the accessibility of web pages and web applications by identifying the role or purpose of an element on a web page (https://www.w3.org/WAI/intro/aria).

We'll focus on ARIA landmark roles in this appendix. A **landmark** on a web page is a major section such as a banner, navigation, main content, and so on. **ARIA landmark roles** allow web developers to configure semantic descriptions of HTML elements using the **role attribute** to indicate landmarks on the web page. It is always preferred to code the HTML semantic element (such as header, nav, main, footer, aside) instead of using the role attribute. For example, to indicate the landmark role of main on a div element containing the main content of a web page document, code `role="main"` on the opening div tag. However, if the main element was coded instead of the div element, the use of the main element alone indicates the ARIA landmark role and the role attribute should not be coded.

People visiting a web page with a screen reader or other assistive technology can access the landmark roles to quickly skip to specific areas on a web page (watch the video at http://www.youtube.com/watch?v=IhWMou12_Vk for a demonstration). Visit https://www.w3.org/TR/wai-aria-1.1/#landmark_roles for a complete list of ARIA landmark roles.

Commonly used ARIA landmark roles include:

- `banner` (a header/logo area). However, when the main header or logo is contained within the HTML header element that is a direct child of the body element, the role attribute should not be coded. The role attribute is needed when the header element contains the website header information and is coded within an article, aside, main, nav, or section element.
- `navigation` (a collection of navigation elements). However, when the navigation is contained within the HTML nav element, the role attribute should not be coded.
- `main` (the main content of a document); However, when the main content of a document is contained within the HTML main element, the role attribute should not be coded.
- `complementary` (a supporting part of the web page document, designed to be complementary to the main content). However, when content complementary to a document is contained within the HTML aside element, the role attribute should not be coded.
- `contentinfo` (an area that contains information about the content such as copyright). However, when this type of content is contained within the HTML footer element, the role attribute should not be coded.
- `search` (an area of a web page that provides search functionality)

465

The code for the body section of sample web page with the banner and main roles configured is shown below. If a semantic HTML header element was coded instead of the div element, the role attribute should be omitted. The navigation role is implied by the semantic HTML nav element. The contentinfo role is implied by the semantic HTML footer element. Notice that while the role attribute will not change the way the web page displays, it offers additional information about the document that can be used by assistive technologies.

```
<body>
  <div role="banner">
    <h1>Heading Logo Banner</h1>
     <a href="#content">Skip to Content</a>
  </div>
  <nav>
     <a href="index.html">Home</a> <a href="contact.html">Contact</a>
  </nav>
  <div role="main" id="content">
      This is the main content area.
  </div>
  <footer>
      Copyright &copy; 2022 Your Name Here
  </footer>
  </body>
```

The same web page could have been coded as follows using only semantic HTML elements to indicate the banner, navigation, main, and contentinfo ARIA roles.

```
<body>
  <header>
    <h1>Heading Logo Banner</h1>
    <a href="#content">Skip to Content</a>
  </header>
  <nav>
     <a href="index.html">Home</a> <a href="contact.html">Contact</a>
  </nav>
  <main id="content">
    This is the main content area.
  </main>
  <footer>
    Copyright &copy; 2022 Your Name Here
  </footer>
</body>
```

Web Safe Color Palette

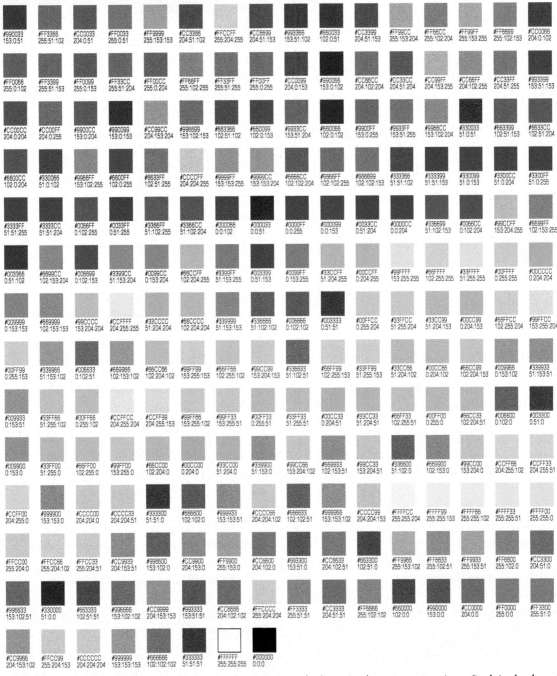

#990033	153:0:51
#FF3366	255:51:102
#CC0033	204:0:51
#FF0033	255:0:51
#FF9999	255:153:153
#CC3366	204:51:102
#FFCCFF	255:204:255
#CC6699	204:51:153
#993366	153:51:102
#660033	153:0:51
#CC3399	204:51:153
#FF99CC	255:153:204
#FF66CC	255:102:204
#FF99FF	255:153:255
#FF6699	255:102:153
#CC0066	204:0:102
#FF0066	255:0:102
#FF3399	255:51:153
#FF0099	255:0:153
#FF33CC	255:51:204
#FF00CC	255:0:204
#FF66FF	255:102:255
#FF33FF	255:51:255
#FF00FF	255:0:255
#CC0099	204:0:153
#990066	153:0:102
#CC66CC	204:102:204
#CC33CC	204:51:204
#CC99FF	204:153:255
#CC66FF	204:102:255
#CC33FF	204:51:255
#993399	153:51:153
#CC00CC	204:0:204
#CC00FF	204:0:255
#9900CC	153:0:204
#990099	153:0:153
#CC99CC	204:153:204
#996699	153:102:153
#663366	102:51:102
#660099	102:0:153
#9933CC	153:51:204
#660066	102:0:102
#9900FF	153:0:255
#9933FF	153:51:255
#9966CC	153:102:204
#330033	51:0:51
#663399	102:51:153
#6633CC	102:51:204
#6600CC	102:0:204
#330066	51:0:102
#9966FF	153:102:255
#6600FF	102:0:255
#6633FF	102:51:255
#CCCCFF	204:204:255
#9999FF	153:153:255
#9999CC	153:153:204
#6666CC	102:102:204
#6666FF	102:102:255
#666699	102:102:153
#333366	51:51:102
#333399	51:51:153
#330099	51:0:153
#3300CC	51:0:204
#3300FF	51:0:255
#3333FF	51:51:255
#3333CC	51:51:204
#0066FF	0:102:255
#0033FF	0:51:255
#3366FF	51:102:255
#3366CC	51:102:204
#000066	0:0:102
#000033	0:0:51
#0000FF	0:0:255
#000099	0:0:153
#0033CC	0:51:204
#0000CC	0:0:204
#336699	51:102:153
#0066CC	0:102:204
#99CCFF	153:204:255
#6699FF	102:153:255
#003366	0:51:102
#6699CC	102:153:204
#006699	0:102:153
#3399CC	51:153:204
#0099CC	0:153:204
#66CCFF	102:204:255
#3399FF	51:153:255
#003399	0:51:153
#0099FF	0:153:255
#33CCFF	51:204:255
#00CCFF	0:204:255
#99FFFF	153:255:255
#66FFFF	102:255:255
#33FFFF	51:255:255
#00FFFF	0:255:255
#00CCCC	0:204:204
#009999	0:153:153
#669999	102:153:153
#99CCCC	153:204:204
#CCFFFF	204:255:255
#33CCCC	51:204:204
#66CCCC	102:204:204
#339999	51:153:153
#336666	51:102:102
#006666	0:102:102
#003333	0:51:51
#00FFCC	0:255:204
#33FFCC	51:255:204
#33CC99	51:204:153
#00CC99	0:204:153
#66FFCC	102:255:204
#99FFCC	153:255:204
#00FF99	0:255:153
#339966	51:153:102
#006633	0:102:51
#669966	102:153:102
#66CC66	102:204:102
#99FF99	153:255:153
#66FF66	102:255:102
#99CC99	153:204:153
#336633	51:102:51
#66FF99	102:255:153
#33FF99	51:255:153
#33CC66	51:204:102
#00CC66	0:204:102
#66CC99	102:204:153
#009966	0:153:102
#339933	51:153:51
#009933	0:153:51
#33FF66	51:255:102
#00FF66	0:255:102
#CCFFCC	204:255:204
#CCFF99	204:255:153
#99FF66	153:255:102
#99FF33	153:255:51
#00FF33	0:255:51
#33FF33	51:255:51
#00CC33	0:204:51
#33CC33	51:204:51
#66FF33	102:255:51
#00FF00	0:255:0
#66CC33	102:204:51
#006600	0:102:0
#003300	0:51:0
#009900	0:153:0
#33FF00	51:255:0
#66FF00	102:255:0
#99FF00	153:255:0
#66CC00	102:204:0
#00CC00	0:204:0
#33CC00	51:204:0
#339900	51:153:0
#99CC66	153:204:102
#669933	102:153:51
#99CC33	153:204:51
#336600	51:102:0
#669900	102:153:0
#99CC00	153:204:0
#CCFF66	204:255:102
#CCFF33	204:255:51
#CCFF00	204:255:0
#999900	153:153:0
#CCCC00	204:204:0
#CCCC33	204:204:51
#333300	51:51:0
#666600	102:102:0
#999933	153:153:51
#CCCC66	204:204:102
#666633	102:102:51
#999966	153:153:102
#CCCC99	204:204:153
#FFFFCC	255:255:204
#FFFF99	255:255:153
#FFFF66	255:255:102
#FFFF33	255:255:51
#FFFF00	255:255:0
#FFCC00	255:204:0
#FFCC66	255:204:102
#FFCC33	255:204:51
#CC9933	204:153:51
#996600	153:102:0
#CC9900	204:153:0
#FF9900	255:153:0
#CC6600	204:102:0
#993300	153:51:0
#CC6633	204:102:51
#663300	102:51:0
#FF9966	255:153:102
#FF6633	255:102:51
#FF9933	255:153:51
#FF6600	255:102:0
#CC3300	204:51:0
#996633	153:102:51
#330000	51:0:0
#663333	102:51:51
#996666	153:102:102
#CC9999	204:153:153
#993333	153:51:51
#CC6666	204:102:102
#FFCCCC	255:204:204
#FF3333	255:51:51
#CC3333	204:51:51
#FF6666	255:102:102
#660000	102:0:0
#990000	153:0:0
#CC0000	204:0:0
#FF0000	255:0:0
#FF3300	255:51:0
#CC9966	204:153:102
#FFCC99	255:204:153
#CCCCCC	204:204:204
#999999	153:153:153
#666666	102:102:102
#333333	51:51:51
#FFFFFF	255:255:255
#000000	0:0:0

Web safe colors look the most similar on various computer platforms and computer monitors. Back in the day of eight-bit color, it was crucial to use web safe colors. Since most modern video drivers support millions of colors, the use of web safe colors is now optional. The hexadecimal and decimal RGB values are shown for each color in the palette above.

Index

Symbols

canvas API, 419

geolocation API, 418

HTML5 API, 418–419

jQuery API, 417

manifest API, 419

progressive web application, 418–419

service workers API, 419

web storage API, 418

area element (`<area>`), 168

ARIA. *see* Accessible Rich Internet Applications (ARIA)

article element (`<article>`), 52

aside element (`<aside>`), 52

assistive technology, 442

attributes

audio element (`<audio>`), 398

check box form controls, 354

date and time form controls, 380

form element (`<form>`), 349

HTML, 17

iframe element (`<iframe>`), 405

img element (`<img>`), 148, 308–309

`lang` attribute in web pages, 19

option element (`<option>`), 359

radio button form controls, 355

scrolling text box form controls, 356

select element (`<select>`), 358

selector, 365

`sizes` attribute, 306, 308

slider, spinner, and date/time form controls, 378–379

source element (`<source>`), 306

`srcset` attribute, 306, 308–309

submit buttons and reset buttons, 352–353

table attributes obsolete in HTML5, 327

table data and table header cell elements, 328–329

table element (`<table>`), 334

video element (`<video>`), 400

Audacity application, 399

audio. *see also* multimedia

and accessibility, 399

element, 398–399

fallback content, 399

file troubleshooting, 396

audio element (`<audio>`), 398–399

automated accessibility testing, 442

B

`background-attachment` property, 153

`background-clip` property, 196–197

`background-color` property, 114, 115, 152

`background-image` property, 152

background images

applying multiple, 156–157

clipping and sizing, 196–197

fluid display of, 304

overview of, 152–153

positioning, 154–155

resizing and scaling, 198–199

tiling with `background-repeat` property, 154

`background-origin` property, 197

`background-position` property, 154–155

`background-repeat` property, 154, 155

`background-size` property, 198–199

Berners-Lee, Tim, 3

Bezier curve, 409

`bgcolor` attribute

configuring tables in HTML, 334

obsolete in HTML5, 327

block anchor, 55

blockquote element (`<blockquote>`), 34–35

body element (`<body>`), 19

`border` attribute, 326–327

`border` property, 188–189

`border-radius` property, 190–191

borders

configuring, 188–189

in CSS box model, 185

rounded corners, 190–191

`border-spacing` property, 335

`border-style` property, 188–189

box model

in action, 185

border area, 185

content area, 184

margin, 185

padding area, 184

`box-shadow` property, 194

`box sizing` property, 228–229

breadcrumb navigation, 94

legend element (<legend>), 362–363

option element (<option>), 359

password box, 372

progressive enhancement, 379

radio button, 355

reset button, 352–353

select element (<select>), 358

server-side processing, 368–369

slider form control, 378

spinner form control, 379

style, CSS, 364–365

submit button, 352–353

textarea element (<textarea>), 356–357

text box form control, 350–351

text form controls, 374–375

fragment identifier, 252–253

free remote-hosted form processing, as alternative to server-side processing, 369

FTP. see File Transfer Protocol (FTP)

G

Generic Top-Level Domain Names (gTLDs), 11–12

geolocation API, 418

GNU Image Manipulation Program (GIMP), 90

Google Chrome Dev Tools, 299

gradient color, CSS

linear, 208

radial, 208

grammar, 81

Graphic Interchange Format (GIF) images, 144

graphics. see also web graphics

applications, 147

applying multiple background images, 156–157

background images, 152–153

canvas element (<canvas>) for dynamic graphics, 419

configuring list markers using CSS, 164–165

favorites icon, 166–167

img element (<img>), 148–149

image formats, 144–145

image hyperlinks, 150–151

image maps, 168–169

lazy loading, 309

placing images in separate folder from web pages, 153

positioning background images, 154–155

reasons why images are not displayed on web pages, 151

use in navigating websites, 95

web design and, 90–91

greater than (>), special characters, 44

grid-area property, 284

grid-column property, 281

grid container

configuring grid columns and rows, 277, 278, 280

display property, 276

grid design, 276–277

grid layout

centered text, 290–291

configuring grid with an empty area, 285–286

configuring two-column layout, 280–281

configuring with grid line numbers, 281–283

and gap property, 278

grid areas, 284–287

overview of, 276–277

progressive enhancement, 288–289

responsive layout with media queries, 296–297

grid line numbers, 281

grid-row property, 281

grid-template-areas property, 284–285

grid-template property, 287

H

header element (<header>), 48

headers attribute, table data element (<td>), 333

heading elements (<h1> to <h6>)

overview of, 28–29

in SEO (Search Engine Optimization), 440

heading tags, search engines and, 440

height attribute, configuring tables in HTML, 334

height property, 183

hexadecimal color values

CSS syntax for color values, 116

web color palette, 82

hidden input control, 372

hierarchical organization, website, 74–75

home page, 57

horizontal navigation, unordered list, 236–237

horizontal rule element (<hr>), 33

:hover, CSS pseudo-classes, 238

markup languages, 16–17
Markup Validation Service, 46–47, 134–135
`max-device-height`, media queries, 295
`max-device-width`, media queries, 295
`max-height`, media queries, 295
`maxlength attribute`, form elements, 351
`max-width`, media queries, 295
`max-width` property, 183
media and interactivity
  audio element (`<audio>`), 398–399
  browser compatibility issues, 396
  drop-down menu, CSS, 412–413
  embedding YouTube videos, 404–405
  fallback content, 399, 401
  HTML5 API, 418–419
  iframe element (`<iframe>`), 404–405
  inline frame, 404–405
  JavaScript, 416
  jQuery, 417
  practice with, 410–411
  `rotate()` transform function, 406
  `scale()` transform function, 407
  `transform` property, 406–407
  `transition` property, 408–409
  video element (`<video>`), 400–403
media query, 294–295
  features, 295
  link element (`<link>`), 294–295
  `@media` rule, 295
  Mobile First, 295
  responsive grid layout, 300–301
  responsive layout, 296–299
  types, 295
  values, 299
`@media()` rule, 295
memory
  designing for mobile web and, 100
meta element (`<meta>`)
  overview of, 19
  in SEO (Search Engine Optimization), 440–441
meta tags, search engine optimization and,
    440–441
`method` attribute, 369
Microsoft Edge, 166, 311
  image processing, 149
  web page display, 23
`min-device-height`, media queries, 295
`min-device-width`, media queries, 295

`min-height`, media queries, 295
`min-width` property, 182
Mobile First, 295
Mosaic browser, 3
Mozilla Firefox, 3, 149
multimedia. see also media and interactivity
  browser compatibility issues and, 396
  use of, 90–91
Multi-purpose internet mail extensions (MIME), 7

N

`name` attribute
  area element (`<area>`), 168
National Center for Supercomputing Applications
    (NCSA), 3
native application, 418
nav element (`<nav>`), 48
navigation bars, 94
navigation design
  breadcrumb, 94
  dynamic, 95
  ease of, 94
  graphics and, 95
  navigation bars, 94
  site map, 95
  site search feature, 95
networks
  client/server model, 6–7
  overview of, 6
Nielsen, Jakob, 85
nonbreaking space ( ), special characters, 44
normal flow
  floated elements and, 223
  nested elements and, 221
  in page layout, 220–221
  practice applying, 220–221

O

Ogg Vorbis, 399
online publishing. see web publishing
`opacity` property, 202–203
  configuring element transparency, 202–203
Opera browser. see also web browsers
  Opera, 73

perceived load time, 92
proximity, 77
repetition, 76
responsive web design, 102–103
target audience, 72–73
text use and, 80–81
visual design principles, 76–77
web color palette, 82–83
website organization, 74–75

W

Web Accessibility Initiative (WAI), 4, 79
web browsers, 6–7
  audio element support, 399
  `background-clip` property support,
    196–197
  background images and, 152–153
  `background-origin` property support, 197
  `background-size` property support,
    198–199
  `box-shadow` property support, 194
  configuring how browsers render
    elements, 236
  CSS flexible box layout, 266–267
  CSS grid layout, 276–277, 288–289
  CSS transforms support, 407
  CSS transition support, 408
  favicon support, 166
  flexbox layout system, 273
  grid layout system, 283
  input element attribute support, 350–351
  media query support, 294–295
  multimedia compatibility issues, 396
  multiple background image support, 156
  RGBA color support, 204–205
  rows and columns, 278–279, 280–281
  server-side processing, 368–369
  size, 271
  testing web page in, 23
  `text-shadow` property support, 195
  video element support, 401
web color palette
  accessibility and, 83
  hexadecimal color values, 82
  web-safe colors, 82
  web-safe hexadecimal values, 82

Web Content Accessibility Guidelines 2.1
    (WCAG 2.1), 79, 442, 463–464
web design basics, 104–105
  usability testing and, 444–445
web graphics
  background image configuration, 152–153
  favorites icon, 166–167
  GIF images, 144
  img element (`<img>`), 148–149
  image hyperlinks, 150–151
  JPEG images, 145
  multiple background images, 156–157
  PNG images, 146
  web page creation, 162–163
  WebP image format, 146–147
web host
  checklist, 433
  providers, 432
  virtual host selection, 432–433
web hosting
  cloud, 432
  colocated, 432
  dedicated, 432
  virtual, 432
Web Hypertext Application Technology Working
    Group (WHATWG), 17
WebM files, 403
web page ranking. see search engine optimization
    (SEO)
web publishing
  accessibility testing and, 442–443
  choosing web host, 432–433
  domain names, 430–431
  file organization, 428–429
  file transfer protocol, 436–437
  hyperlinks, 252–253
  search engine optimization, 440–441
  search engine submission, 438–439
  Secure Sockets Layer (SSL), 434–435
web-safe colors, 82
web-safe hexadecimal values, 82
web servers, 6–7
website organization
  hierarchical, 74–75
  linear, 75
  random, 75
web site submitting, on search engine, 439
web storage API, 418

Credits

Cover image © Nerthuz/Shutterstock

Figures 1.1–1.10, 2.17, 2.19, 2.23, 2.28, 2.29, 2.32, 2.33, 3.1–3.47, 4.1–4.3, 4.19, 4.26, 4.30, 5.1–5.21, 5.24–5.37, 6.1–6.7, 6.9, 6.14–6.29, 6.32–6.40, 7.1–7.14, 7.19, 7.21–7.25, 7.27–7.38, 7.41–7.55, 8.1–8.31, 8.33–8.67, 9.15–9.16, 10.16, 10.18, 10.19, 10.21, 10.37–10.46, 11.4–11.14, 11.19–11.23, 12.2, 12.4–12.6, 12.2, 12.4–12.6, 12.8 © Terry Ann Morris, Ed.D. Reprinted with permission

Figures 1.12–1.14, 4.11, 4.16, 5.13, 12.1, 12.3 © Microsoft Corporation

Table 3.1 © Terry Ann Morris, Ed.D. Reprinted with permission

Figure 12.7 © FileZilla. Reprinted with permission

Figures 2.14–2.16, 4.22–4.24, Copyright © W3C (World Wide Web Consortium)

Figure 10.32 A date form control displayed in the Google Chrome browser. Copyright © Google, Inc.

Figure 10.33 The Google Chrome browser supports the color-well form control. Copyright © Google, Inc.

Figures 1.11, 10.25–10.31, 10.34–10.36, 11.2–11.3 and Mozilla Firefox frames that are used around the OM figures © Mozilla Foundation.